TVA's Public Planning

TVA's Public Planning
The Vision, the Reality

Walter L. Creese

The University of Tennessee Press

KNOXVILLE

Publication of this book was made possible with assistance
from the Graham Foundation for Advanced Studies
in the Fine Arts.

Copyright © 1990 by The University of Tennessee Press / Knoxville.
All Rights Reserved. Manufactured in the United States of America.
First Edition.

Frontispiece: Fort Loudoun Dam under Construction, 1942.
Photo courtesy of Tennessee Valley Authority Archives.

The paper in this book meets the minimum requirements
of the American National Standard for Permanence
of Paper for Printed Library Materials.
∞
The binding materials have been chosen
for strength and durability.

Library of Congress Cataloging in Publication Data

Creese, Walter L.
  TVA's public planning : the vision, the reality / Walter L.
Creese.
      p.   cm.
  ISBN 0-87049-638-7 (cloth: alk. paper)
  1. Tennessee Valley Authority—History. 2. Electric utilities—
Tennessee River Valley—History. 3. Tennessee River Valley—
Economic conditions. 4. Tennessee River Valley—Social conditions.
5. Dams—Tennessee River Valley—History. 6. Electric power-plants—
Tennessee River Valley—History. 7. Tennessee River—Power
utilization—History. I. Title.
HD9685.U7T334   1990
353.0082'3'09768—dc20   89-28129   CIP

To Edwin Rand

Nominally unemployed and in uncertain health because of lead poisoning during the 1930s Depression, Edwin Rand became a thoroughly generous and gifted sailing instructor to me and many others. He built models of clipper ships under full sail, skimming the wooden seas he had carved and painted to receive them. They ended in maritime museums in the United States and Great Britain. For him, the abnormal times signaled an urgent call to colorful restatement, to make something dynamic, yet lyrical, away from the dismal reality. A brilliant intellect with no funds and a powerful physique, he also collected long runs of the Sunday *New York Times* in the 1930s and buried them in heavy boxes, in the event there were humans left alive after the disastrous war he was predicting for the 1940s. Surely they would want to review, step by step, how Armageddon had come upon them in those exhilarating, excruciating decades?

Contents

Illustrations

Preface

For over half a century, the Tennessee Valley Authority (TVA) has stood, noteworthy in scale whether considered in terms of fantasy or reality. Has the TVA actually come to belong in our nation's permanent structure, or was it merely one more emergency improvisation spawned in the face of the Depression? Does it deserve another scan? Whoever the examiner might be, he or she ought to continue to acknowledge TVA as a splendid vision — the glowing medal on the chest of America, as it was imaged in the 1930s. In that decade, TVA certainly merited honor. At the same time, however, an examiner must take into account the underlying political restrictions, inhibitions, and debilities, such as inadequate financing, that tended to impede the TVA's progress during the thirties and thereafter. Such factors determined the subsurface turbulence that in turn has controlled the temporal flow of the TVA. Close interaction between fantasy and reality, not a prolonged separation of one from the other, ultimately renders the TVA story more authentic and illuminating.

To gain insight, the reader must consider four main questions:

1. Does the TVA represent a quick or prolonged, precedented or unprecedented, superficial or profound episode in the American experience?

2. Is the technological or the human side of TVA's history more to be noted and discussed?

3. If the human meaning is as arresting as the technological, do we then have a further obligation to consider human neglect, and the accompanying struggle to recover dignity and self-respect, as a lesson useful to the rest of the country in the 1930s and relevant to its destiny thereafter?

4. Was the TVA finally greater or lesser than the sum of its parts?

In our own era, the limitations on our understanding of the TVA probably reflect dual distractions. The TVA arose partly out of tradition and partly out of a radical experiment. Conceptually, like so many other American undertakings, it was assembled out of spare parts from the public culture — reworked bits of older memory and belief; government, business, and military practice. Often disturbing the clarity of TVA's novel crystallization were several groups of subprophets (TVA directors, scholars, lawyers, or other types of New Deal expediters). Such participants were eager to employ the TVA as a channel for sluicing out the gold of their own preexisting reform notions. As a result, there was often a tendency to move the TVA off the main line to branch lines of initiative. However expertly or efficiently the TVA may have been administered as it moved along its renowned energy "yardstick," it still is necessary to ask, "Toward what end?" Any impression of invulnerable originality, linearity, or monolithicity must be carefully examined; otherwise, the package may look too neat.

Was the philosophy of the TVA truly in the American mainstream in the past? Can it be placed there now? Many TVA enthusiasts seemed to want to deny the organization any previous lineage, apparently in the hope that such novelty would help it attract greater respect and support. But the inclination to seal TVA in its own case prevented it from being seen for what it was — the legitimate culmination of the Progressive and Conservation movements of the 1880s and 1890s. The Progressivist-Conservationist continuity largely explains TVA's alliance with nature and natural resources and indicates why the TVA tried to move beyond the three sequential totems of the Conservation movement — the national park, the national forest, and the wilderness preserve, the last coming up in the 1920s as an environmental possibility.

From the opposite direction, the TVA also was a reaction against what was then taken to be an overripe urbanism epitomized by New York City and by Chicago. As Franklin Roosevelt himself indicated in one of his speeches but no-one has noted since, the TVA arose in response to preoccupations coming out of the Chicago "City Beautiful" model of 1890–1910. The 1930s caused doubts to be raised about social stability, and as a result, postponed conclusions bubbled up from below and spilled over the edge. This motion gave architecture, engineering, and land planning their sudden opportunity — a ferment that showed up in the TVA simultaneously and comprehensively. The three disciplines formed a standing exhibit, a museum without walls. The difficulty was how to encompass it. When the TVA buildings were reviewed in a 1941 exhibit at the Museum of Modern Art in New York, Lewis Mumford complained that lack of wall space made it impossible to display them all, so that one could not grasp the entire landscape with the buildings in it.

Fascination with the possibility of refashioning the surface of a continent in some positive way, to accommodate an entire people better, stimulated a new eagerness to plan and establish more of such mega-units all across the country. But the ambition was nipped in the bud by Senator Taft and his Republican colleagues in 1943, when they abolished the National Resources Planning Board, hoping thus to forestall postwar planning of any substantial kind already exemplified by the TVA. The latter's physical growth had alerted and worried the "saner" circles. What that sequence of events—from the Chicago Fair of 1893, reflecting interest in the City Beautiful, up to the proposal for regional planning on a nationwide basis—ultimately meant was that a show of TVA-like initiatives both began and ended earlier than regularly has been admitted. Those initiatives had a longer and thicker taproot, and a bigger but briefer flowering into adventure, than have been recognized.

As far as this book is concerned, TVA's richer but shorter blooming makes it desirable to ruminate more specifically and at greater length on how the abilities of the architect, landscape architect, engineer, ground planner, and photographer were gathered and utilized, dovetailed together, in a country that had had Army Engineers since the American Revolution but never before had employed teams of landscape architects and architects for such big, ongoing public projects. How did the cooperative efforts of these professionals fit into the operational profile from the City Beautiful up to the time of the National Resources Planning Board (1893–1943)? Were genuine harmony and synthesis ever achieved before the cutoff by Senator Taft and his colleagues? Careful architectural oversight of an extended engineering campaign was sensational enough in itself.

Latterly intruding upon such striving, whether worthy or misguided, was the growing spectre of World War II. By the outbreak of hostilities, the TVA had developed a tempting pool of regional electrical power, and more could be had on demand by building more dams—or so the larger federal government outside TVA had concluded. The original attraction of the possibility of producing electricity in World War I for defense purposes at Muscle Shoals—a plant that never got into production before the war was over — was magnified exponentially. What had been an experimental pilot plant or shelter in a valley to disburse hydroelectricity for the charitable purpose of upgrading the basic living standard of a local people suddenly became a technical plum waiting to be picked for national mobilization in an international action. The first dams had been built too quickly in order to ease unemployment in make-work projects; now defense dams, such as Douglas, Cherokee, and Fontana, were built with an even greater sense of urgency. What time was there for a seasoned consideration of design, in either architecture or engineering? Muscle Shoals had to be converted back to a military purpose in order to generate the electricity needed to produce nitrate for munitions. Electricity also was needed in great quantity to make bauxite into aluminum for airplanes. Finally and most seriously and incredibly of all, great sums of electricity were required to convert uranium into plutonium for the Manhattan Project at Oak Ridge, Tennessee, with the goal of creating, as rapidly as possible, the atomic explosion later heard round the world.

As America's enemies clearly perceived, our national government never plans consistently on a large scale if that can conveniently be avoided. One reason why the exceptional-seeming step of initiating the TVA was hailed so widely abroad by intelligentsias not yet under totalitarian control was that the plan constituted a departure from the usual short sightedness. But once the American government was chivvied into action by external forces, it could sketch things out on a very big canvas indeed. The TVA was big, but World War II was even bigger. By 1955, the Atomic Energy Commission had requisitioned half of TVA's electrical output, so that the latter agency was forced to create coal-fired, air-polluting steam plants, the first of them built during the war at Watts Bar in awkward juxtaposition with the hydroelectric dam being erected on the site. After the war, TVA received no adequate compensation from the federal government for the wartime swelling and sacrifice. The TVA then was simultaneously exploited and taken for granted. To supply enough coal to satisfy its own government's growing demand for electricity, TVA had to resort more and more to ugly strip mining, incurring further public and press opprobrium. The old sense of mission, of being in pursuit of a legitimate dream — a sense that had been ably and imaginatively nurtured by TVA's earlier leaders — atrophied. The happier vision of a renewed arcadian landscape, supplied with abundance, was converted, under the auspices of Cold War psychology, into a singleminded, routine exercise. Those leaders who had looked for the more broadly humanistic, artistic, and sociological opportunities of the Tennessee Valley had either lost power, moved up, or passed on. First principles, badly pitted and dented when Arthur Morgan was dismissed from the board of directors on 23 March, 1938, were further abrogated when World War II broke out on 7 December, 1941. Prewar intention was thwarted by postwar forgetfulness. What has not been well enough understood, in addition, is that, even in the early period, TVA tended to progress in short, staccato, three-year bursts: 1932–35, whereupon the Second New Deal came on line; 1935–38, at the end of which Arthur Morgan was dismissed; and 1938–41, when preoccupation with the coming war took over. Everything happened so fast that longevity was difficult to contemplate. What had been remarkable about TVA before the war, what had made it clearly different from its antecedents, the national park, national forest, and national wilderness, was that people were allowed — even encouraged — to remain to live in it! Exasperation with humanity's treatment of the land had not led to exclusion of individuals. After the war, as the Depression evaporated and the South rose again, increased mobility of families left less room for recognition of the need for inclusion and acceptance of local people. The national attention span for the TVA became preternaturally short. Regionalism now seemed tantamount to parochialism, so there was less willingness to study internal living habits and conditions.

Acknowledgments

Acknowledgments are due first to those who have helped so much with the preparation of the manuscript. They should not be blamed, of course, for any conclusions in the book that the reader may not agree with; responsibility for the content lies with the author. Among students who collected and checked facts have been Malcolm Jollie, Christopher Hudson, Daniel Erdman, Amy Cassens, Dominic Adducci, and particularly Deborah Slaton. Most recently, Roann Barris patiently and competently located numerous books and articles, some so current that they had not been printed yet! Consultation with her was invaluable.

Graduate student Christopher Vernon and Professors John Garner and Robert Carringer of the University of Illinois shared their expertise and their enthusiastic insights into TVA's landscape architecture, land and community planning, and film history, respectively. Paul S. Harris told me about his early study with Arthur Morgan at Antioch College. My wife, Eleanor, traveled with me to the TVA Technical Library in Knoxville, where we assaulted the copious and uniquely well-organized materials. In that vicinity, Professors Marian Scott Moffett and Lawrence Wodehouse of the University of Tennessee generously answered further questions, while Dr. Charles W. Crawford and Delanie M. Ross of the Mississippi Valley Collection at Memphis State University were more than accommodating with their time and their remarkable archive of oral history.

Thoroughly impressive was the splendid command that TVA librarian Jesse Mills, now retired, had of the material that he and his staff brought forth without stint. He made me conscious of other resources as well, including the superb TVA photographic archive, a substantial sample of which is displayed in this volume. It is too often forgotten that the 1930s were a decade unparalleled for black and white photodocumentation, sometimes telling as much as or more than the written documents. Robert E. Kollar, chief photographer of the TVA, filled the gaps I still had at the end of the investigation by searching again through the negatives. Edwin J. Best, Jr., reference librarian of the TVA Technical Library, also quickly, courteously, and most skillfully responded to the several last-minute queries I posed and helped a great deal to add further pictures and population facts. Stephen Cottam of the McClung Historical Collection, Knox County Public Library, gave useful advice on the historical background of the vicinity. J. Wayne Range and Jim Alexander helped with Oak Ridge. Newspaperman Gene Graham and TVA recreation director emeritus Robert Howes shared their long experience. Those photographs, papers, and contacts helped to dispel, as I hope this book may for others, any lingering doubts about the capabilities of government engineers, architects, planners, and bureaucrats in general to do their jobs well — *if* they can be given adequate funds, tools, and moral support over time.

Jesse Mills also kept me acquainted with others investigating the TVA. A fortunate contact was Professor Emeritus William Jordy of Brown University, who kindly commented on my preliminary manuscript and sent his own unpublished study of Norris Village, which I gratefully cite a few times in the text. As would be expected, his grasp of TVA lore was both thorough and profound, and it is much to be wished that the whole eventually will appear in print. I was particularly fortunate, too, in gaining an early interview with Professor Frederick Gutheim of The George Washington University, who was an important actual participant in the formulation of the TVA law, and who illuminated other aspects of American environmental history and architecture for me through our conversations and his writing. Through a lifetime of eloquent essays and activities in behalf of the national and world scene, he has eloquently demonstrated the value of the short-lived Meiklejohn experimental college at the University of Wisconsin, from which he graduated. He has been kind enough to review the mentions of his role in the TVA in this text. Dean James W. Carey of the University of Illinois also checked my lengthy citation of his and John J. Quirk's lively and provocative *American Scholar* article, "The Mythos of the Electronic Revolution" (1970).

Out of his long and very distinguished career as a landscape and town planner, Earle Draper gave me much time and information. His son, Earle, provided a few of the photos of Norris Village. I also was fortunate to have been able to interview, while they were still living, two very unusual and different personalities who told me without hesitation what I asked to know — Charles W. Eliot II and Benton MacKaye. I hope that this book reflects, although it probably does so in too modest a way, the earnestness and dedication I sensed in these several personalities as they sought unendingly to improve the living environment for all Americans.

When I arrived to be an instructor at the University of Louisville in 1946, immediately after World War II, I was struck by how many of the extremely able Louisville leaders were thinking of a new postwar society for the South. Among the many were Mayor Charles P. Farnsley; former mayor Wilson Wyatt; lawyer Eli H. Brown III; public librarian Clarence Graham; symphony conductor Robert Whitney; and newspaper owner Barry Bingham. Their wholehearted desire to provide another, better kind of communal life, an alternate culture for a southern city, brought me to a new and broader awareness. Maybe it had meaning for all of post-war America? As a part of this increasing awareness, I became conscious of the major enterprise of that kind dating from the immediate prewar years and located just to the south, the Tennessee Valley Authority. Subsequently, as chairman of the Louisville and Jefferson County Planning and Zoning Commission, and through attendance at southern conferences, I encountered those who were planning the southern region, including TVA professionals. At the same time, I was attempting to start a state historic buildings survey. These experiences, when rumblings about strip mining were also beginning, further piqued my curiosity as to the type of people who could feel that America still might find its best destiny in regional recognition and definition.

TVA's Public Planning

1 | Introduction

In the 1930s, as the nation was poised and waiting to accept new visions of whatever description, the Tennessee Valley Authority (TVA) had an immediate impact. The TVA was the most practical and visible manifestation of the intoxicating "First Hundred Days" of the New Deal, which ran between March and June 1933.[1] TVA shot up, shone, and spread like a fireworks display against a dark and unsettled firmament. Envisioned as a noble gesture, a means of overcoming all demeaning earthly concerns and deprivations, TVA spread its architecture, site planning, and reorganization of the land as a total public reaffirmation at a time when all other, lesser signs of economic well-being were shrinking and disappearing.

The width of its frame of reference made it indisputably American. Thus, the visitor to the Tennessee Valley first might have noticed the rough, casual, isolated log cabins (1). Throughout the first half of the nineteenth century, the log cabin had been featured as an image paraded to celebrate the humble origins of presidential candidates, but by the 1930s they were regarded as obscure and anachronistic tokens of social decay and abandonment. Nowadays they are appreciated as a skillful "clone" type of vernacular, folk architecture, representing the forgotten frontier settlements that were scattered across the earlier country. From the same symbolic object, in three time slots, we have derived three different meanings reflecting three different degrees of national self-confidence. The TVA planners, however, meant the crudity, uneven texture, and small size of the log cabin to be compared in the mind's eye with the smooth continuity and prominence of a dam towering over a valley and closing it off like a Chinese wall. Just beneath the dam, a spacious power house would linger alone, but in touch, while a giant crane would ostentatiously traverse the dam's top (2).

1. Log Cabin of the Logan Hundley Family, Clinch River, 1934. Such families had to be removed upon construction of Norris Dam on the Clinch River. At lower left is an all-purpose wooden sled called a "skids." To the right is another traditional utensil, an iron kettle. Marshall A. Wilson Collection, Tennessee Valley Authority Archives.

The monumentality of the engineering and architecture of the TVA answered many intangible, gnawing doubts and brought new dignity to a definite locale. The luminosity and flash of TVA's "functional" thinking in an "efficiency" mode were put forward to counteract the brooding, long characteristic of the region. Finely tuned to taut reason, as typified by its later serial "flattop" houses, the TVA could appear to triumph with cleancut thinking over the static, clumsy, independent, isolationist, backwoods style of rumination that the log cabin once had signified. TVA's ambition was to embody a new message on a vast scale. The multiplex of log cabins, crossroad hamlets, churches with rudimentary spires, and one-room schools, along with their poorly clad occupants, at last could be contrasted *en suite* to the smooth, towering concrete dams, whirring dynamos, and high-tension electric towers doing their own singing; and to construction villages, people's parks, paved roads, replanted forests,

2. Traveling Gantry Crane, Kentucky Dam, Tennessee Valley Authority, 1939–44. It was the largest TVA gantry and the most graceful. The contrast in scale, texture, and precision between this crane and the log cabin (1) epitomizes the counterpoint between agency and client throughout the evolution of the TVA. Photo courtesy of Tennessee Valley Authority Archives.

and huge man-made lakes. In the wake of national agitation concerning the plight of immigrants, embarrassment had arisen from the sudden realization that here, half hidden, were some of the most "native" Americans, who were nevertheless among the most impoverished and deprived. Misfortune no longer had to be lonely, whether individual or totally familial. Now it could also be collective and mutual, shared by the whole country. Nationally, the conviction was fast forming that revamped laws, administrative regulations, and moderate social programs — the invisible agenda a republic customarily employs to nudge the habits of its citizens into more approved channels — no longer would suffice as a method of adjustment. A newer age would have to make room for a folk heretofore largely left behind in a hidden and dark pocket as a more aggressive type of American had rolled over them going west in the nineteenth century.

The irony, of course, lay in the intention to overcome this group's marginal individualism and familial cohesion through the leadership of a powerful triumvirate of TVA directors, all strong individualists themselves. That a committee of strong individuals was not an auspicious administrative form with which to begin the TVA was recognized and acknowledged even by the first actors in its creation, the intimates of President Roosevelt. The 1934–35 correspondence between David Lilienthal, a member of the first triumvirate, and Frederic Delano, the president's uncle and his mentor in planning matters, gives plain evidence of this.[2] Roosevelt himself seems to have chosen figures who had previously demonstrated aptitude, originality, and initiative, valuing these qualities over what he saw as their complementary talents in engineering (A.E. Morgan, the first chairman), utility regulation (Lilienthal), and agriculture (H.A. Morgan). The triumvirate pattern appears to have been solely Roosevelt's idea. These three men would give the overall coverage he sought; they could pull the enterprise together as they went along. As an echo of this intention, the term "cooperation" became a watchword among the TVA staff.[3] The triumvirate eventually ignored it.

Roosevelt may have been too optimistic about individualism, just as he expected too much of cooperation between directors having a federalized outlook (A. Morgan and Lilienthal) and one dedicated to the old states' rights orientation of the traditional South (H. Morgan). That latter difference was to be a wedge driven many times among directors with ostensibly common purposes. Larger questions of American history — when individualism should yield to group effort; when practical realism should be transformed into a higher idealism (or vice versa); how often national symbolism should be recorded in objects, and over how great a distance — were about to be rephrased and tested in actual landscape plans and in new engineering and architectural works on a very grand scale. The whole enterprise was to be, in the last analysis, indubitably romantic and visionary, partly because its outline and limits were so indeterminate, partly because it was first implemented by such determined individualists, and partly because it was conceived for such a remote and originally pastoral location. During the 1930s, the TVA exhibited all the qualities of a sylvan court of last resort, much as previous designers had become engaged with suburbs, the public park, the national park, and the wilderness areas as repositories of greater American expectations. In that sense, TVA was a reversion to yet earlier first causes, too, back beyond the pastoral, the picturesque, and the arcadian. Roderick Nash has pointed out that the original American reaction to nature was one of anxiety, a feeling that called forth a fierce moral determination. "Wild country had to be battled as a physical obstacle to comfort, even to survival. The uncivilized hinterland also acquired significance as a moral wasteland, a dark chaos which civilization and Christianity would redeem and order."[4] The extreme condition of the Tennessee Valley during the Depression would reawaken that grimmer pioneer resolution about the wilderness, too, provide legitimacy to the crusade of the TVA as well.

The area's physical remoteness gave rise to the attempt, most often on the part of director Lilienthal, to keep all budgets and programs of the TVA apart from the mother city of Washington, D.C. This desire for total separation from Washington was an additional reason for the more ready acceptance of advanced designs for the dams and their accoutrements, seen in the light of an almost obsessive desire to stand free of Classical ornament such as might be applied to a building in the national capital.[5] For East Coast commentators between 1933 and 1938, when most experts came to see the TVA, the project's great novelty lay in the apparition of so much "progressive" architecture in a setting so largely forlorn, anachronistic, and "deserted," one previously beyond their ken. This association of the new and the removed catered to the romantic supposition that the best American effort, "high-tech" or not, could only appear in the receptive lap of nature, with its softening foliage and scintillating light.

So much was conveyed by distant reference. The hum of electricity would be taken up as a rustic incantation with which to exorcise any inhibiting or diluting, if sometimes picturesque, poverty. The brightness of the huge new buildings contrasted with the dilapidation, dispersal, and smaller, more domestic scale of the older ones. The new buildings stood out against the low-lying hills that sloped up to the ancient, dark, shadowed Great Smoky Mountains. Here once more the longstanding utopian instinct, from Methodism to Mormonism, from New England Unitarianism to Transcendentalism, sought cover from spiritual trouble, early and late, in natural greenery. If these instincts had not existed previously, would any juxtaposing initiatives, so typical of the 1920s and 1930s, ever have been taken up in the format of the TVA? The impulse to make closer and closer contact with the indigenous culture, in nature, was exemplified especially by the "pioneer" Norris Village, rustically half in and half out of the woods. Everywhere in it, we can perceive an assumption that the town was primarily a place of a final gesture: the bucolic, removed isthmus of Land Between the Lakes, terminus of the whole TVA endeavor came again embodying the principle — whether religious, romantic, ecological, or technological — of escape into nature. Initially the TVA might have been thought of as a local emergency relief project, but in the longer run we ought to consider it more significant as a larger, tradition-endorsed pattern (3) being tested for the entire nation.[6] At first TVA's all-inclusive "multiple use" phraseology provided sufficient justification for the monitoring of the project until such time as the region might be able to take better care of itself, until it could become truly the "new model." Thus it was anticipated that the Tennessee Valley would grow more native and indigenous, and more national and timeless, at the same moment. The implication was that if this breeding ground proved successful, the regional cradle in the naturalistic bower could be repeated seven or eight times across the country.[7] Roland Wank, the first chief architect, was still nursing that intuition three decades after the founding: "I believe it is possible the Valley will become an example to the rest of the United States, possibly to the world, as the best way man can live in an industrial society."[8] The ripples were to spread in ever-widening circles. As always on the American continent, Nowhere was intended to become Somewhere as often and as expeditiously as possible — somehow, anyhow.

What jammed this regular message to the hinterland, however, was the too-rapid passage of time and events, particularly the approach of the dreaded World War II and the subsequent postwar affluence. An atmosphere of confidence, permissiveness, and international awareness followed the tight-lipped, nationalistic austerity and scarcity of the prewar and war periods. Ingenuously the people themselves remained convinced that better opportunity was coming up somewhere else. During these years it was hard to get a fix on poverty, because it moved so often. Tramps and individual workers were always on the move. Freight cars, old autos called "heaps," and hitchhiking made such restless movement possible. The unemployed left for cities where there were no jobs. The greatest novelty of the TVA itself was the fact that it would try so hard to hold on to its valley residents, an unprecedented gesture that also at-

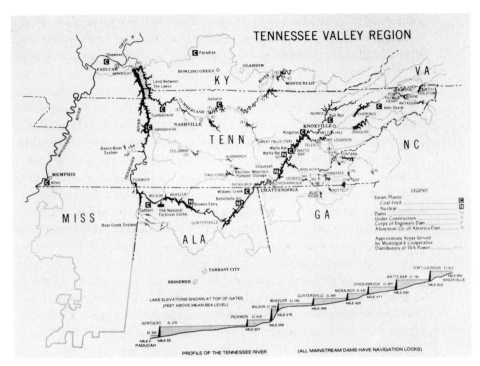

3. The Tennessee Valley Region. The Tennessee River flows southwest, then north-west into the Ohio. To the right are the Appalachian Mountains. To the left is the Mississippi River, not a part of the TVA system but closely related. To the north are the midwestern states, heavily industrialized and richer agriculturally than the South. Finally, to the south lie the states of the once-wealthy Confederacy, simmering with memories. Map courtesy of Tennessee Valley Authority Archives.

tracted footloose outsiders. In 1934, the Florence (Ala.) *Times* noticed the futility of the two-way effect:

> The man who picks up and moves to Detroit in the hope of cashing in on the auto trade revival is apt to get a very painful disappointment, such as was given those who migrated into all parts of the Tennessee Valley when President Roosevelt created the Tennessee Valley Authority. Despite the fact that some of these outsiders have obtained employment in the valley, most of them have not and are being speeded on their way as rapidly as possible, because of the fact that there are still thousands and thousands of persons in the basin who are without work of any kind of a permanent nature. As a general rule it is better to stay at home whether the increased business is at Detroit or Muscle Shoals.[9]

The slow, relaxed, vagabond literary journeys of Thoreau or Mark Twain down the rivers of America were an earlier version of such wandering. The distance always beckoned in America. The rivers' shores, lined with trees and hidden in the mists, would guide the movement along the still waters. Ultimate peace, not often encountered in American life, would reward the sanctity of the voyage. Thoreau had said it: "Time is but a river I go fishin' in." On streams and ponds and in swamps he sought to uncover ancient deposits of meaning, new and more positive rules for an emerging nation. Before the advent of the TVA, the Tennessee River was described by a New York writer as a "sluggish stream, heavy with brown silt, pushing its load down to the Mississippi."[10] This estimate employed a touch of poetic license, since the Tennessee actually fed into the Ohio River. The TVA would clean up and enliven the long run of the river, showing what control by damming and channeling could accomplish. Franklin Roosevelt, like Theodore Roosevelt before him, often indulged in evaluating rivers for their potential for social relief and reorientation. As a young man, he had been a discriminating collector of the lore and prints of the Hudson River, on which he grew up. As governor of New York he maintained a constant interest in the Saint Lawrence River. His generation's wanderlust was much less materialistic, more leisurely and nostalgic than that of the generations after World War II, when the superhighway and jet plane divorced the alert viewer from so much that was of topographic relevance in the areas so quickly passed through or over. The TVA made indigenous sense because it was placed in the waiting middle of the country and made room, at least potentially, for an interlocking network of roads, settlements, rivers, means of transport, and sources of energy that people could see right around them as they went. Of course, the network largely failed to materialize, when later administrators decided that discretion was the greater part of valor.

Rexford Tugwell explained that the first directors of the TVA had came to him in consternation

> because they had not been able to get any answers from President Roosevelt and because they thought someone else might tell them what was in his mind. They had got no enlightenment from him, perhaps because he had not taken the trouble to think out the implication of a valley authority or perhaps because he thought that if he started the thing and gave it institutional shape he could find people who would fight for it and bring it to life.[11]

Open-endedness was for Roosevelt the gist of creativity. Arthur Morgan, the TVA director who later was upbraided for his vague, unwieldy, giant-sized visions, also had a surprisingly unenthusiastic opinion of Roosevelt's reputedly confused outlook. Scrupulously honest, Morgan rendered his opinion after he had been discharged by the president:

> FDR was a man of insight and imagination and was sensitive to the suggestions of others. He threw out ideas and suggestions along the way, but so far as I know, he seldom if ever thought them through carefully. He left his ideas to be filled out by others, which was certainly true of the TVA project. One of his limitations was his inability to give effect to his big, but not clear, ideas.[12]

Some evidence for Morgan's conclusion lies in the fact that Roosevelt fired him for "contumacy," surely not a current and topical word, and one that Roosevelt was rumored to have found in a dictionary and used in the vague hope that the press wouldn't recognize it and would be too indolent to look up. Whether or not he followed sound administrative practice in appointing a triumvirate of such strong-willed directors to begin the TVA, without giving them detailed enough instructions to run it by, he may indeed have confused them enough, as Tugwell and Morgan inferred, to encourage their veering off into self-centeredness and adherence to personal enthusiasms and precepts. Nevertheless, Roosevelt was the only person whose mandate and vision were sufficient to enable him to realize some of the "big dreams" he subsequently pursued; that, even Henry Ford, powerful as he had appeared to be in the 1920s, had been unable to do. Of the TVA, Roosevelt created a compote from which others were invited to draw their favorite conceptual fruit. Thus the history of the TVA could be looked upon less as an outdoor, financial, political, administrative, or engineering tour de force, and more as a grand mental generalization that permitted all kinds of exceptions. In such a frame of reference, the architecture would be less important in respect to its placement, technique, style, and design refinement,[13] and more important as a giant intuitive projection of what American public architecture might someday be expected in general to look like, if it were released soon enough from the customary governmental restrictions on innovation, size, and the frank and open search for excellence. Roosevelt's visualizations may have lacked neatness and certitude, but they did contain promise and élan. The chief executive was not reining in subordinates more imaginative than he; he was urging them on. What has long been acclaimed as a smoothly functioning engineering feat, regulating in electronic synchrony, through flood and drought, the ups and downs of the Tennessee River, may also be important as a singular, monumental embodiment of Roosevelt's, and no-one else's, great good wishes.

The Tennessee River was transformed by a series of retaining dams into long, quiet pools. These pools are epitomized by the last in line, Kentucky Lake, which runs west of and parallel to Barkley Lake (Barkley belongs to the Army Corps of Engineers and not to TVA). The two lakes are large, glistening bodies of water, located in an area that at first had little or no population around to enjoy them. Land Between the Lakes, as one might expect from its name, is held firmly between the two. Kentucky Lake is an impressive artificial body of water 184 miles long. At the northern end of the river, at the official end of the TVA itself, is Kentucky Dam, one and one-half miles wide, the widest of the TVA dams. A photograph of Kentucky Lake, with a lone canoe (4) being paddled close to the uninhabited shore, reminds the viewer of the untoward scale of such projects. In terms of the greater dimension that imagination allows, these projects refer to the other river systems of the center of the country into which the Tennessee flows—the Ohio, the Mississippi, and the Missouri. To recognize these analogies and metaphors the reader would need to be familiar with mid-nineteenth-century paintings such as George Caleb Bingham's *Fur Traders Descending the Missouri* (1844) (5). To view the Tennessee in this wider context brings greater insight because it reveals the static, tranquil, slow-moving tempo, unmatched elsewhere, that is characteristic of mid-continental river life. The people of the central American region were, and remain, hard to understand, because they seem willing to live apart, indifferent to group persuasion, and the fire of competition appears absent from their temperaments. The rivermen in Bingham's painting appear unmoved and unambitious because they look not to have come from anywhere in particular, and not to be going anywhere in particular. Like the native Tennessean in the doorway of a log cabin, they are slightly suspicious of the person looking at them. To the casual observer, they seem almost picaresque, exceptions to the American norm. They travel light and have little equipment with them in the crude dugout canoe. They are not gregarious. With their bear cub, the two are as mysterious and withdrawn as ever were Huck Finn and Black Jim wandering down the Mississippi. But the mystery and enigma progresses, as it does in the Tennessee territory, beyond individual figures to a sense of a huge, moist, and ever-expanding realm beyond, indiscernible in the mist, an empire for democracy, the Louisiana Purchase. It is this impression of languor that most links the Bingham picture to that of the lone canoe on Kentucky Lake (4). What Americans elsewhere do not realize is that in the center of the country there are great hidden distances inhabited by individuals not eager to become part of the pack. Those visiting only the cities there cannot see such folk. When the TVA came, it merely confirmed and enlarged some past assumptions that have been based on loneliness, alienation, independence, and austerity and lived out in very great, mysterious, moist spaces. Regardless of how any particular person might value such a lifestyle, the latter has to be taken cognizance of, because the premises upon which it rests contain not minor, but major symbols of longstanding allegorical significance.

To create a sequence of elaborate, large-scale presentations of native atmosphere, such as the TVA lakes, was very Rooseveltian. Both Roosevelt presidents liked to put on public exhibitions and provide markers along the way. This predilection appears in Franklin Roosevelt's patronage of such art forms as post office murals of rural, historic, and urban life; WPA state and city guides and parks; the founding of the Historic American Buildings Survey; photo essays documenting the people in small towns and on farms; and people's theater projects. Such projects represented a method of visual affirmation, so that the country would wake up, take stock of itself, and rediscover its lost identity as soon as possible, at least within a presidential term or two. The gaze would have to be directed more intensively inward. Stages would have to be set and tableaux organized. The dramatization was another way of returning to the frontier, and to first causes, for positive reinforcement of life on the land and the water. This increasingly inward focus could not be sustained, however. The oncoming need was to be more aware of the rapidly changing political surges of Europe and the Far East, over and above the internal disruptions and subsequent rumblings of the Depression. Attention shifted radically. Everything then came to mean something else — events tumbled over one another in such rapid succession that they appeared to overlap in the actual world. The young men of the Civilian Conservation Corps (CCC) on TVA territory (6) were dressed in coarse khaki uniforms of World War I vintage. And a few of the CCC camps actually were com-

4. Kentucky Lake, Tennessee Valley Authority. The lake traverses Kentucky and Tennessee for nearly two hundred miles. Its peculiar character is suggested by the fact that the shores are uninhabited but not wild. The lake lies in the western part of the TVA system, since there are no mountains and fewer peninsulas in view. It is a typical TVA river-lake, artificial but epic. The lake was built for an America not quite lost but difficult to find again. Photo courtesy of Tennessee Valley Authority Archives.

5. *Fur Traders Descending the Missouri.* George Caleb Bingham, 1844. Mid-nineteenth-century painters provided premonitions of the isolated condition of the Tennessee Valley's people. These traders appear lean and definitely seem separated from others. It was a matter of concern in the New Deal that some groups were in the country but not of it. Rooseveltians wanted to correct that, no matter how reluctant the subjects might be. The setting here is mist laden and extensive, very vague in comparison with the figures. They are hard-bitten. The river is placid. The moment is a selected one. The chained bear cub and his reflection at the bow of the pirogue add an air of mystery that keeps the human figures from becoming comical or picaresque. Courtesy of Metropolitan Museum of Art, Morris K. Jesup Fund, 1933.

posed of World War I veterans who had lost their civilian jobs.[14] World War I innocence and idealism were passed on in a war against economic deprivation and deforestation, in odd contrast to the reverse sequence seen during the Hoover Administration in 1932. At that time the Bonus March of veterans, who paraded from shanty camps on the Anacostia Flats of the Potomac River to Washington, was summarily repelled with tanks, bayonets, fire, and tear gas by active troops under the command of Gen. Douglas McArthur and his second-in-command, Dwight D. Eisenhower.

Some things in that crisis decade slipped too easily out of mind, while others were perceived more clearly. Donald Davidson, a spokesman for the literati of Vanderbilt University, known first as the Fugitives and then as the Agrarians, wished to retreat into previous beliefs in order to sponsor a revival of the older southern culture. At the same time, he offered the briefest and best illumination of the contemporary evolution of the TVA in his 1948 book about the Tennessee River. He saw the initial drama as a long procedural struggle by U.S. Senator George Norris (an Independent Republican from McCook, Nebraska, through which the small river, the Republican, flowed) to rescue the Muscle Shoals development, which, with electricity, had tried to manufacture scarce nitrates for explosives for the federal government during World War I. Norris abhorred the possibility that Muscle Shoals would fall into the "soiled hands" of the private electrical companies.[15] The TVA, he thought, would have "clean hands." Davidson identified the next step after Norris's as the thorough survey by the Army Engineers of the hydroelectric, navigational, and flood potentialities of the river, a survey published in 1930 as *The Tennessee River and Tributaries*.[16] That working study would not have to be redone in order to obtain data when the TVA actually was begun three years later. The Muscle Shoals manufacturing facility and its Wilson Dam in Alabama — a relic that had embarrassed and puzzled everyone during the 1920s, following on World War I — through Davidson's reporting was made to appear a logical beginning, and thus the TVA itself seemed a more reasonable step than it otherwise might have when it eventuated in spring 1933. Davidson followed his interpretation of what had happened to the river up to the early 1930s with the expletive chapter, "At Last—The Kingdom Really Comes!" That long-awaited kingdom (or country) was to be the TVA. The prophet to fling open the gates, the Moses, the catalyst — Roosevelt himself — had finally arrived on the threshold to give all the waiting possibilities body and feasibility. What

6. Civilian Conservation Corps Camp, TVA No. 22, near Esco, Tennessee,
November 1933. Like the Hundley family at Norris Lake or the fur traders on the
Missouri, the CCC enlistees were outsiders to urbanized society. The corps, as the
uniforms demonstrate, was something of an army left over from World War I.
Members did work that was extremely useful to the TVA, including planting trees,
cutting fire breaks, and building roads and parks. The lumber on which the men
are eating will become their winter barracks. A clearing typical for such camps
has already been cut. Photo by Lewis Hine, reproduced courtesy of Tennessee
Valley Authority Archives.

was most typically American in the entire process was Roosevelt's own overwhelming ambition to bring two unlike entities, old and new, the shiny machine and the leafy bough, into close alignment along an extended valley, almost seven hundred miles of navigable distance.

Many have presented the TVA as a cool, professional administrative agency with an almost unlimited capacity for problem-solving. But its capacity for generating hope may have been more important in the long run. That function would have corresponded to Roosevelt's own expectation, at any rate. The TVA took a situation characterized by continuous disappointment and decay and shaped it into something dynamic, inventive, and clean, if too idealistic at times — real but somehow fantastic; practical but straining for utopian perfection; scientific and technological, yet also rolling back into the nostalgic and the arcadian. To contemplate what engineering and the machine might accomplish in an outland setting, within a longer time, was just the kind of reverie that Roosevelt most favored. Protestations notwithstanding, his visions tended to be quite personal, and no-one really could tell what forms they actually would take. Not unlike the TVA, but much less effective, was the Passamaquoddy Bay tidal dams project that, at the same time, he envisioned and activated to no lasting avail for northeastern Maine, another impoverished area so remote that it was almost in Canada. Roosevelt was never lukewarm about such possibilities. In *Franklin D. Roosevelt: The Triumph*, Frank Freidel told how, when Roosevelt was still governor of New York in 1929, he took up a proposal the Republicans had been pushing for a commission "to investigate plans for the development of St. Lawrence Power." He suddenly adopted the study as his very own, "a decision thoroughly startling to the Republicans. . . . Not a person of halfway measures, he did not accept it mildly, but with volleys of skyrockets." It was "a red-letter day," he said, a time for rejoicing.[17] He thought for himself, to be sure, but at the same time he knew a promising idea of someone else's party when he saw it!

Everything that Roosevelt thought appeared to have a penumbra or mandorla about it, "the halo effect of Roosevelt's rhetoric."[18] His domestic effusions, in company with his exhortations, were likely to enchant people, carry them along, sweep them off their feet. The accompanying tragedy, of course, would be the storm clouds of World War II that too soon drew his attention, so that he could no longer sustain his personal involvement with the TVA. This meant that he would have to abandon his incandescent dream of at least one better-organized kingdom within his own country. His sort of courage was based on his view of himself as always ready to set off on a new adventure. When he left on a white steam yacht (another marvel of modern technology) in the gray and discouraging winter of 1940 for the warm blue waters of the Caribbean, he again turned to fantasy, explaining with the greatest good will to the urbane reporters, apprehensively awaiting every word of this leader of the fast-disappearing Free World, that he was leaving for the Christmas Islands to choose Christmas cards, and thence would go on to the Easter Islands to collect colored Easter Eggs![19] As the world situation worsened, his flights of fancy extended even further beyond the continental limits. Such an attitude, assumed so as not to lose zest or courage, would have been called "chipper" then.

Roosevelt's half-shaped, yeasty dreams were always quite pictorial and colorful (pictography was to be the business of the TVA, too). When, at the height of what looked then to be a hazardous war with Japan, he announced a first, token bombing of Tokyo in 1942, he was asked from where the sixteen planes, led by Lt. Col. James Doolittle, had taken off. In reality they had risen from the deck of the carrier Hornet, 650 miles off the Japanese coast, but Roosevelt, irrepressible even in a wartime briefing and determined to add illusions to the news at hand, would say only that they had flown at Tokyo from "Shangri-La."[20] The latter was a mythical mountain-locked lamasery in a hidden valley in Tibet. Shangri-La had been featured in a novel by James Hilton, *Lost Horizon*, published in 1933, the year the TVA was created. In 1937, a film was made from the book, starring Ronald Coleman and directed by Frank Capra, the master of emotional Depression-era escape films. In 1939 came the *ultimate* escape film, one of the first color pictures, *The Wizard of Oz*, starring seventeen-year-old Judy Garland. In black and white film, she was carried up from the drought-plagued plains of Kansas by a tornado to see the wizard in Emerald City. Then, in color film turned on by the wizard's castle, she had to wander "down the yellow brick road" toward it in the company of a tin woodsman, a scarecrow, and an aged lion (an individualistic triumvirate, as within the directorate of the TVA). A short while before, Franklin Roosevelt had come forth with a similar travel itinerary down the long green valley of the Tennessee River, as much inside the country as Kansas. In the Tennessee Valley, as in Oz, gleaming citadels of concentrated power would rise up suddenly ahead of the traveler and seeker in the form of looming dams and power houses. The Witch of the West (Margaret Hamilton) tried to impede the seekers of Oz through the sorcery of electricity. More than other decades of the twentieth century, the 1930s had a sufficiency of images replete with their own adjustability, mythology, color, and animation. They seemed to turn up and come into focus just in the nick of time, as Roosevelt sought to persuade everyone his plans would.

2 | Ideas and Images Behind the TVA

What Kind of a Landscape Was the Tennessee Valley?

There remains an impulse to consider the establishment of the Tennessee Valley Authority as an unprecedented, stellar event, at the outset elevated far above complete subordination to local conditions. Such convictions arose, first, from the project's novelty and boldness. The shock of its impact came, too, from the advent of visible planning and regulation in a society previously loose, open, and thoroughly unused to such draconian measures. The effect was amplified by the extraordinary geographical dimensions of a site that involved seven states. The watershed of the Tennessee River encompassed 41,000 square miles, while the service area with its electrical lines stretched out to 92,000 square miles (3). The size difference between these two domains, each as big as many European countries, occasioned doubts as to just what would be the proper extent of TVA's influence and responsibility. Similarly worrisome were the different characters of the northeast and southwest sections, the mountainous upper section being inhabited by mountaineers and miners while the lower section was occupied by cotton-growing sharecroppers on flat alluvial lands. The ideal had to be imposed on the recalcitrant real.

In a quick, all-encompassing first glance at the map (3), the Tennessee River appears to tie the state and region, east and west, conveniently together. The Appalachian range to the east, with a few mountains over six thousand feet and a high level of annual rainfall, would appear to encourage the river to flow naturally westward to the Mississippi flatlands, at three hundred feet. But the geographical spread takes numerous detours, involutions, and exceptions to itself along the way, bringing in another factor that at times appears to block the accelerated good that the TVA at first wanted to promote. In the Tennessee Valley, resistance to simpler solutions could arise out of the complexity of the setting. The river does not flow easily across Tennessee, but rather moves southwest into Alabama at the outset. Prior to the building of the dams, the river did not flow evenly either, being prone to low water in summer and flooding in January, February, and March, especially around Chattanooga, where millions of dollars of damage was done. About halfway along the river's length after passing Guntersville and Muscle Shoals, at Florence, Alabama, the flow begins to turn directly north toward the Ohio River. So the direction of flow is not essentially from east to west, but rather from north to south, then north again. The river makes, as it were, a gesture first to the Old South and then to the new industrial North. Even at Chattanooga the river turns briefly north and then south as it cuts through Walden Ridge. The Tennessee Valley lay between the two modalities, meditating on what its operational destiny should be, in light of both of them.

The rivers to the west of the Tennessee River, as it ran north, all flowed west down into the Mississippi. Thus the western segment of the state of Tennessee falls much less within the Tennessee River's own sphere of influence. For the last two hundred miles, north up to the Ohio River, the Cumberland River runs parallel to the Tennessee, fashioning the isthmus that runs south and north, Land Between the Lakes. In 1779–80, at Christmastime, John Donelson started his boat, the Adventure, from where Kingsport now stands in northeastern Tennessee. He went south and then north to the Ohio River on the Tennessee, then briefly northeast on the Ohio, and next down the Cumberland River to Nashville so as to bring the families of the first settlers "around." Between the Eastern Valley, next to the mountains and seventy-five miles wide, and the Mississippi bottomlands, lay the Cumberland Plateau, with its thin soil, at two thousand feet; the Highland Rim; and, wrapped within it but three hundred feet lower, the fertile Central Basin, where Nashville is. This Central Basin sweetly was said to be the "Dimple of the Universe." These were the disparate, overlapping geographical elements that the TVA was assigned to recognize and adjust for a mutually beneficial effect.

Breakdown of Other Criteria
Leads to Commitment to the Arts

The TVA policies evolved from so many peculiarly American outlooks that, more closely examined, they appear to have had a historical inevitability. The TVA was unique only in being a sudden culmination of decades of real reform. It could not have come forth in the 1920s, because that was a decade of unprecedented self-centeredness and affluence, accompanied by a relaxation of public standards of morality and responsibility. The TVA was intended to be, above all else, publicly accountable. The options narrowed very abruptly. In the 1930s, businessmen no longer could sustain the preferential positions they had held previously, so the managerial skills of architects and engineers began to be interposed to fill the vacuum. Moreover, those skills, in the TVA, were applied in a cooperative public venture with new ground rules for professional practice involving teamwork. As pioneer austerity returned (along with endorsement of architectural functionalism), freestanding objects needed to be designed both with greater economy and with increased acknowledgment of the human presence around them. The hope was to use the arts to order these utilitarian objects better, as a means of making the experience of life more complete for everyone, for "the People." In the search for this overall, communal satisfaction, the dilemma, of course, lay in how to create common art that would also be high-quality art. Architecture, more even than people's painting, theater, or music, was seen as the medium that most easily might solve the dilemma. Within the huge halls of the power houses (representing the farthest location of the embodiment of artistic dignity) stood the hydraulic electrical generators, like smooth, compact, overturned coffee cans, microsymbolic heart pumps, full of renewing energy and vitality for the national bloodstream.

The scarcity and isolation of certain objects also brought forth a new regional emphasis on minor architectural elements and secondary details to assist in filling the symbolic void. That TVA participated in this disposition caused Frederick Gutheim, one of the more perceptive critics of the time, to remark in 1940, just before the United States reluctantly entered World War II, that TVA's architectural "details themselves are often objectionable: they are often too rich, too fussy, too obviously studied. But such objections are swallowed up in the total composition and the final architectural effect."[1] To emphasize the design of details, the theory went, opened the way for getting directly in touch with people and "upgrading" their circumstances. The slightly overrich, but sincere and purposeful, effect that Gutheim identified was well represented by the visitors' building at Norris Dam (7). The building's imagery was drawn from the small-town movie "palace." The visible surface consisted of troweled marble dust. Over the entrance to "inquiries" was a semi-circular canopy with strings of lights like those on any Main Street small-town theater. Open-ended corners on the building, typical of the structural theory of the 1920s and 1930s, displayed regional crafts. As individualism and private satisfactions became less feasible with the nation's decline into the gray sameness of economic depression, satisfaction from public buildings with elaborated detailing, commonly enjoyed, came to the fore. The bigger world was becoming increasingly dull and monotonous, so the smaller, inside, imaginary world, as represented by this tiny visitors' center, must be made more intriguing. The country was turning into a trinket-and-beads, bartering culture, with increasing emphasis on smaller-scaled satisfactions.[2] Such physical features often would become focal points, as spatial restraint and narrowing inhibition increased due to the expense of gasoline in the Depression and the almost total absence of gasoline during World War II. A saying of the 1930s caught the mood almost exactly: "All dressed up, with no place to go!" The recent monetary constraints and the limited spatial opportunities deriving from them, were well described by Arthur Morgan, the first TVA director:

> Whether under present prevailing methods of planning the recent economic order has reached a limit, whether the mass of the population under that order must settle down to a fairly tolerable peasantry and serfdom, perhaps with a secondhand automobile in every garage, a chicken in every pot, and on rare occasions a week's trip to the nearest national forest or to the relatives, I do not know.[3]

A secondhand car in every garage and a chicken in every pot were what the Hoover administration had promised each family *after* the Depression. Most threatening to individuals such as Arthur Morgan or Henry Ford, however, would have been the possibility that the self-respecting farmer or small-town merchant might sink to the level of a European peasant or serf. America *had* to remain the land of opportunity, because the majority of Americans, including Morgan and Ford, had risen from humble rural backgrounds, a social rise that was less possible in Europe. Similarly, TVA's anxiety over sharecroppers in the South was stimulated by the prospect that they might become serfs, permanently tied to the land. "A week's trip to the nearest national forest" or TVA dam site would be one mode of relief, whereupon the "mass of the population" would receive assurance from the great dam, belonging to them, and a new inspiration from architectural details done well in behalf of the public weal.

The Local Utopian Background

The district of the TVA was so resonant with memories of nineteenth-century utopian colonies as to provide, at the outset, an embarrassment of riches for the reception of this, the latest emerging utopian experiment.[4] From the start of U.S. history, this geographical region had been designated as the one most suitable for acting out European social reforms and building expensive new utopias. In eastern Tennessee in 1880 had appeared the Rugby community for British second sons, under the auspices of Thomas Hughes, author of *Tom Brown's Schooldays*. A socialist community named Ruskin was founded in central Tennessee in 1894. To the west, near Memphis, had appeared the short-lived Nashoba, begun in 1827 for the purpose of educating liberated slaves.[5] Frances Wright, its wealthy patronness from Scotland, had been encouraged to settle in Tennessee by Andrew Jackson and, in the early 1820s, had discussed with Thomas Jefferson her ideas of how she might help blacks. In Kentucky, to the north, there were two Shaker settlements, one east at Pleasant Hill near Harrodsburg, and the other southwest near Bowling Green. Those two colonies represented religious reactions against the early industrialization of that most deplored of English cities, Manchester. Across the Ohio River, at Cincinnati, had been postulated in 1827 the geometric design for the English utopian city of Hygeia for northern Kentucky. In western Illinois, on the edge of the Mississippi River, had been built the city of Nauvoo, belonging to the Mormons and sold by them in 1848 to the French utopians known as the Icarians; just as New Harmony, Indiana, in the Wabash Valley (eventually to be proposed as an auxiliary district to the TVA) had been bought out as a ready-made colony by the Englishman Robert Owen in 1825 from the German Rappites. New Harmony,

7. Refreshment Stand, Visitors' Outlook, East Abutment, Norris Dam. The stand contained a guard booth, refreshment room, and crafts shop, with craft exhibits in the windows. It looked like a small-town movie theater. The walls were covered with local marble dust. Such a countersynthesis of older crafts with new technology would recur in the Tennessee Valley Authority buildings as long as Arthur Morgan was its chief engineer. Photo courtesy of Tennessee Valley Authority Archives.

like the two Shaker settlements in Kentucky, represented a reaction against the specific and growing melancholia of Manchester, England, as did the beginning of Communism in central Europe, since Friedrich Engels had lived in Manchester and also abhorred it. Birkbeck and Flower, two Englishmen acquainted with both Owen and Thomas Jefferson, founded their utopian colony of Albion across the Wabash from New Harmony in southern Illinois. New Harmony also had ties with Nashoba, for Frances Wright was a friend of the Owens and lived at New Harmony for a time. The first manager of Nashoba came from New Harmony. Owenites had founded Arthur Morgan's home town, Yellow Springs, Ohio. So the notion of local geographical opportunity, coupled with utopian colonies, was well established by the time that the TVA arrived in the vicinity. In terms of the previous tradition, the only regrettable aspect of that arrival was that the TVA demonstrated no sustained interest and support for future settlements and ideal colonies. It looked for an older industrialism to be corrected by a newer technology. This made it a less viable demonstration of an alternative manner of sustaining arcadian or rustic residence in and on that land.

Biltmore at Asheville, North Carolina

Even for the inordinate size and spread of TVA's land control, there was a late-nineteenth-century private precedent. The George Washington Vanderbilt estate, "Biltmore," near Asheville, North Carolina, had been begun in 1895 by the architect Richard M. Hunt. It covered 126,000 acres and was also Frederick Law Olmsted's last comprehensive landscaping effort. In charge of the outer forest portion of the estate was the young Gifford Pinchot, fresh from his studies of "scientific" self-perpetuating forestry in France and Germany. He began his long and highly influential career at Biltmore just as Hunt and Olmsted were concluding theirs. For the first time, a great forest preserve received the same professional care and attention as a rich man's private garden. As in the future TVA, the small-holders on what was to become the Vanderbilt estate had exhausted the soil, and therefore had been in a vicious circle, "compelled by economic necessity to exploit fully their scantily productive lands."[6] Vanderbilt and Pinchot's remedy, planting acres of trees in order to revive the played-out land, was a historic "first." But, as with the utopias, the political realities of the 1930s prevented the TVA itself from acknowledging that a model already had been established in TVA territory for future land and forest conservation, since Vanderbilt had been what the New Deal termed an "economic royalist." This is an early instance of the way that refusal to acknowledge precedents encouraged the myth that the TVA was a unique and historically self-sufficient apparition and effort, standing completely free in time.

The notion that there should be an interchange among all the design arts (Biltmore, under the patronage of Mrs. Vanderbilt, became a center for the revival of regional crafts)[7] was common both to Biltmore and the TVA. Harcourt Morgan, of the University of Tennessee's agricultural school and one of the three original directors of the TVA, described his vision of TVA's ideal condition as a "seamless web," and others often used the metaphor as well. His colleague on the board, Arthur Morgan, called for an "all-embracing purpose" for the TVA. However, the possibility of the TVA's reaching that desirable state of unity was greatly inhibited by the fact that there was no one wealthy man to give that purpose the impetus and direction Vanderbilt had given his cohesive aim for the forest at Biltmore. Obviously, too, the pall of poverty and deprivation was much more widespread and indigenous in the bigger TVA territory. Howard Odum, the University of North Carolina social researcher, put his finger on one of the chief causes of the psychological aimlessness and malaise of the whole South when he explained that

> Soil erosion is a problem closely connected with tenancy [not of the Vanderbilt kind, obviously —W.C.]. More than three-fifths of the nation's eroded land is in the South — a national loss of staggering proportions. It is the result of "money crop" farming, absentee ownership, the tenant system, unscientific methods, carelessness of the past, thoughtlessness of the future, ignorance of a better way.[8]

According to Odum, despite the vaunting of interest in the past, the South was careless of it and also thoughtless of the future. It could not be, as Vanderbilt had been, sufficiently cognizant of the long-term future. This chain of indifference and shortsightedness over the generations now would have to be broken by academic intervention, like that made by Odum with his regional research. After that, political initiative would have to be encouraged. Some agency would have to set out to replace the row crops of corn and tobacco with stands of trees, so as to protect the irreplaceable soil that was slipping rapidly down hillsides. Odum recalled that facts such as his finally inspired Roosevelt's New Deal to appropriate action: "From these findings President Roosevelt characterized the South as the nation's Number 1 problem which later changed to the nation's Number 1 opportunity."[9]

Kingsport, Tennessee

Also in the area eventually to be included in the orbit of the TVA was the town of Kingsport, in northeastern Tennessee. It was a model manufacturing town set up by the railroads before, during, and right after World War I to stimulate trade in that location. It was laid out from 1915 on by John Nolen, Sr. With Fredrick Gutheim, John Nolen, Jr., through Senator George Norris, would seek to introduce social legislation into Sections 22 and 23 of the 1933 TVA law. Employed as early as 1915 by the senior Nolen to help him on Kingsport was landscape architect Earle Draper, who was to become head of regional plans for the TVA in 1933. Draper had designed model textile towns in the Piedmont region, running from North Carolina to Georgia, for companies moving down to the South from the declining Northeast.[10]

The Miami River Conservancy and Arthur Morgan

After a bad flood in Dayton, Ohio, in 1913, had been developed the Miami River Conservancy District, the first coordinated flood control system in the United States. It was masterminded by Arthur E. Morgan, engineer, social reformer, Unitarian preacher, and, beginning in 1920, president of Antioch College in Yellow Springs, Ohio. He was a committed educator while there, but he also lived in Yellow Springs so as to keep his eye on the conservancy district. Beginning in 1910, he had developed another engineering headquarters, at Memphis, Tennessee, from which he had directed operations for recovering fertile land lost due to flooding in Arkansas, Mississippi, Texas, and Louisiana by building levees, channels, and bridges, and by draining wetlands.

Morgan was the first director of the TVA appointed by Roosevelt, who said that he had read *Antioch Notes* and liked the train of thought in them. Morgan believed that it was probably Eleanor Roosevelt "who acquainted him with *Antioch Notes* and with my work."[11] Many of the talented employees used in constructing the TVA dams had been educated at the University of Illinois and very early seasoned in Morgan's Memphis and Dayton engineering offices.[12] For the Miami Conservancy District, under Morgan's supervision, five earthen dams had been built, a village removed, and highways and railroads shifted to curb the flooding. He had workers' villages, with provisions for families, built at the various dam sites.

Henry Ford's Initiative at Muscle Shoals

Arthur Morgan, like Henry Ford, was a Midwestern individualist, a self-made character out of the late nineteenth century. Ford was said to have admired Morgan because he didn't have a college degree either. They had a common interest in the revival of old-fashioned village life. Ford was useful in forcing issues to the benefit of the future TVA, although of course Ford's role could not be admitted later, because he, like Vanderbilt, according to the premises of the New Deal, was an economic royalist. Ford joined Morgan in a dedication to the decentralization of industry, so as to blend it with agriculture. Ford's outsized, vivid aspirations made the visions of fictional auto manufacturers, such as Sinclair Lewis's Dodsworth (1929), pale by contrast. In 1921, immediately after World War I, Ford wanted to buy the Muscle Shoals Dam and power facilities from the government. These works had been built around the shallows of the Tennessee River in Alabama and included the Wilson Dam, a steam plant, two nitrate-producing factories (one for the Cyanamid and the other for the Haber process), and a war workers' village. Ammonium-nitrate manufacture required large quantities of electrical energy. A wartime anxiety had arisen that nitrates from Chile for explosives might be cut off. The two plants would have been reprogrammed by Ford to manufacture cheap nitrate fertilizer for southern farmers. Harcourt Morgan would eventually convert the plants to phosphate instead, because he believed phosphate better for southern soil. Contemporary and later critics of Ford have never discussed the fact that he hoped to obtain Muscle Shoals because he had become interested in the devastating affect of the boll weevil on the cotton crop. A theory of the time was that if the cotton plants could be forced to grow faster by the application of nitrate fertilizers, it would interrupt the weevil's life cycle.[13]

Ford envisioned three dams in the Tennessee River. When completed, the older Muscle Shoals Dam of Wilson actually became the middle unit in the TVA three-dam sequence of Wheeler, above Wilson, and, down-river from Wilson, Pickwick. Ford wanted to locate small new towns along the Tennessee River, which would have become lakelike for seventy-five miles. Workers would commute from them to Ford factories. There were to be, in addition, subsistence farms of five to eighty acres, whose owners would be taught to supplement their factory wages with improved methods of food cultivation. Demonstration agents from the company would show them how to operate the mechanized Ford farm equipment, particularly tractors, that could be rented from the equipment pool of the company.[14] Ford estimated that a factory worker might raise five hundred dollars worth of food per year for himself.[15]

Later explanations failed to take into account the fact that Ford had a definite pattern of settlement, in concert with technology, already in mind when he approached the Muscle Shoals opportunity. The balance, grandeur, and all-inclusive atmosphere of Ford's vision never truly would be equaled by the plans of later TVA thinkers, except perhaps those of Arthur Morgan or Franklin Roosevelt himself, who could sense great sweeps of the imagination in others besides himself. Roosevelt wrote to Ford on 8 November, 1934, stating that he had been thinking about getting people out of the cities and into the country through giving more attention to the location of smaller industries in small towns. He wanted Mr. and Mrs. Ford to visit him in Warm Springs to talk such ideas over, although the meeting seems never to have taken place.[16] This invitation was forthcoming only a year after Roosevelt had initiated the TVA. Once again, as with George Washington Vanderbilt's Biltmore estate, the historical, philosophical, conceptual, and capitalist chain of precedence led from a titan of industrial wealth to the public interest of the New Deal; but later chroniclers have failed to acknowledge it, because of a conviction that such parties would have had to be in inevitable opposition to the New Deal, regardless of the intrinsic value of their ideas and actions. That is, everyone appears to have believed that except Franklin Roosevelt and his uncle, his planning mentor Frederic Delano. In the next year, 1935, Delano wrote to David Lilienthal, energetic director of the TVA, pointing out a parallel between Roosevelt and Ford: "When people come and talk with me about the awful things that the President and his associates are doing to the electric power business, I tell them that I think they are doing for the power business what Henry Ford did for the automobile industry."[17]

Since Ford, Roosevelt, and Arthur Morgan exhibited such pronounced views, moved so rapidly, and exercised considerable power and influence, there has been a tendency to oversimplify descriptions of both their personalities and their motivations. All three men had very complex characters and sometimes appeared to contradict themselves. In the Muscle Shoals debate, it sometimes was implied that Ford was a villainous giant out to steal from the government.[18] It was never mentioned, however, that he had, a few years earlier, actually begun to establish small factories in the countryside and on streams within twenty miles of his main industrial headquarters at Dearborn, Michigan. The Muscle Shoals bid was an extension of that initiative. Reynold Wik reported, "With some of these objectives in mind, Ford in 1918 acquired an old grist mill known as Nankin Mills on the Rouge River. Other plants were built to demonstrate that small streams could be used profitably for industrial purposes. As a result, small factories at Plymouth, Newburg, Waterford, and Northville manufactured drills, taps, and valves."[19] Further, explained Wik, "In the early 1920s there were seven of these plants in operation, with the number increasing to twenty in 1934, employing 2,400 parttime farmers."[20] Ford's proposed extensive settlement pattern for the Tennessee Valley around Muscle Shoals, then, was no spur-of-the-moment inspiration, and was not directed exclusively to the potential of the two nitrate factories and Wilson Dam, as his opponents appeared to believe, but was part of an overall blueprint for better physical organization of the American land and its resources to achieve more efficient agricultural and industrial production. Because of the coincidence in time, the early 1920s, Ford's image of a linear city of seventy-five miles along the Tennessee River might well be thought to have derived from the much-praised French linear city drawings of famed architect Le Corbusier (who did come to look into the TVA). But in actuality, if there was indeed such an influence, it is more likely to have come from a native source—Edgar Chambless' "Roadtown" scheme of 1918, wherein business, industry, and residences would be combined in one great linear structure stretching across the open countryside. It is known that W.L. Spillman, of the Office of Farm Management, U.S. Department of Agriculture, had acquainted Ford with the Chambless scheme.[21] With Gifford Pinchot and Sen. George Norris, chairman of the Senate Committee on Agriculture and Forestry, however, Ford's purposes had no buoyancy, no viability, no exchange value, and no integrity; and they saw to it that his attempts regularly were thwarted. In October 1924, upon withdrawing his offer to buy Muscle Shoals, Ford tartly observed that what would have taken a week to wind up in private business had required three years (1921–24) to negotiate with government — and wasn't finished yet.[22]

No Ruhr Region Wanted

Misunderstandings among visionaries were rampant during the 1930s as to reform intentions, and this caused idealizing loyalties and joint reform efforts to break down and diverge. This can be readily seen in the Tennessee Agrarians' opinions of Henry Ford. They believed that he "personified large-scale industrialization."[23] They disdained "his assembly line and the regimentation of his employees," but they ignored (if they ever knew of it) his very similar interest in "small-scale, rural factories" and economic and human decentralization. With Ford, however, they believed that the "ownership of machinery by the individual farmer," when fostered by electricity, "would increase his standard of living and economic independence."[24] At the time, neither the Agrarians nor Ford listened to the other.

During the 1920s a considerable respect had built in America for Germany's Ruhr region replan by Robert Schmidt. It was a program for the reconstruction and development of the steel and iron works of that district in the wake of World War I, a plan that was to prove highly useful in getting ready for World War II. By 1937, Chattanooga editor George Fort Milton, a supporter of David Lilienthal, was demanding that "the Tennessee Valley should become the American Ruhr."[25] Three years before, Arthur Morgan had been traveling along the valley trying to persuade local audiences that no good could come of emulating the heavy industrial development plan of the Ruhr. He, like Henry Ford and the Agrarians, wished for small-scale factories, where individuality and character register on the items produced. "If the people of this Tennessee River region will decide, 'We are going to develop individuality, we are going to put our character into our products, and not make the region the Ruhr of America,' they can win."[26] To Kiwanis and Rotary clubs he declared, "If we can see our way rightly for the future of the Tennessee Valley, we would have some mass production for national consumption, some production of finer goods that the rest of the country is too busy to make; and I think to some extent we would make our own economy and would free ourselves from these deep-worn channels of trade that draw our resources off into the great centers and leave us poor at home."[27] The Tennessee Valley would have to be a *different* place, he said. "We do not want merely to duplicate here on the Tennessee the industrial set-up that has broken down in Detroit and Pittsburgh and the other cities that have sent back penniless the quotas [of workers] they were at such pains to draw from these parts during post-war prosperity."[28] The cumulative message delivered by these key thinkers was that the valley should have had its own alternative kinds of factory, settlement, and housing; but after the hiatus of World War II, such broad questions about the quality of life, and about the texture and disposition of the land and its settlements, hardly ever were raised again. The only exception of any magnitude was the proposal for a new town, Timberlake, in conjunction with Tellico Lake and Dam during the 1970s. It also was to be a linear city on the water.

Franklin Roosevelt and Pinchot's Chinese Village

Ideas, images, metaphors, and similes can never be exhausted as preliminaries for the TVA. They had a cumulative effect. Franklin Roosevelt related that, as early as 1910, when he was a freshman New York legislator and chairman of the legislature's Forests, Fish and Game Committee, he invited Gifford Pinchot to speak before that relatively unimportant group. Roosevelt claimed that he did it to enliven the otherwise dull proceedings. Pinchot, in a burst of enthusiastic interpretation, showed slides of a valley in China, seen both in a sixteenth-century painting, and in photographs taken four centuries later. Where the painting showed a happy, prosperous, tranquil Chinese village surrounded by fertile fields, the photographs pictured utter desolation as a result of overcultivation and erosion. The trees had been cut down, the topsoil had washed away, and the inhabitants had all fled. Pinchot's harsh and disconcerting description of events in China undoubtedly could be traced to an earlier analysis of another very ancient civilization, that of the Biblical lands by Yankee George Perkins Marsh, published first in 1864, under the title *Man and*

Nature, and later revised as *The Earth as Modified by Human Action* in 1874. Marsh's conclusion was that the early peoples of Asia Minor had denuded their land and inadvertently turned it into a desert by cutting their trees (the Cedars of Lebanon) and overgrazing their fields. They had not paid enough attention to irrigation. A report published in 1878 cited Marsh and in turn affected Gifford Pinchot's views. That was Major John Wesley Powell's *The Lands of the Arid Region of the United States.* Powell was a veteran of the Civil War, wherein he had lost an arm. It was in reaction to the destructiveness and dislocation of that war and the subsequent geographical expansion across the continent, much more than as carry-overs from the cheerful Jeffersonian Agrarianism of the pre-Civil War era, that such essays came to be written. Roosevelt liked the imagery of picture puzzles and the irony of anecdotes for arousing people to good deeds, and he retold the Pinchot episode frequently.[29] One can detect its haunting presence in the creation of the TVA. Southern author James Agee connected the same theme to the erosion of the Tennessee Valley that the authority would be attempting, in the early 1930s, to overcome.

> If the U.S. really learns to take care of its land, and really cares to preserve it for the future, it will be just about the first civilization that has done so, and the time is spoiled rotten for beginning. . . . These past four generations, we have wrung the very blood from the land and shipped its health to market and seaward by the sewers and left it exhausted and misplanted for the rain to do the rest. Left to its own devices and the rain's, that whole land could be desert before another century had passed.[30]

The thought of whole civilizations transmogrifying a region into arid deserts haunted Agee, that Knoxville native. Arthur Morgan similarly was troubled by the postscriptions of Marsh and others, and he brought up the unfortunate agricultural decline of Spain and Greece, as well as China, in his initial talks all along the Tennessee Valley.[31]

> Millions of acres [of the Tennessee Valley] are now barren clay hillsides, cut with gullies and abandoned by agriculture. It has taken only a little more than a century to produce this result in the Tennessee River area. If the process continues as it has in some other countries, such as southern Greece, parts of Palestine, and parts of China, great areas will become useless for cultivation.[32]

Contrasting to the Biblically scaled exploitation and erosion of the land, following on the Civil War, came a rapid rise in the material standard of living of the Northeast and Midwest. Upward mobility in economic terms was assumed to be the well-deserved result of an on-going, vigorous democracy. A hierarchy of expectations was being established for the acquisition of private possessions, and that hierarchy could be manipulated to legitimize just about any social result. Nature was to be subordinated to the means of extraction, production, and consumption. In light of this momentum from the 1860s on, it seemed unlikely that the Depression of the 1930s would be regarded, even at the time, as anything more than a passing shock, a temporary hiatus. That assumption was a major reason why it was so difficult for even the federal government to recognize TVA for what, at the time, it claimed to be—a long-range project. Expectations for the TVA then were based on such brief experience that the proposal to manufacture electricity, which at first did not appear at all saleable, was hardly investigated, while the under-consumption of material goods in the South was brought up repeatedly. In that short versus long-term perspective, the TVA dams really were monuments to the aspirations for increased southern consumption more than for increased production, and their visitors' areas accordingly were emphasized and given expanded functions. Leaving nature untouched around such units, to allow for observation and recreation, implied a pledge that the TVA henceforth would — in a leisure consumption frame of reference—work toward the renewal of the ideals and spirit of democracy and the provision of cleancut, invisible electrical units as a substitute for dark, cumbersome, gritty material goods. If the aim had been only to stabilize and save the green valley, it could have looked much more like a national park, forest, or wilderness. The buildings never would have been so much emphasized; they never were in national parks. Those dramatic TVA building and landscape effects in concert sounded a call for the arts of engineering, architecture, and landscape planning to reconcile extant differences among the various social, and more hidden economic, forces.

It was in the Roosevelt spirit to display compassion at the same moment it was assaulting problems with giant, action-oriented programs, vibrant with pictures of power. The Roosevelts dealt with the arts much as they did with the sciences, with a degree of over-application. In the TVA, for the arts this approach would be represented most clearly by the construction of Norris Dam, the Norris parks, Norris Lake Forest, and Norris Village. For the sciences, it would be by the designation and construction of the Clinton Engineering Works at Oak Ridge, which, although not officially a part of the TVA, nevertheless were located where they were because of the availability of power from nearby Norris Dam. The Roosevelts showed their support for conservation by flexing their political muscles and by making their efforts highly visible. Theodore Roosevelt pushed through the Panama Canal (1903–1914) and had his picture taken in 1906 aloft in the cab of a gigantic steam shovel there (8), a great and powerful figure about to do some practical good by cutting an east-west waterway through jungle entanglements by preeminently mechanical means, as "foreign powers" had not been able to do. Franklin Roosevelt's scenario, in the case of TVA, was hardly different. Samuel P. Hays has said about the crusade of Theodore Roosevelt for conservation through the gospel of efficiency:

> That crusade found its greatest support among the American urban middle class which shrank in fear from the profound social changes being wrought by the technological age. These people looked backward to individualist agrarian ideals, yet they approved social planning as a means to control their main enemy — group struggle for power. A vigorous and purposeful government became the vehicle by which ideals derived from an individualistic society became adjusted to a new collective age. And the conservation movement provided the most far-reaching opportunity to effect the adjustment.[33]

The formula was very close to that for the initiation of the TVA: reorganize, give them something to look at, and then place strong, individualistic directors in charge to make adjustments for "a new collective age." If the people could not be tamed, technology and nature could, to set them a better example.

Gifford Pinchot provided a personal tie between the two Roosevelts, for he knew them both, and he used the manifestations of trees and nature as means to solve almost any human problem. His motto of "multiple use" in conservation later was invoked by the TVA staff in its popular publications as often as Harcourt Morgan's slogan of the "seamless web." The revelation of how Pinchot came upon his motto is particularly instructive. It was a motto so unifying, so universal, that one could not resist recognizing it. It pulled everything together. It was typical of that generation that this "mystic" revelation would not take place in an outer wilderness, as might be expected, but rather amid the trees on Pinchot's regular horseback ride through modest-sized Rock Creek Park in Washington, D.C., in 1907. "What had the water in the creek to do with the trees on the bank?" he asked.

8. President Theodore Roosevelt in the Cab of a Bucyrus Steam Shovel from Ohio, Panama Canal, 1906. The Roosevelts believed, after Thomas Jefferson, that governors and presidents had responsibility to the physical as well as the fiscal environment and should let their personalities show in it. Theodore Roosevelt Collection, Harvard University.

> Suddenly the idea flashed through my head that there was a unity in this complication — that the relation of one resource to another was not the end of the story. Here were no longer a lot of different, independent, and often antagonistic questions, each on its own separate little island, as we had been in the habit of thinking. In place of them, here was one single question with many parts. Seen in this new light, all these separate questions fitted into and made up the one great central problem of the use of the earth for the good of man.[34]

Putting it all together, however vaguely or mystically, was to be an aim of the TVA, too.

In that same year, 1907, Pinchot and W.J. McGee, working together, persuaded President Theodore Roosevelt to appoint a commission and issue a report on the inland waterways of the country. This led in 1908 to the White House Conference on Natural Resources, known more officially as the "Conference of Governors." The first Roosevelt's messages about the waterways foreshadowed the second Roosevelt's interest in such ventures as the TVA and the National Resources Planning Board. But the first Roosevelt was, if anything, even more determined to be comprehensive, literal, and thorough, in the uncompromising Pinchot spirit. In his message to Congress of 8 December 1908, Theodore Roosevelt declared that "the plan which promises the best and quickest results is that of a permanent commission authorized to coordinate the work of all the Government departments relating to waterways, and to frame and supervise the execution of a comprehensive plan. . . . We should have a new type of work and a new organization for planning and directing it. The time for playing with our waterways is past." The thought of a "permanent commission" seriously to regulate the waterways may have led to Franklin Roosevelt's appointing a commission of three to run the TVA.

It was McGee who in 1901 had instigated a report that anticipated even more closely the ultimate form of the TVA project. The report was on "the Forests, Rivers, and Mountains of the Southern Appalachian Region."[35] That "Appalachian Report" was a link in a series of events running from the Forest Reserve Act of 1891 to the Weeks Law of 1911 for the purchase of forest land both in the White Mountains of New Hampshire and in the Southern Appalachians.[36] The newspapers praised the acquisition of such eastern forest preserves as a timely balance for what had already been purchased in the Far West. The importance of the Appalachian reserve for concentrating recreation for the eastern United States was also noted by the papers, as would occur later with their remarks about the concentrating character of the recreational facilities of the TVA. In 1930, the Great Smoky Mountain National Park was initiated; in 1933 this Appalachian reserve was joined by the territory of the TVA to the west and south. Nearby Asheville, North Carolina, was originally within one night's railroad ride from everywhere in the Midwest, East, and South, from "Chicago, New York, or New Orleans."[37] This fact was one reason why Vanderbilt, the railroad owner, in the 1890s placed Biltmore at the crossroads of Asheville.

The "Appalachian Report" of 1901 acknowledged the area's heavy rainfall and the sharp downward slope of the mountains that led to problems of erosion of the hill farms where (9–10), "the trees are girdled, and for one, two, or three years such a field is planted in corn, then a year in grain, then one or two years in grass; then the grass gives place to weeds, and the weeds to gullies."[38] Since there were no glacial soils and no natural lakes in the immediate vicinity to absorb the runoff, the necessary step toward flood control in 1901 would be "the building of dams at intervals across the deep, narrow gorges"[39] of the rivers, including the Watauga, Nolichucky, French Broad, Pigeon, Little Tennessee, Tuckasegee, Hiwassee, and Holston — all of which, except the Tuckasegee, eventually had TVA dams built across them. Natural scientists, such as Asa Gray and Nathaniel Shaler, both of Harvard, were quoted in the "Appalachian Report" as favoring the conservation of the locale, the former because he believed that more species of hardwood trees were preserved there than in any other part of America or Europe, and the latter because he felt that, whether or not such legislation could be termed "socialistic," when private enterprise itself could not provide, the government was entitled to step in to shoulder the larger environmental responsibility.[40] Scenic beauty and picturesque views were preoccupations more often in the "Appalachian Report" than in later TVA documents.[41] According to McGee, the demand around 1900 for an Appalachian reservation came mainly from "the mountaineers of the region, a virile and farseeing race."[42] By the Depression era three decades later, much less positive views of the mountaineer and his legendary self-reliance were being put forward.

Frederic A. Delano, Longtime Catalyst

In spring 1933, President Franklin Delano Roosevelt's uncle, Frederic Delano, gave his nephew a then-scarce copy of the "Appalachian Report."

At this time, of course, the legislative message for the TVA founding was being crystallized. Delano admonished his nephew to read it carefully, "as you will find it was a report prepared by your distinguished kinsman President Theodore Roosevelt on the conservation and developement of the Tennessee Valley."[43] Delano failed to note that the report was about Appalachia alone and not the whole valley. Receipt was acknowledged on the flyleaf in Franklin Roosevelt's own hand.

Like his nephew, Delano had a bold and courageous countenance (11). In 1905, when he was thirty-two, he had been elected president of the Wabash Railroad, which ran through Indiana to Chicago. Through his 1904 effort to consolidate twenty-six trunk lines into a union station (an eternal hope in Chicago),[44] he had become increasingly impressed by the architectural and planning ability of Daniel Burnham. He was one of a committee of two from the Chicago Commercial Club and the Merchant's Club who persuaded Burnham to devote the last years of his life to the Chicago Plan of 1908–1909.[45] As a secondary consequence, the "Delano Plan" for the railroads came to be incorporated into Burnham's plan. In 1921 Delano became head of the Committee of One Hundred of the American Civic Association, appointed to promote the 1901 McMillan Plan for Washington, D.C., that had been laid out by Burnham and Charles McKim, working as a team. Next he became chairman of the National Capital Park and Planning Commission, and finally he guided the National Planning Board, set up by his nephew in 1933 and "killed off" by Congress in 1943.

The customary view has been that Chicago's City Beautiful Movement (out of which the Columbian Exposition of 1893, also directed by Burnham, and the Chicago Plan of 1908–1909 came) led nowhere, in fact arrived at a complete dead end in the national design mythology. But the City Beautiful Movement appears to have constituted another of those powerful but hidden inspirations of the TVA. Hence it is only a minor surprise to read a speech that Franklin Roosevelt gave in New York City in December 1931, in which he said he was occupied by the "elements that have developed in regard to planning since the days nearly thirty years ago when Mr. Norton and my uncle, Mr. Delano, first talked to me about regional planning for the City of Chicago." He confided that "from that very moment I have been interested not in the planning of any one mere city, but planning in its larger aspects."[46] This ex-urban initiative caused his vision to brighten and splay. Charles Dyer Norton

9. *Opposite*: Tennessee Farm Erosion. Sheet and shoestring erosion are present. They can occur on relatively flat land. Photo courtesy of Tennessee Valley Authority Archives.

10. *Above*: Erosion of a Steep Hillside with Cows. W.J. McGee used such photographs in his Appalachian report to persuade Theodore Roosevelt and others to establish Appalachian national forests. This picture represents slip erosion. Cows have come to look like mountain goats. Photo reproduced from *World's Work*, November 1901.

had worked with Burnham in Chicago. He had also initiated the Regional Plan of New York City of 1921–31, supported by the Russell Sage Foundation. Delano, as a member of the Sage governing board, had backed Norton's proposal in February 1921. Raymond Unwin, the architect of the Garden Cities, was brought from England as a consultant to the New York Plan in fall 1922. Thomas Adams, the first manager of the English garden city of Letchworth, designed by Parker and Unwin, came in 1923 to be the project manager of the New York Plan.[47] These men were bona fide architects and knew technology, Unwin had started as a mining engineer, but they were less narrow-minded than many Americans of the time and could move from environmental strategy to technical strategy. Franklin Roosevelt was similarly openminded and more receptive to ideas than to procedures. He looked to native "grand schemes" like that of the City Beautiful Movement and didn't want to be too involved with the details. He seems to have been warmed by thoughts and intuitions that made him feel more humane. These intuitions then could be applied outwardly as needed to whatever might appear to be ailing the country. This attitude, so full of good will, might provide too little guidance when systems or proper procedures had to be implemented.

The Round Table on Regionalism at the University of Virginia, 1931

In earlier days, with friends Franklin Roosevelt had formed a private corporation to put up $100,000 to buy outworn agricultural land in upstate New York to replant with trees, to see if that would be profitable many years later. While he was governor of New York, he spoke often and unashamedly of the advantages of decentralizing big cities by means of the wider distribution of electricity, particularly if the Saint Lawrence River, on the northern edge of New York State, could be made to turn out abundant hydroelectric power. Small wonder, then, that in July 1931, while still governor, he was invited to be the keynote speaker at the Round Table on Regional Planning at the University of Virginia, a meeting held under the auspices of Louis Brownlow and Clarence Stein. Other New York regionalists such as Henry Wright and Lewis Mumford were there, as was the southern sociologist Howard Odum of the University of North Carolina, who had a longstanding interest in regionalism. Frederick

11. Frederic A. Delano, 1910. Delano was then in his mid-forties and president of the Wabash Railroad, headquartered in Chicago. Important for the future TVA, he was at the time a great admirer of Daniel Burnham and his City Beautiful Movement. In 1933 he gave Franklin Roosevelt, his nephew, a copy of the McGee Appalachian report (10). His visage resembles that of his more famous nephew. Photo reproduced from *Notable Men of Chicago* (*Chicago Daily Journal*, 1910)

Gutheim and John Nolen, Jr., who would intervene in Senator Norris' preparation of Sections 22 and 23 of the TVA law, were present, too. Radical economist Stuart Chase spoke. In retrospect, some have come to believe that this meeting was the authentic seedbed of ideas about the TVA.[48] By way of contrast, Lewis Mumford regarded the meeting as the last important gathering of the New York-based Regional Planning Association of America, which he and Stein had been instrumental in founding.[49]

Everyone at the Round Table took an individual approach to regionalism. Gifford Pinchot, by then governor of Pennsylvania and inordinately busy, had been billed to elaborate on an "Industrial Power Policy for the South," but he did not arrive. Howard Odum discussed the subject he was already well into, "Sociological Aspects of Regionalism." Benton MacKaye—who was to carry out studies of the "folk" of the Tennessee Valley, was affiliated with Stein and Mumford's Regional Planning Association and had contributed first input to the New York State Plan prepared by Henry Wright and Clarence Stein for New York's Gov. Alfred Smith—discoursed on "The Cultural Aspects of Regionalism." MacKaye said that he wanted to rescue the "town form" from the "fatalistic," noncreative, unimaginative school of planning. Raymond Unwin was his hero, and the New England village his ideal community, his "emphatic type" of community. Besides an overall plan on paper, what future deliberations concerning the TVA were most to lack was a continuing interest in experimentation with model towns such as MacKaye, Mumford, Stein, and Arthur Morgan would have liked to see in the TVA region.

Charles W. Eliot II spoke on "Historical Considerations in Regional Planning." Despite the generic title, he discussed only the historical area within a twenty-mile radius of Washington, D.C.—his and Delano's main interest in their work with the National Park and Planning Commission. Eliot, in a surprising extension of his thinking about the capital, asked in an aside if "perhaps the next step in regional planning is national planning?" He wanted to have as "a first step toward a National Plan the formation of a research staff with authority only to collect data to place the facts of the situation before the interested agencies and to keep the story before the public until Congress and the Administration adopt some coordinated program which can really be called a National Plan."[50] How much did this query reflect his own view and how much that of Delano or perhaps even Franklin Roosevelt?

Lewis Mumford, "critic of Architecture and Letters," called his paper simply "Regionalism." He endeavored to define the difference between metropolitan planning (the Russell Sage Plan) and "genuine" regional planning (his, Stein's, and Henry Wright's ideas for New York State). Mumford's last sentence was: "Regionalism is only an instrument: its aim is the best life possible."[51] Electricity was not mentioned by him.

The preceding meeting in 1930 had featured talks just as prophetic as those of 1931, but they are even less known. The general topic for 1930 had been "The Economic and Industrial Development of the South." Preston S. Awkright, president of the Georgia Power Company, spoke on "Industrial Power Policy for the South." He noted "There is a hue and cry over the nation for socialization of the electric industry. It did not originate in the South but it has unquestionably affected the attitude of the public toward us in the South."[52] Awkright also quoted from Anne O'Hare McCormick's recent article in the *New York Times*, in which she caught the luminosity of the ubiquitous bodies of water that graced the South: "The South is islanded in water. The mountain torrents are at its back, the sea spreads at its feet and all the ground between is raked with creeks and rivers and treacherous with marshes, like a landscape in moire. It is inevitable that progress should be related to water power, and it is, so closely that it is hard to determine which is the cart and which the horse, individual initiative or overabundance of hydroelectric energy."[53]

There it was! All the potential power packed into the soft and lyrical landscape of the South, before anyone had a word to say about the structures and practicalities of the TVA! It would be only a short time before the lesson of McCormick's aqueous ecology would reappear in places like the visitors' hall of Pickwick Dam, the second TVA dam built.[54] There, in the checkerboard visual style that became typical of TVA exhibits, Harcourt Morgan made sure that what he thought of as his personal message was communicated by beautiful photographs and telegraphic captions: "50 inches of rain in the Tennessee Valley — controlled — speeds protective cover growth — makes deep river channels [legislated at nine feet for the TVA — W.C.] — generates cheap electricity." Below that line of illustrations was another strip with a contrasting message: "50 inches of rain — uncontrolled — erodes naked lands — floods town and country — makes people poor." The last phrase "makes people poor," was framed in a manner typical of the whole New Deal, as the caption of an eloquent photograph showing above an isolated elderly woman (12) in an old dress looking out from her rickety porch at an entire hillside of erosion. It was apparently necessary for the viewer to be told that such TVA people were both fragile and vulnerable. Even in this locale of unrelieved desolation the poetry of the place and moment could still be rescued by the eye of the truly sensitive TVA cameraman. Harcourt and Arthur Morgan had both been teachers, academics (one reason why Roosevelt recruited them?), and one sometimes feels that their messages are a little pat, too condensed, putting the emphasis on saying rather than doing. However, long before Anne O'Hare McCormick and Harcourt Morgan penned their aquatic salutes, the first overlyrical and the second somewhat cryptic, the liquidity inherent in the TVA region had been acknowledged in the "Appalachian Report" of 1901: "Upon these mountains descends the heaviest rainfall of the United States, except that of the North Pacific Coast. It is often of extreme violence, as much as 8 inches having fallen in eleven hours, 31 inches in one month, and 105 inches in a year."[55]

James C. Bonbright, professor of finance in the Business School of Columbia University, in 1930 spoke on "The New Role of Public Ownership in the Power Industry." Bonbright had been appointed to the New York Public Service Commission in 1929. Felix Frankfurter of Harvard had recommended him to Governor Roosevelt.[56] One of the commission's purposes was to consider possibilities for the state ownership of power houses and lines along the Saint Lawrence River. Bonbright did not specify in

12. Woman on the Porch of a Farm on the Old Sevierville Pike near the Smoky Mountains, February 1939. This picture at Pickwick Dam was in one of H.A. Morgan's didactic shorthand displays. The caption of the picture was: "Rain Makes People Poor." Only lava could be more inexorable in its progress than this earth erosion. Often, as here, the Depression was presented as "one against the world." Photo by Charles Krutch, reproduced courtesy of Tennessee Valley Authority Archives.

his talk how this public ownership was to come about, but it is obvious from his text that he strongly favored the public development of hydroelectric power because of its need for larger capitalization over day-to-day operating expense. He indicated that water rights, which had been conferred by the government on too many private companies, were being used to obtain higher rates from consumers than they could have had if that conveyance had been more openly recognized. Bonbright used the example of rates levied on the American, as contrasted to the Canadian, side of the Saint Lawrence River as the worst example he knew of. On the American side, the legally permitted annual return on investment was 8 percent, but, because of the government gift of water rights, actually was closer to 10, 12, or 14 percent. The contrast between U.S. and Canadian costs to the consumer for electricity on this river was what caused Senator Norris to become so exercised that he finally wrote the TVA bill. Bonbright referred to Boulder Dam in the West, the Saint Lawrence River, and Muscle Shoals on the Tennessee River as being the "super-power" sites still available to public use. These sites also were being indicated in the writing of Governor Roosevelt at the time.[57] Bonbright warned in 1930 that, while it was possible to build dams and power plants, it was much harder to arrange the distribution of the resulting electricity, a process that could be "highly unstable."[58] Successful distribution could come only from heavy promotion and expert management. Indeed, profitable distribution was to become a big and unadmitted problem for the TVA until World War II, with its unprecedented demand. The problem of balanced distribution was what gave David Lilienthal some of his superior status within the TVA operation too, for he was in charge of distribution from the beginning.

At the key 1931 Round Table on Regionalism the following year, R.D. McKenzie, the human ecologist from the University of Michigan, referred to the social concerns that loomed larger as the Depression wore on. The question of what to do with people had become more pressing. McKenzie did not approve of the population's current state of instability. During the last decade, more than a third of American localities had declined in population, he said. He also deplored the "tidal shifts of population from one region to another." Thus:

> The increasing mobility of population has disrupted the traditional bonds of security and control and created new problems of administration and welfare

organization. The life of the individual is becoming increasingly insecure. He lives in an eternal present, shorn from the supporting traditions of the family and the neighborhood and forced to change his domicile in response to the dictates of a highly sensitive economy over which he has no control. It is a serious indictment of our present economy that the masses of our citizenship have recently been precipitously tossed from the highest level of individual income known to our country into a condition of destitution the gravity of which surpasses former record.[59]

Any new effort ought to shore the human condition up. Stuart Chase, the economist and charter member of Mumford's Regional Planning Association, offered a more comprehensive alternative answer for this instability that surely would have attracted Donald Davidson of the Vanderbilt Agrarians, Mrs. Roosevelt, and Arthur Morgan, if they had known of it at the time. Chase observed that "in the regional idea, furthermore, things of the spirit receive more consideration: the beauty and order of one's own countryside; a rebirth of craftsmanship aided by cheap electric power; a local literature, a local art."[60] Everything was to be brought back to regional sources, because in that return would lie greater stability and closer identification. Too, Chase spoke of the desirability of a national plan, to be enacted by Congress, but he wanted first "to take a specific region, study it in detail, note carefully the problems involved in the technique, and prepare and publish a series of blue prints."[61] Chase's direct association of electricity with the more slowly-paced crafts would appear incongruous to many, and especially to David Lilienthal later on. The concept was closer to Arthur Morgan's hope for the humanistic integration of the Tennessee Valley, a first and last cultural harmonization. The essence of the arts and courtesies were what Agrarians Frank Owsley and Donald Davidson had wanted to reestablish; life in the 1930s was badly cramped and incomplete. Still, technology in and of itself ought not to displace the rural values. Rather it should stabilize and reinforce them. There was a lot of emptiness in the social receptacle left to fill.

The "opportunity" ostensibly would be fulfilled in the postwar era by the translation of electricity into other forms of energy — at the atomic center at Oak Ridge, Tennessee; the engineering experiment center at Tullahoma, Tennessee, where artificial wind was a major feature; and the space rocket center at Huntsville, Alabama. There was disagreement, early and late, over what electricity ought to "do." It was really Frederick H. Newell of the Research Service, Inc., of Washington, D.C., who pro-

vided an unlikely linkage, as early as 1931, among the thoughts of Herbert
Hoover, Henry Ford, and Franklin D. Roosevelt on this issue. Unfortunately the linkage did not become sufficiently visible at that time as to
be recoverable by every liberal intellectual thereafter. Newell was something of a maverick engineer. Known to Brandeis and Frankfurter, he
had worked closely with Gifford Pinchot. He was also the first chief of
the Reclamation Bureau of the Department of the Interior, which had
built Boulder-Hoover Dam. In his 1931 Virginia Round Table talk, entitled "The New Industrialism and Power Development of the South,"
more clearly than any of the others, he set the stage for the TVA. He
began with a quote from "our sober-minded President [Hoover] some
years ago at a time when he was Secretary of Commerce." Hoover, an
engineer, had observed that Americans were in "the midst of a great
transformation in the development of electrical power" that would reduce
the burden of human toil, increase productivity, and bring "increased
comfort to our people." So the vision of the beneficence of electricity at
that time was not Roosevelt's and the Democrats' alone. But it was the
nonpolitician, Henry Ford, that Newell pointed to as the genuine in-
novator, having first suggested for the Tennessee Valley a program of planning policies, better land management, and fresh use of new crops for
manufacturing purposes, as well as more efficient use of electricity.[62] Ford,
he said, had wanted to reinstate the value of the home, so that the country residence, if reorganized as a "mode of life," would better support
the family. Newell, like Ford, endorsed the soybean as the crop of new
hope.

> The successful cultivation of this in enriching the soil and in producing forage,
> food, and raw materials for manufacturing, may lead to improved conditions
> in rotation and crop diversification. It seems a far cry from hydro-electric
> power and the development of the resources of the South to the somewhat
> prosaic subject of soybeans, but the connection can be traced through . . . the
> larger use of electric power and its full enjoyment by the people of the South
> must rest largely upon the ability of the people in the country to pay for the
> power utilized by them in their farm homes. This ability in turn rests upon
> the development of a better type of agriculture and this in turn upon the in-
> troduction to new kinds of crop and the crowding out of those which must
> compete in the world market. . . . The particular contribution to the subject
> made by Henry Ford is his insistence that on these farms of diversified crops,
> there should be produced not merely the foods needed for home or city con-
> sumption, but also the raw materials which may be manufactured locally,

such, for example, as fibers, meal, oil and substances hardly yet known or recognized but which through careful research and practical experiment may become important factors in the industrial world.[63]

This was Newell's pre-TVA assessment in 1931. Ford believed that industry could rest directly on agriculture, that the two should be interdependent. His penchant for experimentation to modify lifestyles through land and settlement change also would characterize the early TVA. Ford's investment in soybean plastics was not to provide a permanent return, however, because by World War II it was evident that plastics could be made more conveniently from mineral than from vegetable matter.[64] Most of Ford's laboratory investigations of the possibilities of soybeans took place in 1932 and 1933, ending after yielding a fluid for shock absorbers, a superior enamel for Ford cars, and soybean meal "horn" buttons, gearshift balls, distributor cases, window-trim strips, and electrical switch assemblies.[65] Ford used soybean meal to form a trunk cover that he unfortunately cut through, when he applied the blade rather than the handle of an ax to it when testing the surface. He had timing gears and other essential parts manufactured from compressed corn husks. From soybean threads Ford had a suit made. His instinct to return industry directly to the soil and to agriculture was a strong, consistently American one.

Although at Virginia Franklin Roosevelt seemed of all the speakers to have the vaguest notion of what "regionalism" ought to be, eventually he was to prove its most powerful advocate. His vagueness appears to have been due to his initially open-minded, improvisational approach to social problems of whatever kind; and to his preference for leaving all options available until the last possible moment. Whatever the cause, inevitably his approach would be, on the surface at least, circuitous. In Virginia, he announced at once, in his typically insouciant manner, "I did not come here with any prepared speech this morning." He followed that with a brief tribute to Thomas Jefferson as an architect-planner (although he was speaking in a building designed by Stanford White, at the end of Jefferson's Mall, that had thwarted Jefferson's original intention to keep the view open). He followed that unpromising warmup with another totally characteristic, open-ended statement. "I do not know what 'Regionalism' means."[66] In the face of that "frank" disavowal, it comes as a surprise to read the account of his reception in the local newspaper, which reported that, while his talk was "impromptu," nevertheless "he was extended perhaps the most thunderous applause that any speaker has received from an audience here in a number of years."[67] To gauge this cordial reception, one must recall that here was a very powerful politician, sending friendly signals toward intellectuals having absolutely no secular power, as rare an extension of goodwill from a politician to scholars then as today. It was already known, too, that Roosevelt soon would become a candidate for the presidency.[68] Today it is easy to forget his personal charm and magnetism. But maybe the chief underlying reason for the audience's unequalled display of enthusiasm was the accurate feeling that here at last was a person who could take a constructive role in relation to the environment and was not too timid to do so. Roosevelt's own rhetorical questioning always carried the implication of "too much talk, too little doing." In his speech at Virginia, he returned to his favorite themes: false urbanization, rural overtaxation, deserted farms, and the need for trees. As the governor of the state having the biggest and most influential city in the country, New York, he was never afraid to speak out against cities in general and New York City in particular. "The state of New York is not merely the city of New York," he said.[69] He noted that many would be surprised to hear that his state was at all important agriculturally. He reported that, out of twenty-two million acres that had once been under cultivation, four million had been abandoned. Four million more ought to be returned to trees. Then he moved to a position very similar to that of Henry Ford and the other speakers. He asked rather plaintively, "Isn't there a third possibility, a possibility for us to create by cooperative effort some form of living which will combine industry and agriculture?"[70] This search may have been TVA's chief initial raison d'etre, although it has been little acknowledged since. Roosevelt was looking for a new breed of hyphenated men, he said—"factory-farmers." He made himself appear educable to this audience, even though by slow degrees and in a ruminative—and to some probably exasperating—way. His ever-present sense of theater worked even among the intellectuals. He asked such simple questions as "Why did milk and cream have to come to Florida from faraway Wisconsin?" He could hear the milktrain, "a very noisy long train," at five minutes to three every morning going by his property at Warm Springs in Georgia, and that noise not only woke him up, it set him thinking! Why couldn't Georgia supply Florida with milk more quickly and cheaply? He could not understand why milk for New York City could not be shipped directly from New Jersey, Pennsylvania, and Vermont, rather than from faraway Indiana, Illinois, Iowa, or Wisconsin. He looked upon the distribution of food as a geographical problem. On that problem he now hung the sobriquet of "regional planning." He didn't like the fact that peaches and pecans, including some of his own in Georgia, were overabundant in the South and so had to be either sold at a loss in the North or discarded in the South. He preferred to speak in warm human terms and not to revert to impersonal bookkeeping, but he concluded that "the waste that has come about on account of lack of planning in this country has run into billions of dollars."[71] Distribution should have been better managed in the United States. Through better management, the country could be made more scientifically efficient, less wasteful. The ultimate benefit however, should arrive on the "social side." He was in pursuit of the general welfare as it depended upon the agricultural and aboreal environment. As he reverently explained in "Growing Up by Plan," an article in the *Survey Graphic* the next year, "Land is not only the source of all wealth, it is also the source of all human happiness."[72]

Morris Lewellyn Cooke

Gifford Pinchot introduced "scientific" forestry to the United States at Vanderbilt's estate of Biltmore near Asheville, North Carolina, in the 1890s, and stirred Franklin Roosevelt's Forests, Fish and Game Committee into life in the New York Legislature in 1910. In the 1920s, Pinchot, by then governor of Pennsylvania, hired Morris Lewellyn Cooke to be his advisor on electrical energy and water rights. These issues began to be more important after 1905, when at Niagara, on the Saint Lawrence, it became possible to transmit electricity over greater distances by high transmission lines. By the 1920s, it could be carried for 250 miles and more, by the 1930s up to 350 miles.[73]

Much of the planning for the TVA was based on this method of conveying power for two hundred to four hundred miles. The valley's long, narrow geographical shape lent itself to such an extended, vertebrate reach. Cooke, an engineer, was a disciple of Frederick W. Taylor, the champion of "Scientific Management," a movement that later turned into "Management Engineering." The key word in all the various improvement endeavors during the Progressive Era of 1890–1920 was "efficiency."[74] Taylor looked upon "a campaign for national efficiency"[75] as a logical successor to the Conservation Movement. There would be a convergence of these two concepts, an attempt at knitting them together, in the TVA planning. Cooke paused at various stations along this line of thought and policy. He had worked with Brandeis and Frankfurter in the Public Utilities Bureau in 1914, and they helped him to rise, as they did later with Lilienthal and many others. Through the press, Brandeis gave national prominence to the terms "scientific management" and "greater efficiency systems" in hearings in 1910 before the Interstate Commerce Commission in a fight to lower railroad freight rates.[76] Brandeis wished to place "professional" managers like Cooke and Lilienthal in utilities in order to replace "incompetents," those bent on bilking the public from various vantage points in private utility companies. Pinchot summed up the Progressive drive between 1890 and 1920 when he declared in his gubernatorial inaugural address of 1922, "The movement which resulted in my election is the direct descendant of the Republican Progressive movement of 1912."[77] Three years later, in 1925, Senator Norris considered making Cooke director of a government corporation to run Muscle Shoals, but he was not able to get the proposal past Calvin Coolidge, who rejected public ownership.[78] The next step was to have Franklin Roosevelt hire Cooke in 1931 to be his consultant on the Saint Lawrence Riverway problem. Finally, in 1935 Roosevelt appointed Cooke as the first director of the New Deal's Rural Electrification Administration, which was to be a most important instrument in the formulation and realization of the TVA. Roosevelt and Alfred Smith, in their plans for New York State, always had associated the availability of electricity with land reform. Rural electrification was one large reason for the anti-Ruhr attitude in TVA planning, with its emphasis on the decentralizing possibilities of cottage and light industries and, until World War II, less emphasis on major installations or heavy industry. During the 1920s, industry had boomed, but the farmer had been in distress since 1922, when the overseas agricultural market collapsed with the recovery of European countries after World War I. It was a later version of agrarian life, rather than a simple electrical distribution system, that was being encouraged by the first TVA enthusiasts.

Behind Cooke's career trajectory lay the conviction that electrical cartels and holding companies, first made possible by the wider transmission of electricity, should not be allowed to obtain a monopoly on this spreading source of wealth, drawn from natural resources and the downward flow of waters — from the very body and soul of the original, pioneer country. In the 1920s Cooke thought up for Pinchot a motto big enough to embrace the growing public opportunity: "Giant Power." Giant Power never came to anything materially, but the catchy slogan did express in a preliminary way what was to occur when the TVA eventuated in the next decade. Even then, Giant Power did not have to imply giant industry, giant commerce, or giant urbanism, as was later assumed. Expressing quite typical early aspirations, Pinchot said to his General Assembly of Pennsylvania in 1925, "Giant Power may bring about the decentralization of industry, the restoration of country life, and the upbuilding of the small communities and of the family."[79] "Big" was intended to have "little" implications around it. The huge concrete dams were supposed to cause the countryside to become more habitable, symbiotic, prosperous, bucolic, and receptive. It was this hope of gaining "little" from "big" that generated Norris Village out of Norris Dam, both under the inspiration of the thoroughly competent original engineer, Arthur E. Morgan.

The Character of the Initial Dream

All the emblems of the TVA and the New Deal tried to say something eminently useful and forceful, but often they could not because they were constantly being reviewed and revised. The short-lived National Recovery Act (NRA) of 1933–35 was so permissive, fragile, and ambiguous that it soon broke down under the bulldogging tactics of its tough-talking director, ex-Marine General Hugh Johnson, and under the hesitancy of the Supreme Court. Like the TVA, the NRA looked much bolder than it was or ever could be. The logo of the NRA was a taloned blue eagle holding three thunderbolts. The TVA, founded in the same year, 1933, was represented by an upraised fist holding a single thunderbolt. The raised fist may have been an awkward reflection of the aggressive symbols of European fascism, but Roosevelt himself had another explanation he liked to give, derived from Woodrow Wilson.[80] Wilson had pointed out to him that the Conservatives stuck together and so had "the striking power of a closed fist," while the Progressives (the associates of Wilson and the Roosevelts) often approached their problems with their "fingers spread out stiffly" and thus were likely to break them when tackling a problem. While American symbolism was not as self-assured as that of fascism or communism, it achieved an improvised "whistling in the dark" bravura and optimism that they did not.

The efficiency expert and professional manager were early on the scene to try to dominate the new situation. Another odd involution of assumptions, that surely would not have been entertained in a more confident era, was the persuasion that through applying normal legal methods bureaucratically, the material accomplishments of more conventional businessmen in a better set of decades could be reproduced and surpassed. Roosevelt declared in his speech proposing the TVA to Congress in April 1933 that he wanted a new type of organization that combined the flexibility of private corporations, "clothed with the power of government." Back of this interest in a new combination of elements drawn from older value systems was Brandeis' faith in Taylorism and its "Scientific Management" concept, to be implemented through "Management Engineering" and the earlier "Gospel of Efficiency" that had originated in the administration of Theodore Roosevelt.[81] Frankfurter told Raymond Moley that Brandeis' views of promoting "public works, reforestation, and control of waters, including the Tennessee Valley project," should be conveyed to Roosevelt as soon as possible before his 1933 initiating message to Congress.[82] Thus the language of an abstract and elevated legal assault would become at least as important as many of the acts of the Progressives in using conservation measures to set the physical stage for the TVA.

When David Lilienthal graduated in 1923 from Harvard Law School, where Frankfurter had been one of his teachers, he went, on Frankfurter's recommendation, into the Chicago office of another Frankfurter protégé, Donald Richberg. Richberg in turn recommended Lilienthal to Progressive Gov. Philip La Follette of Wisconsin.[83] Lilienthal stayed with the Wisconsin Public Service Commission, fighting the utilities, until July 1933. It was La Follette who coined the term "energy yardstick," which Lilienthal later tried to apply to the output of the TVA. Brandeis had employed the same technique for measuring the efficiency of the railroads.[84] The Progressive La Follette was not reelected governor; consequently Lilienthal at thirty-three felt his position in Wisconsin to be untenable, despite the state's previous dependence upon expertise such as his. In the spring of 1933, Brandeis and Frankfurter, respectively, recommended him to Arthur Morgan and Franklin Roosevelt for a position on the TVA board of directors.[85] Within three short years after his appointment, he would become Arthur Morgan's rival for the lead position within the TVA triumvirate.

New York Regionalism

The second half of the 1920s, following the era of "Efficiency" and "Scientific Management," was the period of "Regional Planning." The concept had found recognition within New York intellectual circles in the first half of the decade, following World War I. In 1926 Henry Wright had offered Gov. Alfred Smith a visionary nonstatistical brief for the reorganization of New York State.[86] The outlines of it had been prepared in 1924 by Benton MacKaye.[87] The opportunity for reorganization depended upon the fact that, since 1920 and the improvements in transmission of electricity, there had existed at least a theoretical possibility of human resettlement along a major electrical alley or "runway" from Buffalo east to Albany and then south down the Hudson Valley into New York City, based on hydroelectric power from the Niagara River, the Saint Lawrence, and streams in the Adirondacks. However, as with the TVA, "the aim of the State should be clearly to improve the conditions of life rather than to promote opportunities for profit."[88] The professed aim was more eleemosynary and sociological than technological or managerial. The patterns ultimately were based on those of the pioneer Edinburgh proponent of regionalism, Patrick Geddes, who set great store by the "Valley Plan of Civilization." The settlements along the "larger belt" in New York State were seen as catchment basins surrounded by rims of reforestation. Decentralization away from major cities was to be made possible through trucks, buses, and easier transmission of power. According to the projections, 75 percent of the electricity for the northern half of New York and 25 to 50 percent of the electric demand in the Greater New York City area could be supplied by hydroelectric power.[89]

In the 1920s data and thoughts about New York City and New York State abounded. The 1926 study for Governor Smith was a reaction to the much more statistically-based 1921–31 one on the New York Metropolitan Region, prepared for the Russell Sage Foundation. That foundation was directed by Thomas Adams, who was very familiar with Garden City theory because he had been manager of Letchworth, England, the first Garden City. In 1932, while terminating as governor of New York, Franklin Roosevelt wrote a sympathetic foreword to Adams' classic treatise, *The Outline of Town and City Planning*, published by the same Russell Sage Foundation Roosevelt's uncle was interested in. Roosevelt's foreword stressed the issues of redistribution of industry and population, "the wasteful consequences of haphazard growth," and increased interest in "a high social purpose" for planning.

The Regional Planning Association of America, founded in 1923 and carried forward by Lewis Mumford, Clarence Stein, Henry Wright, Frederick L. Ackerman, Stuart Chase, C.H. Whitaker, Tracy Augur, and Benton MacKaye, organized the state study for Gov. Smith because the association felt that Adams' Russell Sage plan would only make New York City more congested and dense, since his plan excluded consideration of anything beyond a fifty-mile radius of the city. Mumford declared in his foreword to Clarence Stein's *Toward New Towns for America* (1966) that the group was "freely co-operative in act," with "neither factionalism nor desire for priority or publicity"; consequently, it is difficult to single out particular contributions. Stein became chairman of the commission appointed by Governor Smith. Then he hired Henry Wright to direct the study. Wright was to be Stein's partner in the subsequent design of Radburn Garden Suburb in New Jersey (1927–29). The New York State study, with its one brief booklet, turned out to be considerably sketchier, more general, and far less statistical than the Russell Sage one begun in 1921, that finally ran to eight volumes.

New York Regionalism As It Affected Tennessee Regionalism

Two personalities, Tracy Augur and Benton MacKaye, bridged from the New York Regional Planning Association to the Tennessee Valley Authority. As assistant to Earle Draper, the director of the TVA Department of Regional Studies, Augur mainly supervised the buildings of Norris Village. MacKaye was hired from 1934 to 1936 to study Tennessee folk culture and suggest how to keep it intact.

Any further involvement of MacKaye with the TVA was cancelled with the eviction of Arthur Morgan. Nevertheless, his cogitation about regional plans remains intriguing. With it, ideas that first appeared to belong to one value system can easily be seen as fitting into another. One prominent instance of that viability was the Appalachian Trail, first proposed by MacKaye in 1921.[90] Today it is looked upon as a path for therapeutic hiking in the wilderness from Mt. Katahdin in Maine to Mt.

Mitchell in Georgia. However, the original vision was never as restricted, was never merely a thread in the wilderness. Instead, the trail would have broken down the sense of total isolation and total removal from coastal cities by means of "highwayless towns" and "townless highways" that would have been protected by "wilderness reserves" down the mountain slopes (13–14). What was new and positive for a further development of the national park and wilderness preserve philosophies was the notion of a close and more continuous interweaving of an improved human settlement within nature.

A new energy source would make everything possible in a new planning methodology. MacKaye wanted to "stretch the power belt" along the Appalachian range, he said. This potentiality demanded conceptualization at a much greater geographical scale.[91] Contemplating the increased potential of power transmission in the 1920s in relation to his vision of better usage of the mountain belt, MacKaye asked himself a basic planning question: "What shall we do with our power when we get it? We are harnessing up a team of giants — a black giant called 'coal' and a white giant called 'water-power.'" The "giganticism" of power was recognized at about the same time by MacKaye as by Cooke and Pinchot.

> We have the power of a giant wherewith to turn the wheels. . . . In which direction shall we head our giant — toward becoming our master or our servant? We can turn him loose in the superpower zone to add more wheels and chaos to a headless commercial tangle; or we can conserve his strength to up-build a hinterland in the original molds of a purposeful life. We can use him to huddle our people further in the grimy congestion of our urban wilderness; or we can get him to start us on a new frontier.[92]

The ultimate result was intended to be a better distributed and balanced form of community, strung out on a "quanta" basis in the "rural wilderness." The older idea that the American settlement should have spread through the wilderness was combined with the new imagery of the linear city, all to be made possible by the recent extension of power lines from water sources falling down mountainsides. This new type of community would be "self-comprehensible." It would utilize industrial mechanics and technology, and its population would come in "democratic proportions." It would be compact and have access to the "natural attractive environs as much of our rural countryside provides," with modern sanitation, including smokeless factories, and cooperative unity.[93]

This futuristic vision of an electronic strip (13) along the entire Appalachian Ridge was largely formulated between 1921 and 1924 and before the 1926 Governor Smith plan for New York State. New York City itself became more and more the locus to be detached and weaned from, as Manchester, England, had been for the older utopian colonies. It only took one city to incite a revolution. The skyscraper was its one-of-a-type incubus, just as the dam would be the one-of-a-type architectural affirmation of the TVA at the opposite end of the symbolic scale. The twist was that, although New York State gave premonitions of what might happen next in the TVA, almost all of the effort of the New York intellectuals themselves was engaged in untangling their plans from the city as a central, dominating, phenomenon. That prejudice appears to have been the main generator of feeling against the solely urban introspection and real expertise of the Russell Sage report, so exemplary in its own way, on the validity of New York City.

MacKaye's 1928 book, *The New Exploration*, was once more full of such observations, turning away from New York City and pointing attention down country toward Appalachia. Inevitably, five years later, he pointed further on down to the Tennessee Valley Authority, which had been founded in 1933 (13-14). MacKaye declared that a New Eden would be made possible through an improved population movement that he called "folk flowage," as speeded up through readier access to electricity by means of wires stretched farther and farther out. "The factory need not go to Niagara, Niagara can go the the factory: that is, within a radius of some three hundred miles."[94]

Against New York City, MacKaye now charged that "New York is the outstanding example of a broken-down reservoir of population,"[95] and "The titanic turreted skyscraper viewed through evening lights from New York harbor combines rational power in repose and majesty with imperialistic affront and sinister contemptuousness."[96] Mumford, who gained much of his publishing success from being associated with the intellectual power brokers of New York City, likewise was repelled by that city. According to his 1925 article, "The Fourth Migration," New York City depended solely on "business efficiency" for its importance. Thus, "the forces that are concerned solely with business enterprise lose no opportunity for stressing the necessity of continuing the third migration [into the cities]; and city planners [no doubt Thomas Adams, Delano, and their Russell Sage Foundation] who fall in line with it plan for the agglomeration of ever-greater urban regions."[97] It was the next, gener-

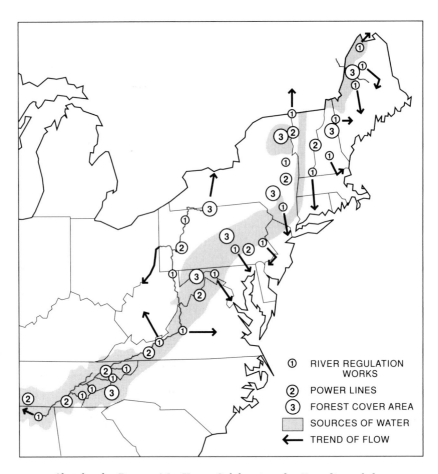

①	RIVER REGULATION WORKS
②	POWER LINES
③	FOREST COVER AREA
▨	SOURCES OF WATER
←	TREND OF FLOW

13–14. Sketches by Benton MacKaye, Celebrating the Founding of the Tennessee Valley Authority. The map (13) on the left shows dams (1) and power lines (2) in a forest cover (3). The plan on the right (14) represents MacKaye's way of settling the Appalachian Trail. Contrary to later interpretation, he did not wish the trail to be a wilderness without towns. The string of A's in the right diagram stood for a limited-access highway down off the hiking ridge. B's were "highwayless towns," along this highway but not on it, while the C's were wilderness areas to be reserved on the slopes. The large dots with arrows indicated the principal cities that would yield population on the coming of the Fourth Migration. The idea was remarkable, proposing to link the themes of the TVA of 1933, MacKaye's 1921 Appalachian Trail scheme, and his 1924 New York Regional Plan in a linear city program stretching two thousand miles or more from the TVA to

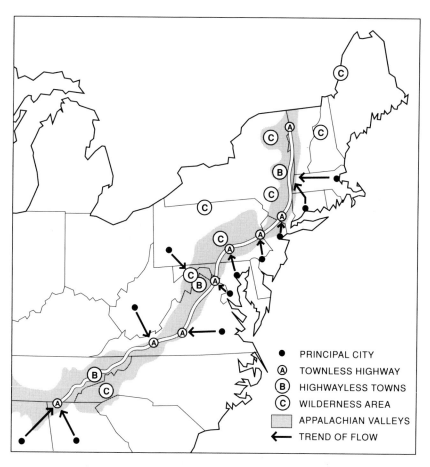

PRINCIPAL CITY
Ⓐ TOWNLESS HIGHWAY
Ⓑ HIGHWAYLESS TOWNS
Ⓒ WILDERNESS AREA
▢ APPALACHIAN VALLEYS
← TREND OF FLOW

Canada. The long, narrow linear shape would depend on the new transmission possibilities of electricity. Linear settlement would be a subject
of recurring discussion within the TVA district; in 1922, Henry Ford had
suggggested it for seventy-five miles along the Tennessee River at Muscle
Shoals. MacKaye's scheme resembled the future TVA, too, in that it could
be like a national park, forest, or wilderness, but also would have human
residents. It would be a merging of the big city and the unpeopled
wilderness preserve. Like so much the Regional Planning Association put
forth, the MacKaye diagrams were at once too simple and too complicated,
but they promised an alternate (and better?) kind of America. Illustrations
adapted from MacKaye "Tennessee — Seed of a National Plan," *Survey
Graphic* 22 (5 May 1933), redrawn by University of Tennessee Cartographic
Services, 1990.

ously redistributing Fourth Migration that really was wanted now, because "only here and there have we even fitfully attempted to utilize the land intelligently, relate industry to power, resources and market, and provide an adequate 'human plant' for the community at large. To affect this union is the task of the fourth migration."[98]

The question that MacKaye and Mumford were posing was: "Where ought Americans to belong? Could they find a better adjusted, more proper milieu in another locale?" MacKaye turned to his old backbone of the mountains for an ultimate answer: "Appalachian America promises to be a strenuous battleground. It looms large on the map of the world. Here is the nucleus of what seems to be, potentially, the mightiest industrial empire on earth." Previously, all that America had been able to come up with by way of alternative earth and forest forms was a series of remote and unusually beautiful national parks and national forests. This Appalachian reserve, not too far from New York City, ought now to become activated, implemented, and inhabited. MacKaye's most daring innovation was to people his proposed wilderness, a radical departure flying in the face of the wilderness enthusiasts, whose cause he had seemed at first to be championing with the Appalachian Trail. Ultimately, MacKaye wished for a physical system that found its greatest significance in being morale-building, uplifting, spiritual. People should be immersed in "'spiritual form' in our society. . . . The immediate job of the regional planner is to prepare for this awakening—not through unconstructive and chimerical efforts on the metropolitan 'Bottle Neck' [the Russell Sage study of New York], but through a synthetic creative effort back on the crestline sources [his Appalachian Ridge plan] where an indigenous world of intrinsic human values (and specifically an indigenous America) awaits its restoration and development as a *land in which to live*."[99] To fulfill that purpose, linearity would take over from outmoded urban concentricity.

MacKaye knew what conditions his new synthesis would require. One requirement would be more "space" to take up the slack caused by the failure of historical time.[100] Two late nineteenth-century developments had put time out of phase. "Through such developments as the railroad in the sixties, and motor transportation and electrical transmission of power, since the nineties, a series of sudden jumps has taken place in this country. The leisurely growth of an American indigenous culture, which promised so much before the Civil War, has been burked or diverted into purely metropolitan interest and concerns."[101] The gradual evolution of the evergreen landscape, from which to take spiritual as well as physical sustenance (MacKaye was indeed a Transcendental Yankee in outlook and tastes), had been diverted and "burked" by the Civil War and the postwar attention to northern cities. The lyrical unfolding of the prewar romantic republican landscape had been thwarted, leaving only a deep and hurting void, especially in the South. There had been an overconcentration on an artificial form of denser settlement in an unnatural place — the city. What was needed instead was an *"indigenous, innate, symphonious environment,"*[102] MacKaye said, where all the arts could be deployed and examined while "working all together."[103] MacKaye's gift was for moving easily up in cultural stages by means of an imagined set of societal steps. At last, well-designed architecture (not including the skyscraper) would show up and come into its own, providing a spiritual shelter and relief in concert with the evergreen background.

The assessments and visions issued by MacKaye and others from 1921 to 1933 echoed and reflected off the harder surfaces of pragmatic politics as well. In 1932, before the TVA was proposed in 1933, Governor Roosevelt was saying, in his ruminative, dissembling way that often was right to the point,

> I am wondering if out of this regional planning [talk] we are not going to be in a position to take the bull by the horns in the immediate future and adopt some kind of experimental work based on a distribution of population [by which he meant redistribution, the Fourth Migration in MacKaye and Mumford's terms]. We know from the economic point of view that every skyscraper that goes up puts a dozen older buildings out of business.[104]

Even David Lilienthal, in a speech he made at Columbia University in January 1940, appeared to accept the reform-minus-skyscraper thinking of Roosevelt and MacKaye:

> There are few sights so impressive as the magnificent towers of New York City. They stand as evidence of the might of our metropolis, the wealthiest and most secure city of modern times. But the towers of New York rest not upon the rock of Manhattan but upon the soil of the whole United States. The great structures and the millions of men and women who work and live in them are no more secure than the soil of this nation.[105]

The differences in the appearance and scale of city and country intrigued Lilienthal also. He too wanted to know where the real strength of a vision for the future, after — or even in — a war, would lie.

The New York Port Authority

The gap in concepts between New York City and the Tennessee Valley was so marked that the attachment of the title of "Authority" to the latter, deriving as it did from the 1921 creation of the New York Port Authority, appears slightly out of order today. A credibility gap develops because the Port Authority, covering a radius of scarcely twenty miles out from the Statue of Liberty, embodied little or no sociological implication and was in actuality merely an urban administrative unit to promote commercial and transport efficiency. The TVA, in contrast, at the outset bristled with sociological ramifications. As with MacKaye's Appalachian colony proposal, the rub with the TVA would come from the introduction of the uncertain factor, people, into a sylvan setting. Raymond Moley, former New Deal brain truster, wrote that he had not liked the application of the New York term "Authority" to the people of Tennessee, because it connoted the imposition of an outside "dictatorship" on them.[106] Douglas Yates, in his *Bureaucratic Democracy: The Search for Democracy and Efficiency in American Government,* called attention to the same dichotomy of human needs versus governmental efficiency when, much later, in 1982, he described the tug-of-war between "pluralist democracy" and the "efficiency model" of administration.[107] Although he identified the New York Port Authority as being of the "efficiency" type, Yates only briefly mentioned the TVA and did not allude to the fact that Lilienthal had chosen the efficiency approach for internal operations, while publicizing the first, or "pluralist democracy," theme in pursuit of his public-relations "grass roots" imaging. Although the National Resources Committee (Delano's group) reported in *Regional Factors in National Planning* (1935) that "The Port of New York Authority is perhaps the most important regional administrative agency yet developed,"[108] it also regretted that, while the New York Authority had flourished up to 1935, it had also defaulted on its bigger responsibility of forming a more efficient railroad system around New York City. Given Delano's own past dream (with Daniel Burnham) of a railroad integration plan for Chicago, such a failure certainly would be a demerit in his eyes. Nevertheless, his report added, "A similar plan for a port authority for development in metropolitan Chicago has been proposed."[109] The report compared the New York Port Authority to the TVA, but rather unfavorably in terms of steps that ought to be taken with both in the future. As he indicated to Lilienthal in letters that same year, 1935, Delano continued to hope that a more vested and focused management style might be assigned the TVA. But he liked the picture of an organization formed by the "compact" method, on an interstate basis, even less than he liked the model of an "authority" with its faults.

> When the problem is a continuing and complex one, the [interstate] compact method is not only ill-adapted to the planning function, but it leaves much to be desired from the standpoint of effective administration [his reaction reflected his firsthand knowledge through a visit to the scene of the multistate Colorado River compact, that had caused a lot of dissension—W.C.]. It is a well-known fact, applicable to other forms of administration as well as business management, that successful administration requires adequate authority and opportunity for initiative, flexibility, and even experimentation [this last FDR's cardinal rule for how government ought to be conducted—W.C.]. The more complex the economic or social problem, the more necessary are the last-mentioned administrative attributes. But autonomous administration is far from characteristic of the compact method; it is the principal difference between the compact [the Colorado River agreement of 1922 or the New York Port Authority of 1921] and the regional development authority represented by the T.V.A.[110]

The Tennessee Valley "Authority" had adapted its title from the immediate postwar New York Port Authority, but it proved difficult to establish a more functional bridge between those two cultures in the dislocated 1930s. MacKaye did not encounter the robust, justice-loving, apple-cheeked rustics that he had been anticipating. New Yorker Moley reacted candidly: "In the Tennessee region, most of the people engaged in agriculture are the products of long residence. To be frank, their urge to progress is not marked."[111] A critical issue was whether groups of Americans had been driven too far apart, become so different from each other, that even when no ethnic or racial issue was pressing they no longer could communicate or even see anything positive in each other. In the context of such a climate, TVA could amount to little more than a colorful display of increasingly divergent hopes which America, in developing its material and technological wealth, had made possible. TVA was one more desperate attempt to recapture a common cultural language and a common set of goals — an American social Esperanto or Basic English (a popular notion during the 1930s). In addition, TVA became a further exploration of a uniquely mid-American utopianism.

The Electronic Revolution and Technocracy

The TVA was a complex, supposedly organic, enterprise, fitted into a waiting natural setting. The organization, however, seemed constantly to revert to simpler and more elementary definitions. H.A. Morgan's mottoes of the "seamless web" and the "common mooring" were evidence of this craving for increasingly simplistic formulae, as was Lilienthal's image of the "grass roots." The drift was always toward one quantity or mode, and that search for the single value system finally came to rest in electricity alone. As a latter-day commentator put it, "Lilienthal wrote of TVA's generating capacity as 12 billion genii, and measured democracy in terms of kilowatt-hours."[112] The compelling expectation was for an "electrical utopia" where "America's redemption from the past was to come from 'Nature' and a 'Virgin Land,' a new scene of human society filled with unique possibilities," as James W. Carey and John J. Quirk were to describe "The Mythos of the Electronic Revolution" abiding in all America. "America was to realize, through a marriage of nature and mechanics, an unprecedented solution to the problem of industrialization, a solution that would rejuvenate all immigrants who ventured into the new world and allow us to transcend the typical evils of industrial society."[113] This newer, fresher vision was supposed to supplant the older one of the Industrial Revolution in England and Europe. Similarly, the Shaker religious colonies in Kentucky; Owens' utopian New Harmony, Indiana; the Mormons' Nauvoo, Illinois; and Engels' continental Communism were all in some way and to some degree reactions to one industrial city, Manchester in northwest England, whose manufacturing presence appeared so degraded and so overwhelming. The movement toward purifying utopianism in the middle of America bore some similarity to that of the TVA toward the same area and away from New York City. Yet, by the last third of the nineteenth century and the end of the Civil War, some doubts had to be raised as to whether America at large would, after all, be able to avoid those urban and industrial forms that had been the bane of Europe. So, from 1870 on, Americans turned their hopes toward electricity, creating through a "rhetoric of the electrical sublime" an exalted picture of what could be accomplished with technology on this continent. The outcome was that this rhetoric has invested electricity [in America] with the aura of divine force and utopian gift and characterized it as the progenitor of a new era of social life, which somehow reverses the laws and lessons of past [European] history. Despite changes in vocabulary, the idea of an electrical utopia possesses a common rhetorical tendency whenever it has appeared over the last century: it invests electricity with the capacity to produce automatically, on the one hand, power, productivity and prosperity and, on the other, peace, and a new and satisfying form of human community and a harmonious accord with nature.[114]

In that connection, electricity was another, and surprisingly durable and powerful, utopian elixir. The interest of the TVA in the "human community" diminished as World War II approached and the interest in not losing the hold on "peace" became greater, so that the TVA was looked to as an alternative and an exception, not only to the older industrial disadvantages of Europe in general and England in particular, but also to the hostilities just beginning there in the late 1930s. Lilienthal himself could see clearly that electricity brought on beneficial "social change." In the in-house film, *The Electric Valley*, he said, "Electricity happens to coincide more than almost any other service with a change in the standard of living. . . . [Electricity] had a profound social effect. In many respects it is the charge that makes for social change."[115]

This was a peculiarly American invocation of faith in electricity, to which the intellectuals paid as much tribute as the politicians. Lewis Mumford, with his "neotechnic" imagery of dams and power houses, was followed in time by other North American intellectuals with greater faith in technology to solve most problems, such as Buckminster Fuller and Marshall McLuhan; and by American presidents such as Dwight D. Eisenhower, with his interstate highways and promotion of nuclear power; John F. Kennedy, with his New Frontier; and Lyndon Johnson, with his Great Society. The latter two appeared to believe that they were elaborating on basic social and energy visions of Roosevelt's First New Deal. According to Carey and Quirk, such politicians and intellectuals were forever predicting a brighter future after the passage of a dark and confused stretch in the national history. Every other consideration got squeezed out by the enthusiasm for electricity and technology. The later personalities "always proclaim the future in word and then desert the future in fact. Technology finally served the very military and industrial policies it was supposed to prevent [for the TVA, represented by the

Dixon-Yates controversy in President Eisenhower's term—W.C.]. At the root of the misconceptions about technology is the benign assumption that the benefits of technology are inherent in the machinery itself so that political strategies and institutional arrangements can be considered minor."[116] There was a residual innocence and impulsivity in taking up technology as an unquestionable good that could only be American. The serious indictment made by Carey and Quirk was that advocacy of technology was often used, in addition, to plaster over ineffective governance. The TVA went to war in 1941. Afterward, with its confidence in and thirst for electrical output only confirmed, the TVA had difficulty recovering its original interest in the "human community and a harmonious accord with nature." Much that was too narrowly specialized came to be expected of the TVA technology, which had already given signs of definite limitations, such as the end of hydraulic potentiality or limitations of political activity. The free word sketches of rural settlement and lighter industries that Pinchot, Ford, Roosevelt, MacKaye (13-14), and Arthur Morgan had once submitted were entirely forgotten. The promise of recreational resorts, assuming increased importance through the return of affluence and leisure time, might have become a design or managerial mission for the TVA to respond to; however, even though the new TVA lakes had created a regional magnet, the recreational function was largely ignored. The Upper South and Lower Midwest were areas that, after World War II, could not offer the same scenic and recreational opportunities that both coasts did. The glamour and color simply were not there. Instead, in the middle kingdom, electrical energy became the hard, bright beacon, the artificial element, to pursue, because it was the easiest to think about and carried with it a cachet in the society at large, as Carey and Quirk discerned. The visions of a more balanced core country-within-a-country, "a land in which to live," as MacKaye had put it; the hope for a successful amalgam of agriculture and manufacture; the need for a fresh infusion of homesteads and villages to support an improved family and community life; and the wish that light industry and vernacular crafts might take up suitable positions in the natural background were ignored and went glimmering.

For the TVA the hope of organizing geographical space in a balanced manner, in order to compensate for the previous distorted era of the late nineteenth and early twentieth centuries, which MacKaye had pointed to, showed up most conspicuously in recurring debates over why, as governing units, regions would be preferable to the present states. The paraphernalia of the power plants then became symbolic stakeouts, flag plantings, for this newest conception of legitimate territorialism. "The myth of the powerhouse is implanted into the pathways of theory and practice and acted out upon the real environment."[117] The TVA became "great" because its electrical technology could turn up in the form of terse and compact structural and visual statements in every corner of its farflung territory. The initial status of its technology depended upon the opportunity to affect real situations and genuine territories. Carey and Quirk evidently doubted that such a "marriage of ecology and technology" would be either easy or permanent, because politicians were always focusing on the next moment, rather than on the more distant future, and trying to enlarge their spheres of personal influence; and because American intellectuals were forever inventing new vocabularies to exalt technology and make it appear more beneficial than it ever could be in actuality, thereby generating a "contradictory image of humanized technology." The glamour of electrical technology outstripped human nature to such a degree that the projected sequel became impossible to fulfill, since it neglected and pushed aside "the values of the arts, ethics and politics where man finds fulfillment."[118] Technology allowed less and less room for the other, less tangible human activities and concerns that Ford, Roosevelt, Gutheim, Nolen, and Arthur Morgan had recognized and tried to incorporate.

Technocracy

During the 1930s, the desire to evoke the most vivid technological imagery possible led to some bizarre manifestations. Engineers especially felt that, through electricity, they had an inkling of some more profound truth, an X factor through which a new and better social system could be structured. The least diluted of such mechanisms was the cult of "Technocracy."[119] Sometimes Technocracy is described as the inspiration for the TVA, but in fact it was a parallel development that also began in 1932–33, in the same general milieu, New York City. This notion had currency as a mesmerizing revelation, a blinding insight. In essence, Technocracy was a vehement complaint over what disproportions in the society technology had recently wrought. However, what was exactly to be done by way of productive correction after a more accurate inventory of natural and human resources was taken, seems never to have been spelled out. George Norris did attend a meeting in February 1932, at which Howard Scott, the advocate of Technocracy, spoke. While noting that the ideas expressed were "very interesting," Norris was not in agreement with them.[120] But major magazines like *Harper's* and *Fortune* paid Technocracy the flattering tribute of articles. Wall Street pundits gave it a hearing at their banquets and dinners. Figures already legendary on the national technical scene — Charles F. Kettering, president of the General Motors Research Corporation; Carl Compton, president of MIT; and Dr. Nicholas Murray Butler, president of Columbia University; among others — directed attention to it through talks and articles. Technocracy went straight into Butler's 1932 Columbia University report. There was such a dearth of masterplans during the 1930s that any sudden, prophetic sign in the intellectual or academic sky was likely to be taken seriously. To find one sure formula, based on pure energy, meant that other, more nebulous formulae could be ignored or discarded. Bearings and beacons were being lost in Europe and America. On 23 December 1932, comedian Will Rogers wrote a brief letter to the *New York Times* containing the best summation of the situation. He noticed that people, in the year before the TVA was undertaken, easily entertained the most absurd ideas, since "people right now are in a mood to grab at anything," even though Technocracy might "go out as fast as Eskimo pies or miniature golf courses." In desperate times, he observed, even the wisest were susceptible to the enticement of "strong" ideas.

Technocracy was as oversimplified as the "Electronic Revolution." Imbedded in all its slick and shiny body of theory was an elemental belief in the efficacy of "energy determinants." Wattages or "ergs," rather than square or cubic feet, were to be the prime measurement. All was to be marshalled by alert and efficient engineers. In the previous two or three decades the possibilities of extracting and distributing mechanical force had increased by leaps and bounds. During the last two centuries, a more intensive combination of energy transfer devices such as levers, axles, belts, and wheels had been introduced into textile factories and mines, there to be powered by the newly developed steam engine. Unexpected side effects were that, in building factories and railroads, too large a capital debt was incurred, and a great abundance of material goods was produced and accumulated. Yet this abundance was not sufficiently well distributed. On the one side had arisen an increased efficiency in production, while, on the other, no adequate system had been set up for the wider distribution of all this new and substantial "wealth." The technologies, with their machines, had produced so much so quickly, had become so efficient, that by overproducing, they had partially destroyed their own reasons for being. One of the subtler reasons why the first TVA structures attained such high visibility was the wish to show a nation whose confidence was shaky that such a technology, if lined up with one kind of "cleaner" energy, could sustain — by cheap production — a last-minute, last-ditch struggle toward a different kind of productive efficiency and a more proper type of distribution.

Part of the fascination with the electrical units that the TVA was to furnish arose from the fact that those units would not constitute unused inventory to be stored in warehouses. One could be detached about their significance and worth, if not their distribution yet. The Depression really was the period when businessmen first realized that they had created so many objects that social purchasing power was no longer capable of absorbing them, taking them into the fabric of society, except through some redistribution of the means of acquisition. If a proper new system of acquisition and distribution could be arranged, Technocracy said, the "adult population (25–45) need work but 860 hours per individual per year to produce [in 1932] a standard of living for the entire population ten times above the average income of 1929. The present political system is wholly incompetent to effect the technical coordination necessary to handle the problems, which today, involve the whole continent."[121] The opportunity to factor out the whole continent was as important to Technocracy as it would be to that ever-changing macrocosm of the TVA, called at various times the National Planning Board (1933), the National Resources Board (1934), the National Resources Committee (1935), and the National Resources Planning Board (1939). The standard of living would, as a result, be measured by "energy determinants," a concept that was as much an abstract reaction to material goods as the cleanliness and purity of electrical production was to the grime and smoke of burning coal. Despite the disclaimers of any deeper aesthetic concern, and despite the thought that all the problems of society could be solved by invisible kilowatt hours, symbolic assertions were as important for Technocracy as for the TVA.

The Industrial Revolution was said to have overproduced in three ways — heavy capital debt, too many factories and goods, and lots of energy. It followed that social problems could be attacked better by the cooler and more objective minds of newly educated engineers. These new technicians would work better on teams. For the TVA, such a group would be created in the Sprankle Building in Knoxville, Tennessee, and for the Technocrats on the second floor of the engineering building of Columbia University in New York City. At Harvard it was the team of the Bauhaus group, replanted from Germany in 1937. Each enclave would be made up of a reassuring number of "professionals," turning out a stream of "informed" reports. At Columbia, thirty-six such experts were engaged in producing "The Energy Survey of North America." Three thousand charts were planned on that subject alone. An earlier enclave was Howard Scott's Technical Alliance of 1919, that listed on its "Temporary Organizing Committee" such names as Benton MacKaye; Frederick L. Ackerman, New York architect and planner; and Robert H. Kohn, another architect later prominent in public housing. All three became charter members of the Regional Planning Association of America, when it was founded in New York in 1923. Charles H. Whitaker, another member of the Regional Planning Association and editor of the *Journal of the American Institute of Architects*, was listed on the temporary organizing committee, too. He had sent Ackerman to England in 1917 to investigate Unwin's war housing.

The nominal backer of the National Energy Survey was a professor of industrial engineering at Columbia University, Walter Rautenstrauch. His talks projected the same image of engineering as a discipline able to accomplish comprehensive changes for the whole society. He picked the power station as the indispensable symbol of the lighter, braver, cleaner new society: "The modern power station is a possibility [of showing how the newer society ought to be run], because the many pieces of apparatus to be operated in combination to generate current at varying load demands are integrated and controlled by properly designed control devices. The social mechanism presents the same picture to the technologist and he can see no possibility of uniform and stabilized economic society if the control devices of the systems of regulation which it employs are not scientifically designed."[122] The destiny of the society could be perceived better through the simple act of visiting a power house, as TVA eventually strongly encouraged the public to do. The scientific design of its material equipment would offer a more precise control model for American society, by now too open and eclectic. "All social problems of North America today are technological." Rautenstrauch emphasized manipulation of a whole culture through more particularized technical devices, in a manner that in Germany and Japan shortly would serve to assemble tanks, planes, rockets, and gas chambers for aggrandizing purposes — for example, a blitzkrieg or a surprise attack on Pearl Harbor.

The expectation was not too different from that shown in the 1936 science-fiction film, *Things to Come*, whose futuristic buildings in an underground city were modeled by Moholy Nagy and whose "technocratic credo" had been adapted from H.G. Wells' book *The Shape of Things to Come* (1933). In the film an enlightened scientist (Raymond Massey) brought technical order out of social chaos after World War II, to last until the year 2036. The appearance of the utopian city, derived from a newer, more orderly, and more logical approach, oddly paralleled what some were formulating for the TVA itself at the time — except that in *Things to Come*, the setting was underground, genuinely subterranean and not just in a valley far away. "The lines of the new subterranean city of Everytown begin to appear, bold and colossal. Swirling river rapids are seen giving place to a deep controlled flow of water as a symbol of material civilisation gaining control of nature."[123] And really of humanity too. As people grew more powerful in means, but less controlled in their use of them, the temptation was to look for more arbitrary and elementary solutions. Water control represented social control.

The actual director of the "Survey of Energy" at Columbia University, Howard Scott, was a type quite different from his patron, Dr. Rautenstrauch. Scott was an amateur propagandist who developed a highly personal vision. Even in his own time, a newspaper described him as the "Baron Munchausen and Marco Polo" of engineering. He claimed to hold a doctorate from a prestigious German university, but in 1932, when he had his greatest standing as the guru of Technocracy, it was discovered that he had no degrees at all, German or American. His inspiration and guidance came rather from self-education and the theories of economist Thorstein Veblen, translated by Scott into engineering terms. The most pertinent of Veblen's books, as far as Scott was concerned, was *The Engineers and the Price System*. Scott spoke to meetings of the Taylor Society, the earlier organization promoting the scientific management that Brandeis favored, and he had links to the International Workers of the World, the so-called world "Wobblies." Truth being strange in the

1920s and 1930s, Scott also had a real connection to the future TVA through Wilson Dam and other constructions at Muscle Shoals during World War I. He described himself as having been a "leading technician" there, though in actuality he appears to have been "an incompetent workman in a cement pouring gang," assigned to Nitrate Plate No. 2. He was discharged from that "sensitive" position because his fellow workmen "insisted that he was a German spy." His later view was that "technologically trained people must form the nucleus of a new revolutionary movement."[124] However, an equally important ingredient in any of the Technocrats' declared goals was the hope for increased efficiency through the proposed cleanup of the economic value system, which they felt had been made wasteful through overcompetition. Economists Stuart Chase and Paul H. Douglas and poet Archibald MacLeish felt that there was a logic beneath the superficial glitter of Technocracy. Despite lively discussions during the 1950s and 1960s over postwar, postindustrial, postmodern society, experts in the 1930s and 1940s even more arrestingly juggled prospects for a new and different future society, because they actually visualized what such a future milieu might be like in organizational terms. The TVA staff had more group or team density than anything before, except perhaps an army, and had more ordered monuments. All up and down the Tennessee River, the power houses were meticulously synchronized, as Rautenstrauch said power houses should be.

University of North Carolina Research

More based on local circumstance was the far-flung sociological research carried on at the University of North Carolina by Howard Odum and Rupert B. Vance. They were not partisans of New York metropolitanism, large industrial complexes, or highly sophisticated engineering, since the South did not yet possess any of those to use as models; they would come only after the war. At the time, there were no substantial graduate schools in the South. Odum and Vance looked more to hardscrabble agriculture. In 1932 Vance observed that his and Odum's approach to "Regional Planning in the South will be forced to face sooner or later the task of salvaging marginal highlanders."[125] This rescue of humans would inaugurate "A Folk Renaissance for the South." There is a Shakespearean timbre to the sounds emanating from southerners in relation to their mutual destiny. Historic dignity had to be maintained in all future transactions: "Here is nature and there stands the folk. Behind the folk stands a tragic history. What we need to know is that, in spite of its tragic history, the mold in which the South is to be fashioned is only now being laid."[126] Such pre-TVA statements by Vance and Odum looked frankly for long-term solutions to come from history rather than technology. The difficulty with this method, of course, was that change would be presumed to be gradual, while the TVA was geared toward solving an emergency. How different the outcome of the TVA project might have been, had it been begun in a period of confidence and prosperity, with authentically long-range plans and goals readily in view, and taking into account the historical premonitions generated out of previous utopian, Progressive, Agrarian, and Conservationist experiences and the Civil War? How different would the result have been if, in the first place, the ideals had not been so high and so multiple, and if, in the second place, the actual means to support those goals had not been so minimal and shortlived?

Arthur E. Morgan; Frederick Gutheim; John Nolen, Jr.; and Charles W. Eliot II

Outside the groups already described, four key individuals held no special allegiance to either the New York intellectuals or the southern regionalists. Those four were especially important because they actually influenced the writing of the TVA law, through its author, U.S. Senator George Norris. Arthur E. Morgan was chief engineer for the TVA dams (15) and the earliest appointed director of the TVA. Together with his wife, Morgan was deeply interested in the surviving folkways and crafts of the remotest, most mysterious and challenging area of Appalachia. He was not preoccupied exclusively with engineering. His approach to all phases of the TVA was highly experimental, which of course followed the original inclination of Franklin Roosevelt. That approach lasted until 1937–38, when open discord arose among the three directors, ending with Morgan's discharge by Roosevelt. A report in *the New York Times* of 4 December 1933, headlined "Tennessee Valley Wakes from Dream," gives an excellent summary of that first approach. TVA would be a model social experiment after which the national system might be patterned. Dr. A.E. Morgan went about from city to city preaching . . . that the Authority was not thinking of a Ruhr of America at all. 'Go to Pittsburgh, go to Detroit and look on the rows on rows of hovels occupied by the workers of the big mills, and you will say, "I don't want to see anything like that in Chattanooga." We are looking to a valley inhabited by happy people, with small hand-work industries, no rich centres, no rich people, but everybody sharing in the wealth.'

15. President Franklin Roosevelt, Mrs. Roosevelt, and Arthur Morgan in the Tennessee Valley, November 1934. The main figures appear to be completely themselves — the president as genial master of ceremonies and great communicator, with his radio microphone always at the ready; Mrs. Roosevelt looking socially concerned; and Arthur Morgan so earnest and eager to get on with the next challenge that he almost comes out of the car. Photo courtesy of Antioch College Library.

According to Morgan's guidelines, even though he was an engineer, what was wanted was not so much greater investment in electrical facilities as it was rethinking of actual living conditions. Earlier, Muscle Shoals had been spoken of as a seed site for an American Ruhr. But Morgan, Roosevelt, and MacKaye all had said plainly in their writings and speeches that another Pittsburgh was not wanted anywhere in the United States. In the TVA law, finally pulled together by Senator Norris, the phrase at the end of Section 22 dealing with the right of the TVA organization to cooperate with state and municipal agencies derived from Arthur Morgan, as perhaps also did the call in Section 23, Item 6, for developing "the economic and social well-being of the people living in said river basin."[127]

Two other individuals having input into the TVA law were Frederick Gutheim and John Nolen, Jr. The former had been a student in the Experimental College of the University of Wisconsin under Alexander Meiklejohn, when a discussion of differences between ancient Greek civilization and American culture was carried on annually. The curriculum had no interest in professional education as such, but Wisconsin's John Gaus, professor of public administration and the "voice" of regional planning there, had considerable influence on Gutheim's thought. Lewis Mumford had been a visiting lecturer. Moreover, "each student would be required to conduct a regional study, an investigation of his own home region, modeled on the recent study by Robert and Helen Lynd, *Middletown*,"[128] a requirement that linked the curriculum with one of the most original and influential social documents of the Depression decade. In this college, which lasted from 1927 until the critical year of 1932, only the year before the TVA was instituted, "the regional study proved to be the most effective and stimulating part of the Experimental College's curriculum."[129]

Gutheim was a prodigy, active in organizing important early shows on architecture and planning at the Museum of Modern Art in New York City. He was also at one time assistant to Catherine Bauer, the housing expert and lobbyist. She had been the youngest member of the Regional Planning Association of America, and in turn had worked as an assistant to its leaders, Clarence Stein and Lewis Mumford. All three advised Gutheim briefly during the creation of the TVA law.[130]

Nolen Jr. was the son of the leading town planner of the time from

Cambridge, Massachusetts, who laid out Kingsport, Tennessee, with Earle Draper.

Gutheim's wishes for the future TVA included:

1. Use of Civil Service regulations for first employment to avoid political patronage (a precaution that eventually was gotten around by having TVA set up its own employment rules).

2. Avoiding land speculation, a problem created at Muscle Shoals by Henry Ford's ill-fated proposal for it in the 1920s. This drawback was brought to Gutheim's attention by Bauer, Mumford, and Stein simultaneously, in connection with the lack of interest in supplying housing and community planning apparent in the forthcoming law.[131] The reluctance to extend responsibility for new housing and community plans was to become a chronic blind spot and gap within TVA programming, if the Regional Planning Association definition of regional planning was to be seriously taken.

3. Some method of interlocking the efforts of TVA personnel with the officials of the seven states to be served by the TVA.

4. Avoiding concentration of the enterprise upon hydroelectric power exclusively. Gutheim wanted to witness an "enrichment and stabilization of life," as compared with the "more simple technical questions alone."[132]

Gutheim and Nolen approached Norris in 1933, while he was preparing the TVA legislation to be introduced in March, and asked him to revise Section 22 by acknowledging responsibility to make "surveys and plans" and endorsing projects that would improve the quality of life in all its aspects, based on "an orderly and proper physical, economic, and social development of said areas." The aim was environmental enhancement through planning in the Tennessee drainage basin and "adjoining territory."[133]

Charles W. Eliot II was the grandson of the famous president of Harvard University and had been named after him. He conferred with Gutheim and Norris while they were trying to enrich and broaden the Norris law. For seven years, he had been director of the National Capital Park and Planning Commission and later had directed the National Resources Planning Board, both under the regular guidance of Frederic Delano. At the time, it was thought that Eliot was about to be appointed Under Secretary of the Interior and therefore that he would be in a key position under Harold Ickes to give initial shape to the TVA.[134] That didn't happen.

Eliot's uncle had been the landscape architect Charles Eliot, pupil and partner of Olmsted, who had created the "emerald necklace" of regional parks around Boston, the first such string of parks and parkways in the world. The uncle had died of meningitis, at age thirty-seven, in 1897. Charles W. Eliot II then was nominated by his grandfather to fill the role of landscape architect and civic planner, left vacant by the deceased uncle.[135] President Eliot had a deep interest in the arts and in environmental conditioning, seeing these as desirable means for the positive evolution of the country. Scientific intellectuals, of which he was one as a chemist and mathematician, were supposed to care about the environmental arts. This dedication of dynastic talent to the shaping and handling of the land also was carried on by the Roosevelts (even Nicholas Roosevelt, the family maverick, was a conservationist all his life).

The perhaps inevitable outcome of the insertions in Sections 22 and 23 was evident in a very early assessment report in 1940 from within the TVA itself: "As far as the record can be followed, no complete, balanced, and long-range program of surveys, studies, experiments, and demonstrations directed toward the purposes set out in Sections 22 and 23 have been achieved, and the Authority has too little to show for the two million dollars expended on Sections 22 activities up to June 30, 1939."[136] The report came from the office of the general manager of the TVA himself in August 1940, only a year before the outbreak of World War II. After the war, programs to relieve the impoverished and alienated at home would be directed toward urban areas, and overseas toward Third World constituencies.

George Norris' Degree of Awareness

Senator Norris (16) appears originally not to have had a broad social view in mind for the TVA. The breadth was suggested to him by Arthur Morgan; Frederick Gutheim; John Nolen, Jr.; and Charles W. Eliot II in connection with Sections 22 and 23. Norris fixed his attention rather on the hydraulic power distributed from the Saint Lawrence River, where kilowatt hours coming from the same dam customarily had been sold on the American side at International Falls for five to six times the price on the Canadian side. The company on the Ontario shore was publicly owned, but that on the American side was private. Norris' hopes for

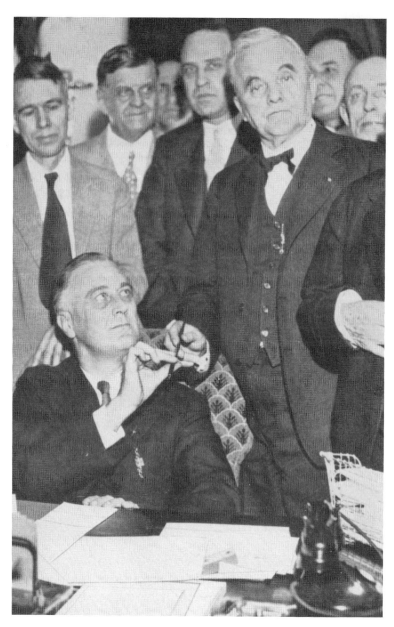

16. Franklin Roosevelt Presenting Senator George Norris with the Pen
Used in Signing the TVA Bill in 1933. Norris, like Arthur Morgan (15),
was a self-made man of the Old Progressive school. Hence he was able
to confront any camera with an unapologetic, individualist, uncompromising stare when doing good. Photo courtesy of Wide World.

public power from Muscle Shoals had been dashed twice in the 1920s. Previously he had helped to thwart the hopes of Henry Ford in regard to the same site. President Coolidge had kept Norris at bay with a pocket veto in 1928, and President Hoover, when he used the veto in 1931, had been very vocal (for him) about interference with regular business investment. Norris boarded Roosevelt's train, at the latter's request, on its way to Muscle Shoals in January 1933, just before the inauguration but after the election, in the so-called "First Hundred Days." Roosevelt wanted to know from Norris what should be done about Muscle Shoals. When Norris got off the train, he was elated, sure that a bill of some kind, dealing with the whole Tennessee Valley, would soon pass.[137]

Who can know, even today, what truly happened and what the secret dreams of the first TVA participants were? Even the chief players, the Roosevelt group themselves, seem never to have been entirely certain of what had transpired. A quaint incident suggesting this uncertainty occurred after Frederic Delano had retired from the National Resources Planning Board. He wrote to George Norris, also retired, in McCook, Nebraska, asking if *Norris* knew what had *really* happened in the creation of the TVA! On 5 October 1943, Norris replied that he long had been interested in the efforts of Adam Beck of the Ontario Power Commission to bring down the perunit cost of electricity in Ontario:

> The Commission, of which Sir. Adam Beck was at the head until his death, had power to purchase any private system, their generating and transmitting. The Commission generated and transmitted electric current to the municipalities,—sold it to them at cost. They in turn sold it to their consumers at cost and it resulted in revolutionary changes in the price of electricity. . . . It expanded and expanded until it controlled all of the electricity in the Province of Ontario.[138]

Norris then prepared a speech contrasting the price of electricity on the Canadian and American sides, circulating many copies. In a two-sentence statement, Norris summed up for Delano how he saw his role in the TVA. "The T.V.A. is to a great extent the result of my study. I followed the Ontario system wherever it was possible."

Delano, in a followup typical of him, then wrote to David Lilienthal on October 9, four days later, to ask how the present rates of the TVA compared with those of Norris' chief inspiration from Ontario. Lilienthal's reply of 12 November 1943 stated in a noncommittal manner that the Ontario rates were being approximated in the urban areas of Tennessee, while the TVA's own schedule of rates was being applied in the rural areas. He couldn't give a full answer. The vaunted "power yardstick" had never quite become a precision instrument. A major figure in the whole American planning movement, Frederic Delano, whom his nephew, Franklin Delano Roosevelt, had credited with fanning his own enthusiam for planning, now deferentially was asking later arrivals on the scene, Norris and Lilienthal, how "they" had brought it all about!

There was always this spirit of generosity, a sense of high-mindedness, in whatever the Roosevelt circle undertook. Also there was likely to be a vagueness about where some of its principles had come from, and a lasting question as to whether those navigational aids for the whole country, developed through so much effort, would ever be realized. Perhaps even a four-term president would not have had time enough? So much had to be undertaken in a confused moment and compromising manner in the United States. Over a longer period of time, a certain cohesiveness of intention might inadvertently have built up, however partisan the opposing declarations were at times. The realization may never have been exactly on the mark, but there appears always to have been a certain buildup of premises and convictions as to the preferable social goals, all to be disregarded in the prosperity of the postwar period. Harry Slattery, the third administrator of the Rural Electrification Act after Morris Cooke and John Carmody, in 1940 wrote a book for the National Home Library Series. It was published by a foundation organized "to promote and inculcate in more people the desire to read good literature; and to these ends to provide for the delivery and holding of lectures, exhibits, public meetings, classes, and conferences, calculated to advance the cause of education and promote the general culture of the nation."[139] This series was one of the pedagogic enterprises of Louis Brandeis and was organized in 1932. Arthur Morgan had written *The Long Road* (1936) for the series without knowing that Brandeis was his patron.[140] Slattery's book reveals that "in his efforts to decentralize industrial activity, Justice Louis D. Brandeis early saw the need and potentialities of electricity in small villages, towns and on the farm."[141] So Brandeis would have to be placed in the company of Ford, Franklin Roosevelt, and Arthur Morgan as a rural decentralist? Slattery likewise credits Gifford Pinchot as being the true initiator of rural electrification, since he had begun in 1923 "an electric farm survey of extensive character" while he was governor of Pennsylvania.[142] Beyond that, the foreword by Senator Norris makes clear that the TVA of 1933 and the REA of 1935 really were parts of the same grand scheme: "the bringing of low-cost electricity to all the farmers of the Tennessee Valley, not only for their own good, but as an example to the nation."[143]

The Conservationist-Progressive tradition, beginning in the 1890s, had been criticized as too often "devoted to forests, water and wild life but rather neglectful of humanity."[144] The Asheville, North Carolina, *Citizen* of 17 March 1934 wrote that naturalistic means were being used for humanitarian purposes within the TVA: "The main object back of the whole program is to better the social and economic condition of the people in the mountains here. In order to do that it will take cheap power, good roads, flood control, reforestation and a dozen other objectives in the program." Franklin Roosevelt showed similar optimism in the book he brought out in 1934, purporting to be "a history of the first year of the New Deal." Like the newspaper, FDR was looking for multipurpose means to achieve broad results, and he gave Senator Norris credit for a more diversified interest than the senator perhaps ever had had. "In enlarging the original [Muscle Shoals] objective so as to make it cover the whole Tennessee Valley, Senator Norris and I undertook to include a multitude of human activities and physical developments. By controlling every river and creek and rivulet in this vast watershed, and by planning for a highly civilized use of the land by the population of the whole area, we believed that we could make a lasting contribution to American life. . . . It touches and gives life to all forms of human concerns."[145] Splendid thoughts, but where would they go and what did they actually mean? What would that "highly civilized use of the land" really consist of? How far did "all forms of human concerns" stretch? Who really would be in charge of the effort? Franklin Roosevelt's penchant for improvisation, pay-as-you-go empiricism, novelty, and surprise may permanently block any thorough understanding of the TVA as an outcome more of evolution than revolution. We may never come to know who was the true progenitor of the TVA — Norris,[146] Franklin Roosevelt, Theodore Roosevelt, Gifford Pinchot, Henry Ford, Benton MacKaye, Arthur Morgan, David Lilienthal, Frederic Delano, or even Louis Brandeis and Felix Frankfurter — it could have been any one of them, or any combination thereof. All we can say surely at this point is that they all had input. However, in reevaluating the original circumstances today in the light of further knowledge, it would appear that Frederick Gutheim, in his succinct 1933 essay, already had the clearest idea of what the TVA could and should accomplish:

> A well-developed regional plan for the Tennessee basin would treat of power production and its use, transportation facilities, the conservation of natural resources, the control of industries and their location, human habitations and settlement, and the control of land uses. When the elements of the Tennessee Regional Plan are re-grouped, they fall naturally into three main divisions: (1) conservation and development of natural resources; (2) physical equipment of the region, including transport, highways, communities, and recreation areas; and (3) all other appropriate considerations which make for "social and economic well-being," including broad provisions for public health and education.[147]

Gutheim had taken up the unorthodox subject of regional planning in his undergraduate days at Wisconsin, under the unorthodox Alexander Meiklejohn. Now, in 1933, he thought long and carefully about the TVA mission. Many others took action concerning the TVA with much less contemplation or definition than he had applied.

3 | The People's Reactions to the TVA

Frustrations of a Cultural Group

The technical competence of the TVA staff and their early pledge to carry out tasks boldly and thoroughly never could smooth out all the bumps along the way for the TVA, nor conceal its failures of omission or commission. The local people, or at least the segment that should have benefited most from association with the authority, had been inflexibly connected to the land over a long period. They did not rest easily upon it, for a variety of reasons. First, they never had been large-scale agricultural planters. As the *WPA Guide to Tennessee* (1939) indicated, "The census of 1850 showed 118,941 farmers [in Tennessee], none of whom were classifed as planters. In significant contrast are the census figures of South Carolina for the same year: 8,407 planters, as against 32,898 farmers."[1] Tennesseans were traditionally small-holders. They thought of themselves as yeomen, as independent agrarian proprietors, and therefore as able to speak one-to-one with any possible employer or government official.[2] But by the 1930s their economic position no longer could support such a self-image. Jerrold Hirsch's 1986 introduction to the reprint of the same *WPA Guide* describes how pervasive and poor the rural population had become by 1939: "When this guide was published, 62 percent of Tennessee's people made their living on the farm." The farms often were quite small in acreage, and "the guide appeared at a time when almost one in five Tennessee farmers were sharecroppers. Almost half did not own their own farms."[3] They were underprivileged, but obviously a majority, not a minority.

While these people were being cast in the mold of ancient pioneers on a frontier that had lasted and lasted, they also had used up the land at a frightening rate. By the time of the publication of the *WPA Guide*, fourteen million acres of Tennessee needed reclamation.[4] Richard Kilbourne, forester in charge of erosion control for the TVA, reported that erosion (9–10) showed up everywhere in the system, with West Tennessee having the most and the deepest gullies, and East Tennessee displaying hillsides washed down into the Clinch and Powell rivers.[5] Like their Virginia predecessors, who had exhausted their soil within a few decades by one-crop cultivation of tobacco, the Tennesseans had managed very early to produce a great deal of erosion in their fields:

> As early as 1854, the State agricultural bureau warned that excessive "mining" or one-crop cultivation of the soil would finally lead to economic disaster. Farmers following this practice grew one crop year after year without letting the land lie fallow or rotating crops to build up the soil. . . . However, the period was one of prodigal shortsighted waste and little was done to check the menace. . . . In 1935 the State planning commission made a survey which revealed that three million acres had been practically ruined for cultivation by deep gullies, and that 75 to 100 percent of the surface soil had been taken from eleven million acres by sheet erosion. . . . The severity of sheet erosion, which occurs on all lands not protected by a crop of heavy-rooted vegetation, depends on the degree of slope to the terrain and the type of soil. Shoe-string erosion is also responsible for much damage. Gullies, started by little rain rills, are gouged out to such a depth that it becomes almost impossible to fill them. Slip or landslide erosion, Tennessee's third principal type, is frequent where shale soils predominate.[6]

The cutting of trees had added to these negative results. The people were thought of by themselves and others as being of an ancient lineage, but they prematurely had exhausted the land they depended upon. Moreover, geographical differences between the mountainous East and the bottomland in the west, plus internal variations in height, utility, and soil quality in Middle Tennessee, had caused political separation as well. "Divided into three 'grand divisions' of East, Middle, and West Tennessee, the state has been plagued throughout history by sectional bitterness."[7] The mountaineers of the East were Union sympathizers and early Republicans, while the cotton growers of the West were Confederates and later Deep-South Democrats. As late as the time of the demand for coal during the Korean War, with the new steam plants coming into being, a similar bitter rivalry built up between eastern and western coal miners.

Alternative Type of People Proposed

There was a rising urge to replace the present inhabitants with new human types, most notably with a sage, responsible, middle-class breed of
farmer. Frank Owsley of the Agrarians pursued this theme assiduously
in his book, *Plain Folk of the Old South*. His analysis was of cause-and-
effect: "Because of the great increase in population, together with the
enclosure acts and other similar monopolistic trends, the freeholds and
leaseholds of the yeomen farmers of Western Europe, especially Great
Britain and Ireland, became too small for comfortable support and in
many cases too small to furnish subsistence." Many of those Scotch-Irish
and other nationals were driven forth from their homelands because of
too little land and had consequently settled in the South, a "boundless
domain" of "genial climate." Though some became "landless tenants" and
"squatters," most eventually acquired the entitlement of the "freeholder"
(a title of great worth in the South, in accord with the notion of holding
land "in fee simple"). Large planters and members of the professions came
from those freeholder ranks, "but the greater part remained landowning
farmers who belonged neither to the plantation economy nor to the des-
titute and frequently degraded poor-white class."[8] So a substantial, stable
enough, well-to-do middle class had once existed in optimal contact
with the land. Owsley was looking for a new majority of that type that
would exhibit "integrity, independence, self-respect, courage, love of
freedom, love of their fellow man, and love of God."[9] But, in the book,
his quest remained an antebellum one, at least outwardly and by his own
definition.

Oswley, however, was stung into proposing a similar formulation, the
most distinct one put forward by the Agrarians, for the post-Civil War
present (1934–35). In an attack on the Agrarians, and particularly on
Donald Davidson, the Baltimore critic H. L. Mencken asserted that the
Agrarians were "intensely uncomfortable in their brummagen Zion, but
they lack the skill and resolution to undertake its reform and sanitation,
and so they seek relief for their troubled minds by discovering enemies
over the fence."[10] Mencken felt that the southerners no longer were capable of rescuing themselves, and that "the rise of those lower orders [in-
cited by evangelists and revivalists] in the South, following the crippling
of the aristocracy by the Civil War, is probably to be blamed for the pre-

sent state of affairs."[11] For both Owsley and Mencken, circumstances were bad because of a demographic vacuum, a missing pivotal or keystone class. Mencken also recalled what the Agrarians were most anxious to forget, an event that constituted a kind of red flag for them, the 1925 Bryan-versus-Darrow Scopes trial concerning the teaching of evolution in the schools of Tennessee.[12]

The chain of pictorial evocations that Owsley put together in retort to Mencken's challenges culminated in his "five great pillars" of Agrarianism. Corporations, banks, power holding companies, and big industries for Owsley were wholly northern inventions, that had exacted a tremendous toll from the southern farmer.

> Most of the white tenants were once landowners, but have been thrust to the bottom of the economic and social order by the loss of their lands through [northern] high pressure salesmanship of radio, automobile, and farm machinery agents. Industrialism has persuaded, or created a public opinion which has virtually driven, the farmer to accept industrial tastes and standards of living and forced him to mortgage and then lose his farm. Battered old cars, dangling radio aerials, rust-eaten tractors, and abandoned threshing machines and hay balers scattered forlornly about are mute witnesses to the tragedy of Industrialism's attempt to industrialize the farmer and planter.[13]

Oswley wrote with pungency and conviction. He saw a northern culture that created detritus rather than artifacts.

The ideal figure that should take control of such physical decadence — the unpainted architecture, rusty field equipment, and gullied hills — was still imaginable as a "small or large" farmer who should be protected by the government and augmented in numbers by the return of planters from above and sharecroppers below to the ranks of the middle class. "Actually, most of the planters are without credit, and are no better off than the tenant or share-cropper. . . . The fate of the small or large farmer is much better. As a rule he is thrifty, owes less than the other classes, and lives to a great extent off his farm. Sometimes he sends his children to college, especially the agricultural college. His house is usually comfortable and sometimes painted."[14] Those willing painters should be added to by the state and federal governments through the implementation of "five pillars": (1) purchase of eighty-acre homesteads by the government for the ever-increasing numbers of landless; (2) restoration of "a modified feudal tenure where the state had a paramount interest in the land and could exact certain services and duties from those who possessed the land," including measures for its restoration and preservation; (3) reduction of money crops such as corn, cotton, and tobacco, in favor of subsistence foods; (4) leveling up of southern agriculture by the federal government to the status of northern industry, finance, and commerce; and (5) creation of regional, in preference to state, governments. Following on those five fundamental steps, further governmental intervention would be unnecessary, and an ideal life would ensue in its several topical dimensions. The proposal certainly was not made with material or monetary gain in mind; it was reaching more for social stability, an old American aspiration in the face of negative charges. The poignancy of recovery would derive from a subsequent renewal of the old humanistic values. So the spectator would end up relieved and disarmed, taken up in an atmosphere in which everything and everyone *belonged*. "The old communities, the old churches, the old songs would arise from their moribund slumbers. Art, music, and literature could emerge into the sunlight and from the dark cramped holes where industrial insecurity and industrial insensitiveness have often driven them. There would be a sound basis for statesmanship to take the place of demagoguery and corrupt politics. Leisure, good manners, and the good way of life might again become ours."[15] What the Agrarians saw as missing was the profile of better character and the sense of interpersonal contact and loyalty that could be identified with earlier times. They wanted intellectuals to matter in the social fabric. The very adherence of the Vanderbilt University professors to Agrarian consanguinity probably could not have been achieved in the northern Ivy League universities of the time, however much better equipped or more distinguished for their scholarship the latter may have been. The Agrarians felt that a fullness, congeniality, and wholeness had gone from American life because of too much emphasis on objectivity, efficiency, and the making of money. They had spotted a genuine fissure in the concrete of American culture, that intellectuals should know better than to widen further. Owsley wanted for southern people the same thing the TVA did — a new dimension and a greater range — but for him, it had to come out of the past and out of a tradition and could not result solely from new machinery pushed by a new technology.

Agrarians and others frequently conjured up such alternate images. Andrew Lytle presented "a highly idealized verbal portrait of an actual farm in the upper South" in his chapter in *I'll Take My Stand*, the first catechism of the Agrarians, as Paul Conkin noted.[16] From nearer the outer edge, others images could be tested as well. Walter M. Kollmorgen tried the ingenious theorem that the English settlers at the utopian colony of Rugby had not succeeded because "the absence of professional opportunities abroad brought them to Rugby, the site of an idealistic experiment in farming which failed."[17] They were misplaced upper-middle-class persons. It was the peasantlike Poles, Germans, and Swiss who succeeded more readily in Tennessee agriculture, because they were better able to adjust to frontier conditions; knew or learned more about the cultivation of grapes, berries, and vegetables in the given soil; carried on better stock and fertilizer practices out of Old World traditions; and "introduced better barns and were the first, as a group, to paint their houses."[18] They kept up their farm machinery, did not borrow money recklessly, and were interested in cover crops such as alfalfa and crimson clover. They definitely were not "land skinners." Therefore, after some qualification, "it seems reasonable to assume that if the South had received a more substantial number of immigrant peasants from abroad, the change toward a more diversified and constructive farming program would have been accelerated."[19] This slightly different argument, pitting seasoned peasant against useless aristocrat, also depended on the assumption of a missing class. So many "lost harmony" or "other way out" situations could be produced as models, so many contrary-to-fact conditions could be invoked! It made for an exciting exercise in futility until industrial employment increased via the TVA. By January 1971, David Halberstam could report in *Harper's Magazine* that "the TVA worked, brought factories, raised the income, and made some Republicans here."[20] Would Henry Ford have thought that *his* first dream had come true when a Ford plant was located in Nashville in 1956, in which three-quarters of the employees commuted from rural counties and drew almost a third of their income from government soil-bank programs and the rest from the Ford assembly line? That agro-industry income enabled them to go "on buying sprees for new houses, . . . two cars, new furniture on the installment plan, color television, and boats to fish from,"[21]— the kind of boats, incidentally, that they might use on TVA lakes. The pay at the Ford factory was "the best in the area." Otherwise these rural employees would have had "to go to Detroit or Chicago" for employment. The industrial intrusion that the Agrarians so feared, finally had penetrated to the heart of the valley. The "factory-farmer" (Roosevelt's term), finally had gained a foothold, however accidentally or incidentally.

Glimpses of the Inhabitants in the Crucible Years

A curious enough inquirer would have to preface or qualify his observations on the desolate (or unusual?) look of the Tennessee homesteads and their inhabitants in the 1930s with several considerations. First: "It should also be pointed out here that in general the soil fertility in the northern states, particularly in the Middle West, was much greater than it was in most parts of the South and so land was not worn out so quickly. Climatic conditions in the South are conducive to the leaching of soil and also to the destruction of organic material. As a result the pioneer in the South usually did not have as fertile soil to begin with as the pioneer in the North."[22] From the beginning, the fields were substantially inferior relative to the Midwest, and the improvement of the South out of that frontier condition consequently was slow. "The frontiers did not disappear as rapidly or as completely in Tennessee as in some middle western states. . . . Geographic conditions account for this slow encroachment on new lands because areas with poor soils and unfavorable topographic conditions have retarded development for varying periods of time."[23] The farmers who had endured entered the 1930s hampered by a second unfavorable condition, poverty. The price slump of the 1920s, following immediately on World War I, affected agricultural prices nationwide but struck more deeply in the South, precluding any savings or even establishment of sound enough credit before the Depression set in. The third precondition, lying heavily on the psyche of the Tennessee farmer, miner, and mountaineer, was the disruption, inhibition, and hurt remaining as effects of the Civil War and Reconstruction, periods that in the 1930s were no more than five or six decades back. In these eras, initiative and self-confidence had withered while pride and a need for a sense of dignity and self-worth had grown. Over and over, attempts were made to place a platform halfway down the pit of chagrin, and then to build back up from there. Conceptually, the TVA was one more such emergency structural device designed to effect rescue.

But the social and economic buildup would have a long way to rise. The sudden arrival of the TVA in the valley was taken by some, even by some of the educated, as a setback. Such an immediate reaction found voice in the erudite *Sewanee Review* (edited during 1944 by Alan Tate and from 1961 to 1972 by Andrew Lytle, two Agrarians). The commentator, James R. McCarthy, took particular umbrage at the "patroniz-ing" tone employed by the first director of the TVA, Arthur E. Morgan. He called into earliest question Morgan's interest in "reviving" Tennessee folklore, when it didn't appear to need reviving, and what he took to be an egregious effort on the part of Morgan to introduce a new Tennes-see currency. He assured his readers that the mountain and valley people already were self-sustaining, remained pioneering, and came "of a stock, largely, that goes back to the old immigration of stout English, Scottish, and Irish settlers who went to Tennessee and made their homes there and asked for nothing else but to be left alone to achieve their own happi-ness." He identified the New Dealers, just intervening, as "utopians" and reported that the natives "resented the implication that Washington was the seat of a government which regarded the valley as a colony."[24] By 1934, the year following the enactment of the TVA law, resistance to collective outside governmental authority had already hardened that much.

The less educated natives were largely indifferent to the imperceptible (at least by them) future offered by Big Technology. They had acquired radios and cars, small technology, but the roads often were not solid enough for steady driving, and in a few instances they had no outlet. The highway infrastructure was almost nonexistent in modern-day terms. For all-weather travel, some settlers preferred the wooden sledge (1), known locally as "skids," pulled by a mule. Skids were used to bring crops down from the hills, for hauling lumber, and occasionally as an ambulance or hearse. Consequently the territoriality of these in-dwellers was circumscribed, which was another reason why they could be fiercely defensive and introspective about the few possessions they still had left in a particular place. Feelings of isolation, loneliness, neglect, and ali-enation were bound to be strongest in this landscape. In other regions, such resentments might be sublimated and concealed more easily. *Scrib-ner's Magazine* of New York City observed that "at almost every step you are among people who look on any intrusion from the outside, any seri-ous regulation of conduct with suspicion and resentment."[25] But the TVA had resolved to rejoin these citizens to the rest of the United States, so extreme tension was generated between the two poles. The irresistable force of ideas from the North had met the immovable object, a body of nationally unacclimated people.

The fervor with which Franklin Roosevelt and his associates sought to readjust human poverty and eliminate soil erosion stemmed in part from the hyphens they constantly placed between folk and environment. Physical afflictions that brought on lethargy — among these, hookworm, pellagra, and malaria — were endemic in the region. Black lung was frequent in the mining districts. That situation was regularly discussed in premedical and sociology courses in northern universities, and young doctors planned to make whole careers of eliminating them. The general condition of widespread illness was sometimes blamed for what was occasionally termed the "analgesic subculture" of the Appalachian area itself.[26] Indeed, commentators likened life there to existence in a somnambulistic state. While the rest of the country became angry after taking the first blows of unemployment and bank failures, the Tennesseans remained passive and numbed, or became more so. What is so instructive in retrospect is not only how much suffering and distress America allowed to pile up within a segment of its own population — how extreme that could be — but also how apathetic the suffering then could cause its victims to become. That degree of psychological exhaustion appears to have been unmatched before or since:

> Many of the behaviors commonly cited in descriptions of the folk culture can be construed as examples of regression induced by frustration. These characteristics would include the lack of esthetic appreciation, anti-intellectualism, the preference of anecdote over abstraction, the insistence upon a literal interpretation of the Bible, the entanglement of religious fundamentalism with deep superstition, the improvident squandering which often accompanies "pay day" or a welfare check, the tendency for self-pity, and the conversion of the "sick-role" into what local physicians sometimes half-seriously term a "chronic passive-dependency syndrome."[27]

Too much distress, too much detachment, appeared to snap the cables. Plainly, some kind of environmental intervention of a major kind was needed.[28] But what the scholars, engineers, and politicians of the 1930s often forgot was that the slow tick (or drip) of time was apt to shape people and their settings as much as, or more permanently than, any sudden radical intervention. These individuals looked less likely to be able to rise to any challenge, more likely to retreat from any official confrontation. Their remote locations gave them special opportunity to withdraw. In the nineteenth- and twentieth-century scramble for opportunity and gain, they had been held on the sidelines, pushed up against the hills and coves of the Appalachians, from the Piedmont area of the East Coast on the one side, and from the fertile meadows of the Blue Grass regions around Lexington and Nashville on the other.

In the late 1920s and early 1930s, a new wave of awareness began to arrive for these largely forgotten people. But in the crisis there was little time left for making substantial analyses. David Cushman Coyle was interested in the connection between the devastated land and the people and buildings dependent upon it:

> What did this spoiling of the land do to the people? You would see if you could go to one of the mountain schools in a badly eroded section and take a look around. There are panes broken out of the windows in the two rooms, and the cold northwest wind is leaking in around the rags stuffed in the holes. The children are barefoot and some of the boys have nothing on but a pair of overalls; they cluster around the stove trying to keep warm. . . . It is not a good place to be born, where the land is washing out from under you.[29]

Odette Keun, a Frenchwoman, had in 1937 what she regarded as a much less parochial point of view. These inhabitants did not measure up to what she felt should by now be a more internationally-oriented cultural standard: "The TVA calls it culture, I'd call it an obsolete form of Americanism that America needs no more than Europe needs the ideology of the Corsican or the tribal pattern of the clans of the Caucasus, to speak to peoples I know — and yet teach them responsibility to society and give them a significant and useful role in the community: this is the task the TVA has undertaken."[30] In contrast, the native journalist from Knoxville, James Agee, later renowned for his *Let Us Now Praise Famous Men* (a book in which Agee's text was paired with documentary photographs of sharecroppers by Walker Evans), emphasized the durability of this strong sectional conditioning:

> TVA knows that it is, among other things, a passel of smart Yankees descended to improve a tetchy people; . . . What TVA also knows but may not realize seriously enough — such realization may be impossible — is that generations of poverty and habit breed a quite indescribable inertia; that hopeful and faintly skeptical apathy and an almost childlike dependence are in general very possibly the liveliest attitudes that can be hoped for without very considerable guidance from above.[31]

Commentators felt that the TVA would have its work cut out for it, if any indigenous group attempted resistance. But Lorena Hickok, a former newspaperperson, traveling across the country in 1934 in order to report back to Harry Hopkins and Mrs. Roosevelt what was actually transpiring nationally, saw the problem as an opportunity for an imminent rescue, in the New Deal way. She expected the longstanding difficulties to be corrected within a decade: "But then — there's TVA. It's coming along. My guess is that, whatever they do or don't do about rural rehabilitation down in Tennessee, in another decade you wouldn't know this country. And the best part of it is that here the Government will have control. There's a chance to create a new kind of industrial life, with decent wages, decent housing. Gosh, what possibilities! You can't feel very sorry for Tennessee when you see that in the offing."[32] FDR's people never took "no" for an answer. Progress, according to Rooseveltians, had to be rapid and inevitable. The big defect was that their glance was so quick, like the snap of a newpaperperson's camera shutter.

TVA Initiatives in the Norris Lake Setting

Since the condemnation of properties for the building of Norris Lake and Dam was a first effort for the TVA, the conditions prevailing then, and TVA's handling of those conditions for the benefit of the local inhabitants, deserve closer scrutiny too. The pattern of settlement in eastern Tennessee and Kentucky, instead of consisting of fixed villages, more often was made up of loose clusters of households connected by bloodlines. Among such groupings in the Norris vicinity was a collection of twenty-five Stooksbury families (17) in what was then known as the Big Valley, fifteen miles long and two wide. The removal of the families from the Big Valley for Norris Lake was hardly a Biblical Exodus. At the outset, the people ignored the notices to vacate by 1 January 1935.[33] Next they indulged in fantasies of get-rich-quick schemes or became jealous of their neighbor's purported levels of reimbursement. They had difficulty making up their minds where they wanted to resettle, in spite of TVA's scrupulousness in keeping them informed of available farms. Government agents drove whole families time after time from their original farms to look at others. The average distance for a family to be moved was sixteen miles. The average education level there was fifth grade.[34] Less than four in a hundred houses in the Norris area then had electricity.[35]

Musical instruments, furniture, tools, and guns were highly prized as cultural artifacts, but they were too few to be totally believable as a body of objects supporting a full cultural tradition. A tiny item like a 104-year-old ten-cent piece was considered a true talisman.[36] The scale at which

17. Mrs. Jacob Stooksbury Family, Loyston, Tennessee, 23 November 1933; The close-knit family, nuclear and extended, was typical of this area. Stooksburys lived throughout the Norris district. The figures express a dignified self-respect. The grandmother holds what is perhaps the "Good Book." Around the family circle is a typical miscellany of small objects — the clock, the calendar partially hidden by it, the almanac pinned at the side of the mantel, the bottles and kerosene lamps, the spinning wheel, the ladder-backed, rush-bottomed chairs — all in a very loose aggregation but carefully contained. The windows do not match. The oversized wooden mantel has intimations of a Federal-Neo-Classic chimneypiece, but the components are out of proportion. The wood of all such rooms was genuine, as, in their particular ways, were the families and their furnishings. Photo by Lewis Hine, reproduced courtesy of Tennessee Valley Authority Archives.

these natives arranged or disposed of property, and the manner in which the TVA approached the same tasks, were ages and worlds apart. One elderly woman refused to have her picture taken unless her antique table was brought out into the sunshine to be photographed beside her. This woman's loyalty to the small-scale and miscellaneous, in its original location and not be be disturbed, was replayed, much later, in the 1960 film *Wild River*, directed by Elia Kazan and starring Lee Remick and Montgomery Clift. Clift was the third TVA agent sent to persuade Miss Ella Garth to leave her condemned island. Her sons threw him in the water. A sign on the island read, "TVA Keep Off!" Garth couldn't stand being hemmed in and preferred that people and animals be left "running wild." Her sons were depicted as being untamed, beneath a veneer of civility.

It was difficult for the poorer people to tell what objects properly belonged in any given time, there were so many times to cope with; and the 1930s, with their increasing hardships were not reassuring. That decade's suddenly increased lack of certitude could lead to physical disorientation and actual loss of contact with reality. William Henry Hawkins, who owned the land on which Norris Dam was built, laid a circle of brush around his house and ignited it when the TVA agents drew near to confer with him, burning his house to the ground.[37] A series of similar anecdotes were told about older people becoming confused and resorting to suicide and other extreme measures when their land was taken by the TVA.[38] A related story was told by Jack E. Weller in *Yesterday's People: Life in Contemporary Appalachia*.[39] A son asked if he could move into his parents' former dwelling. The parents would not answer "yes" or "no." The house had remained vacant for years because they had been unable to decide which offspring ought to inherit it. The son decided to move in, anyway, whereupon the vacant house was set on fire. The crucial moment of James Still's novel, *River of Earth*, was the instant when the wife applied the torch to her own cabin in order to cause her husband's long-resident great-uncle and two cousins to move on to other relatives. Thereafter, her more immediate family resided in the smokehouse.[40]

Differences in scale and in tempo of operation caused a lot of the misunderstanding. An opposite, more constructive option was taken up by an old man who had suffered a fire in his house on Island F (which was being turned into an ecology center at the behest of Arthur Morgan). The property owner then carefully rebuilt the house so that he might

18. House near Baker's Forge on Cedar Creek, Island F, Removed for Norris Dam. The man and his friends are dismantling a neighbor's abandoned house, salvaging materials to repair his own, damaged by fire, so that it can bring a better price from TVA assessors. These people's dealings with such materials were intimate and firsthand, their exchanges small in scale and based on the ritual of barter. Photo from Marshall A. Wilson Collection, Tennessee Valley Authority Archives.

get more for his acre-and-a-half "homeplace" when the TVA did come around. He obtained his new materials from friends' houses (18), already abandoned. His commitment to crafting was painstaking, down to the last handsplit chestnut shingle. After the TVA extension service had properly evaluated his home, it was as cheerfully and painstakingly torn down.[41] The New Deal "problem solvers," who despite their sophisticated techniques sometimes couldn't solve the larger problems, had come face to face with those who thought in a much smaller and more compartmentalized fashion and who responded to the problems they did recognize by ignoring or erasing the original question.

The building sequence on Island F was to be less innocently, but just as advantageously, exploited at Land Between the Lakes, at the northwest terminus of the TVA. The number of families to be relocated for Kentucky Dam there was approximately the same as it had been at Nor-

19. Okolona Baptist Church, Two Miles Below Miller's Bridge in Union County, Removed for Norris Dam, 1935. Graves are being moved prior to the flooding of Norris Lake. Cedar trees were thought to be appropriate for memory of the deceased. Some retention of dignity *has* to result. This example shows especially well the loneliness and isolation of many southern church buildings. The church would be the most important building in the community, but the congregation cannot afford paint, the steeple is short, and the roof is made of metal. Photo from Marshall A. Wilson Collection, Tennessee Valley Authority Archives.

ris — 2,607, as against 2,899 at Norris[42] The appropriations for moving 949 families after 1964 to clear the isthmus of Land Between the Lakes itself were much slower in coming through. This circumstance encouraged some enterprising owners to move the same building about in order to achieve "a plant," thus setting the stage for "a full return," since the structure then would have to be purchased by the TVA twice or thrice over. The simple faith was that the government assessors would not have long enough memories to realize that they had seen the building under consideration previously, on some other land.[43]

Local opinion resisted the idea of consolidating farms into larger, potentially profit-making units, as recommended by federal advisors, who claimed that nothing less than fifteen acres was needed to support a family. This local recalcitrance was abetted by TVA personnel who had grown up in the area, such as John C. McAmis from the University of Tennessee's Agricultural Extension Service (TVA director Harcourt Morgan's old agency). The service held the contract to buy the land for the Norris Dam site.[44] McAmis' permanent conviction was that families needed a small-hold, a "homeplace," that they could return to when employment ran out in the northern manufacturing cities of Cincinnati or St. Louis, Detroit or Chicago. This "farm" would be held for them in absentia by kinfolk.[45] Connected with that desire for a possible retreat to the homeplace was the custom of bartering the tools of cultivation with kinfolk. In McAmis' view, the extant web of family interdependence and permanence should not be rent and snatched away. The people around the Norris Lake site mostly were Pennsylvania Germans who had come down the Virginia Valley and then over into Tennessee. When the TVA was begun, two-thirds of Tennessee Valley residents were engaged in farming, and the annual income was $168 per capita. The main cash crops were corn, cotton, and tobacco.[46]

This sequestering impulse took on spiritual overtones in connection with churches (19) and their gravesites. Through those associations it could be shown why "the families simply did not want to go." The church meant more in this setting, because it was the only community center. There were twenty-nine churches on the Norris ground, only six of which were moved out of the flood area. Some of the cemeteries had served particular families for over six generations, but action had to be taken to remove the graves, after it was realized that water would rise two hundred feet and more above them. The newest kind of people (from Washington, D.C., and previously unattached to the immediate locale) soon would be trying to help the oldest kind of people (very much attached to one site for a long while) to relocate.[47]

Was the Norris Experience Unique?

Was the Norris chronicle merely a first-case occurrence? Did the later displacements to make way for the other dams represent a continuity or a different experience? Later observers from outside the South might consider the writings of such novelists as William Faulkner, William Terry Couch, Tennessee Williams, and even Erskine Caldwell, with his 1932 *Tobacco Road*, as exaggerated in their cast of characters and situations, leading up to what might be regarded as caricature. The cool and objective documentation of case studies by TVA agents, however, depicts people almost as exotic, they possessed so little well-being. What first becomes apparent in perusing the official reports is the insularity, the exclusivity, the dilution of means and hope, that turns life in on itself. America still pinned its hopes on a mobile society, but this segment was fixed or downwardly mobile. There was no indigenous capital pool from which to build a new hegemony. Most unsettling of all was that the region's general state of health reflected financial, psychological, and spiritual difficulties of long standing. More ominous but less acknowledged were the Calvinistic thought that maybe these people were destined to suffer so, and the political apprehension that American agrarianism now might be degenerating into conditions even worse than those of the European peasantry and serfdom.

If one concluded that Miss Ella Garth's instinctual resistance to moving for the TVA in the film *Wild River* was an exaggeration, to check that with "reality" one would have only to read the TVA report on the attempt to remove Miss Sadie J. and her sister, Miss Katie, aged 61 and 65, from their farmstead in 1940–41.[48] These sisters were relocated to make way for Cherokee Dam in East Tennessee on the Holston River, in the wartime readiness program. "The sisters do not seem to be able to grasp the facts of the situation in which they have been thrown. . . . While they seem anxious to get the advice of the [TVA–Agricultural Extension] worker, they do not seem to grasp the importance of carrying out his suggestions." They were taken on two auto trips to look at other farms, not having been off their homeplace for three decades. They "thoroughly enjoyed the opportunity to see the changes which have been brought about in the period of their seclusion." Still, they could not be hurried along by any psychological means then known, at this moment of impending national "peril." In the end, the unaccustomed stress of the situation caused Miss Katie to refuse to cooperate any longer with Miss Sadie. To some, these sisters' narrow circle of awareness and slow tempo of life appeared definitely un-American, or at least unnorthern, at the moment.

There was a certain southern gentility, fragrance, and lack of anxious awareness from the sisters' case, as there was in the acquisition of the homestead (20) of Jack W. for the same dam.[49] At age fifty-nine, he owned five hundred acres and engaged in stock dealing, so in several respects he was the prosperous "small or large farmer," the middle-of-the-frame figure, the missing link, that Professor Owsley of the Agrari-

20. Antebellum house of the Jack W. Family, Removed for the Cherokee Dam. The trees in front have been cut preparatory to demolition. The house is that of a more affluent middle-class farmer, a member of the class, neither sharecropper nor planter, that the Agrarians hoped to augment in order to make agricultural practices in the South, along with the whole society, more systematic and stable. The appreciation of Classical detail remains, but while the dentil molding on the porch is delicate, the Victorian entablature above, with its brackets, is heavier. Two eras of Classicism overlap here, as happened in the South even before the Civil War. Photo Courtesy of Tennessee Valley Authority Archives.

ans wanted to make central in the South. Mr. W.'s wife Bessie had inherited their substantial home, built in the 1860s. It was not a historical type rendered purely, but it held together adequately enough in terms of its tradition. It was Georgian-Federal in its form concentration and symmetry, but slightly untrue to that style in its Victorian cornice brackets and its later, blanker windows of rolled glass without white muntins. So much of the tincture came from unconscious chronological overlaps. The house showed as obviously Tennessean with the contrast between red brick and white trim (always a little heavier in Tennessee and Kentucky), and also with the larger size of the window-door of the second-story hall, now partially filled-in by painted glass. Inside, the structure communicated the same Tennessee-Kentucky message with eight rooms, *all* of the same dimensions, eighteen by twenty-two feet (Tennesseans and Kentuckians liked to approach the square as often as possible), with twelve-foot ceilings. Each room downstairs was of a different wood — pine, poplar, oak, and cherry, attracting attention to the once surrounding hardwood forests and lending a note of versatility to offset all the Classic space and trim, in case visitors dropped in.

For the Hiwassee Dam, also in the eastern mountains but farther down into North Carolina, evacuation took place during 1937–38, two or three years earlier than for Cherokee Dam. The houses near Hiwassee Dam "as a rule were unpainted, one-story, frame structures of three or four rooms." The people who lived in them were "given to a fierce love of personal liberty and a strong attachment to home and family, hesitating to contract an obligation but likely to fulfill one, and usually courteous but slow to make friends." The case of Will S. from there (21) demonstrated again that the results of the TVA removals were disconcerting yet not all unfortunate.[50] Mr. S. was a tenant farmer to begin with, but the precipitate action of the TVA made it possible for him quickly to acquire twenty acres and to build a five-room "box" (sawn boards as distinguished from logs) house on it. "This was made possible by the securing of TVA employment for his son, Howard, and help from his brother Sam." Howard did not "fully meet the standards of merit and efficiency required by TVA of its employees," but Mr. Sharp, his foreman, was giving him friendly counsel, and it looked as if the situation could be retrieved. The whole family took "a new interest in life" as landowners. The TVA agents never tried to destroy the family members' dignity, innocence, or individuality. Despite TVA's commitment to technological and mechanical innovation, Mr. S. was enabled by it to acquire a spanking new turnout of archaic oxen in an archaic yoke (21), pulling an ancient wagon in which to drive slowly up the dirt road to his newfound home, where he soon would be cultivating "a normal crop."

There were other, longer case histories to indicate how thoroughly miserable, tragic, and grinding daily existence could be in Tennessee in the 1930s and 1940s. Because of the start of the Watts Bar Dam in Middle Tennessee, the Gene Mc. family had to be told that it would have to

21. The New Ox Team of Mr. S. A former sharecropper, Mr. S. was removed for Hiwassee Dam. His new house, farm, and team were made possible by TVA benefits. Photo courtesy of Tennessee Valley Authority Archives.

22. Family of Bert R., Removed for Hiwassee Dam. They were beset by several illnesses — the father with silicosis, tuberculosis, and heart failure; the mother with pellagra; and the children with chronic malnutrition — but they kept together and struggled through to better conditions, the children helping. Photo courtesy of Tennessee Valley Authority Archives.

move early in 1941, in response to a faraway war.[51] On this first notice from the TVA agent, "Mrs. Mc said they were both in good health and had no physical handicaps." They lived in a two-room box, board-and-batten house, without any "modern conveniences." They had been married when he was fifteen and she thirteen and expected their first child in April. But over the winter, Gene Mc. contracted influenza with pleurisy, which led quickly to galloping consumption. He had put in a crop in the spring but no longer could tend it. His wife became anemic. She needed an operation to relieve pelvic inflammation caused by the birth of her child. Finally, the Pine Breeze tuberculosis sanatorium would not admit Mr. Mc. because it saw him as having only a few months to live. By then he was twenty-four years old. That fall, on 10 November 1941, the family's possessions were removed from the house standing alone in the now-neglected fields. The Mc.s had lived in the Sagebrush community since birth. Besides helping to disrupt this family unit, World War II played a large part in obliterating the memory that in prewar days such Third-World conditions had existed within the continental limits of the United States itself, where the lives of even the youngest could be profoundly touched by adversity while residing on native soil.

Bert R. (22) did not perish when abdicating for Hiwassee Dam, but he already had been afflicted by silicosis, tuberculosis, and heart trouble.[52] Originally he had lived with his wife and five children (with one more on the way) in a house where he paid no rent because it had been condemned. He went to work for the TVA in fall 1936 but had to be laid off because of his illnesses. Mrs. R. (22), his age but looking older, suffered from pellagra and had to go into the hospital. The children were often without food and took odd jobs to keep the family going. The father finally was able to gain admission to Black Mountain, North Carolina, Sanatorium, but reemerged from it, claiming the doctors had told him that he didn't have tuberculosis after all. Three months later, in June 1938, he was still improving and could be employed again with light work on the construction of Hiwassee Dam. The TVA regularly tried to make such humane adjustments. The health of the whole family improved, and the oldest son gained nearly twenty pounds. Bert R. became "decidedly more self-reliant and optimistic." At the end of 1938, he built a three-room box house on land purchased by him on Bumpus Creek. He cleared eight acres for cultivation. The TVA concluded that "this can be regarded as a very remarkable relocation and adjustment."

Could it be that, after all, health came from a state of mind? Were these the "average" Americans with whom organizations normally prefer to deal, or were they made more real than that by their acute suffering and anomalies? Despite the compassionate attitudes of the New Deal, at the time no food stamps, housing subsidies, child welfare benefits, or health grants for poorer families existed.

The areas within which Sadie J., Jack W., Will S., Gene Mc.'s family, and Bert R.'s family tried to stay and recover their equilibrium could be described as outer fringes and fastnesses: in the mountains, out on the plains, or up the rivers and away from the major cities of the valley. Yet in 1943, at the Fort Loudoun Dam directly adjacent to Knoxville, similar or worse conditions were in evidence. Some people were driven off their houseboats on the edge of town. An even more excruciating situation was that of Mrs. Mattie X.; her daughter, Martha X., later Y.; Frank Y., the son-in-law; and the sons of Mrs. Mattie—Blane, Condon, and Hugh, respectively thirty-five, thirty-two, and thirty years of age; of Unitia, Tennessee.[53] The X.'s were "an old and prominent family in the community," and highly respected, but they had suffered reverses. "Mr. Y seems to be a poor worker and a poor manager but is a willing talker." All three sons of Mrs. Mattie were mentally retarded and physically infirm. There were also two grandchildren, eight persons in all, living in. Once a handsome Georgian-Federal house (23) on a ferry road running right down to the river, the X. manse had an ell and a two-story, Palladian front porch of a house type frequent in Tennessee, that James Patrick has traced to the famous Miles Brewton House in Charleston, South Carolina.[54] The house by now was no longer being painted to keep up appearance, and the floors of the two porch levels sagged badly. The Classical piers had lost their equilibrium too. "The inside is dark and gloomy." A new house was found (24), but the occupants were reluctant to vacate for the X.'s and Y.'s in a time of great housing shortages. Mrs. X. was driven to her bed for a month by anxiety and high blood pressure. Rationed gasoline had to be begged from the Office of Defense Transportation for the three moving trucks. The combined annual earnings of the X.'s and Y.'s for the family of eight were around $1,200, from corn, livestock, poultry, pears, milk, and the Soil Conservation Service.

Charges against the TVA of indulging in wild extravagance to assist such families don't appear to hold up when the actual records are closely examined. The feeling remains that the TVA could have built new houses and villages for them without undue expenditure, especially in light of the unusually fine roster of architects and planners it had assembled, probably the finest of the twentieth century for a team project. Much thought and care had gone into the dams, to establish a tandem series such as had never been seen before in the whole world, functioning integrally. However, even for dams, money, labor, and materials were extremely short during World War II. Everything had to be done by fits and starts. So the X. and Y. family had to start over in a secondhand, too long, too narrow bungalow that vaguely resembled an American Classic temple form in wood, overlaid with the recollection of a New Orleans double

23. The Mrs. X. Mansion at Unitia, near Knoxville, Tennessee, Removed for the Fort Loudoun Dam. The two-story, three-bay, temple-type porch is characteristic of middle- and upper-class houses in pre–Civil War Tennessee. Mansions had more difficulty surviving. The road in front originally led down to a ferry landing. The X.'s had been "an old and prominent family," but they had fallen on hard times. Photo courtesy of Tennessee Valley Authority Archives.

shotgun house. The Classical form was present, without any refinement being contributed by Classical detail, but at least the structure was painted a grubby white. (24) It gives another reason to regret the thinning of the Classic tradition as it moved inland from the Atlantic and up from the Gulf. Yet, at the very end of the Tennessee River, in the log cabin of Kenneth M. (25), which previously had belonged to his father, greater substance and dignity can be discerned.[55] Even more than the X. mansion, this cabin nevertheless exhibits a satisfactory set of ancient proportions and a solidity and balance that seem beyond the grasp of the twentieth century. It is as if the well-balanced frontier, and then the Colonial-Classical presence, grew fainter and fainter, fading away in the distance of time. The architectural tradition was sound, but it had been let go too long.

Only by returning to the main street of Unitia (26), the village of the

24. White Painted Southern Bungalow into which the Tennessee Valley Authority Relocated the X. Family. It is another neoclassical temple form, influenced by the New Orleans double shotgun house type. There is no longer any Classical ornament. Photo courtesy of Tennessee Valley Authority Archives.

X.'s, can we come to a fuller realization of where these transformations came from, and where they might or might not lead. The men standing quietly on the porch of the Unitia store, with their felt hats and railroad caps, were living on borrowed social time in a soon-to-vanish place. Before TVA, the village street was an exceedingly dim folk memory of the Duke of Gloucester Street in Colonial Williamsburg, where one rejoices in the street's Baroque axiality and width, its wonderful compromise of domestic, commercial, and civic scale. After TVA, Unitia would become a miniature Lost Atlantis under the waves of the lake. A relatively minor

25. The Kenneth M. Log Cabin, Removed for Kentucky Dam. Several claddings have been tried on this cabin, such as board-and-batten and clapboard overlays, and verge boards with teeth on the raking cornices. Also, the house is being torn down. Nevertheless, it still appears to be out of a better-proportioned, more-contained, sturdier tradition than the X. family bungalow which was provided by the Tennessee Valley Authority (24). It was not easy to keep a hold on the folk tradition, and after the Civil War, except in the major cities, there was also little hope of continuing the "educated" architecture that Tennessee had had in good measure before. Photo courtesy of Tennessee Valley Authority Archives.

Somewhere would disappear suddenly into a hidden depth of Nowhere, contrary to the usual sanguine American expectation. As attractive entities, such hamlets were never much more than dim emulations of the traditional southern county seat, with maybe (as at Unitia) a crossroads; a supply store, seen in the photo; a gas pump; one or two signs; and sometimes a church.[56] The wealthy and well-read class of the South had been eliminated by the Civil War, so that its contribution of architectural discrimination and taste was no longer available. That type of

26. Main Street, Unitia, Tennessee, Removed for Fort Loudoun Dam. This village, destined to disappear, was the home of the X. family. The last house visible down the street had the same type of two-level temple porch as the orginal X. manse. Otherwise, the long narrow oblong form prevails, with little differentiation between house and store. The only ornamental or textural effect comes from the white paint on the narrow clapboards. Even the porch columns were thin tree poles. This was a minimalist architecture derived from what was once a richer eighteenth-century tradition. The memory grew fainter and fainter. Hence the native vitality has to come from the figures with their caps and hats. Photo courtesy of Tennessee Valley Authority Archives.

input Thomas Jefferson had thought absolutely indispensable for a burgeoning republic. Money was pumped into this region by the TVA, acting as a good right arm of the federal government, but never enough money to give its dwellings and their locations new and permanent forms, such as easily could have been limned by its talented architects, engineers, landscape architects, and land planners, working in concert. The Tennessee Valley was a national laboratory where not quite enough experiments were to take place. Refined and uplifted in some respects by the TVA, the valley was left too crude and impoverished in others. The need of the clients was then pressing, the potentiality of the talent to answer that need manifold and ingenious, but actual responses never were sufficiently well financed.

Nevertheless, an examination of "the homeplace" and the people in it during the 1930s is an instructive exercise, primarily for two reasons. The shanties of the Depression-generated Hoovervilles in the gray areas of the cities, the jalopies being used as homes on wheels in Farm Security Administration photos and in the pages of Steinbeck's *Grapes of Wrath*, and the camps of the Bonus Marchers on Washington all exhibited degrees of deprivation equivalent to that of valley residents. None of them, however, showed a similar desire to stay attached for generations on the same plot of land and in the same house; none was that heavily ensconced. In the valley a different kind of struggle went on, one more tied to a given landscape. Moreover, peculiarly hollow vibrations were set up in that landscape, vibrations that responded only to singular manifestations. The tiny unpainted churches (19) in the valley had a unique function, often being the only focal point uniting the community. The arcadian setting meant fewer buildings, each more potently symbolic because of its isolation. Lonely houses up a creek or in a cove often emitted similar symbolic meanings. The same type of psychological echoes or reverberations over architectonic status could be communicated by the power houses of the great dams that, once built, also would vibrate and hum on alone as isolated objects. Then, in World War II, came Oak Ridge, bigger yet, but also a unique settlement—top secret and off by itself preparing the new nuclear power of vast dimensions. Size, scale, texture, and usage differed for each of these building types and places, to be sure, but each could generate a powerful allegiance and close attention from its adherents as a singular object, that might induce hope, delight, and admiration, or as much indescribable pain and anguish. The depth and singularity or experiences were greater and more vivid in the Tennessee

Valley than in the rest of the country during the 1930s and 1940s. It was one of the most interesting places in the whole world. It could be demeaning and pain badly to remain in the Tennessee Valley then, as in the rest of the United States, but it also could be a heady and exalting place to be present in with its exploration of the extremes.

The Agrarians Start a Backfire

By more recent standards, any analysis of social problems and conditions in the Tennessee Valley undertaken during this period of 1930–45 is apt to appear naive and unscientific. Where are the armfuls of diagrams, the reams of statistics? There was less time for painstaking analyses during the 1930s, and much more affinity for traditions, sentiments, and pride. And in the South there was a stronger, more enduring pull toward retrospection, while, among the New York intelligentsia in the 1920s, there had been continuous calls for more rational future use of land and water resources. The initial impulse for the southern intellectuals had, of course, come from the Civil War; while for the New York intellectuals the fuse obviously was lit by World War I, although the motivational forces also went farther back. It was significant that the southern model for regionalism originated primarily with professors of English literature, since they represented the more genteel tradition in the South of a literacy and a leisure always at risk. The Agrarian movement of Donald Davidson, John Crowe Ransom, Alan Tate, Andrew Lytle, Lyle Lanier, Frank Owsley, and Robert Penn Warren began in the English Department of Vanderbilt University. Southern concern over regionalism, doctoral-level literature, graduate social science studies, and industrialism centered in two institutions: Vanderbilt, and the University of North Carolina at Chapel Hill (UNC).[57] These two enterprises were bootstrap operations in every sense. In 1932, the year before the TVA was founded, Rupert Vance, a sociologist long engaged with Howard Odum in regional studies at UNC, had indicated the future battleline between the Vanderbilt Agrarians and the not-yet-appointed directors of the not-yet-formulated TVA:

> Just as the post-war bourbons, the Confederate Brigadiers and the Mossbacks were being abandoned by southern liberals, the traditional agrarianism of the South became vocal. Driven to utterance by the encroachments

of promotion and industrialization [not by the TVA, but by a presentiment
of it—W.C.], the writers of *I'll Take My Stand* [the 1930 Agrarian proclamation of Donald Davidson and eleven other southerners subtitled "The South
and the Agrarian Tradition"] have given in the first coherent presentation of
regionalism for the South a plea for return to agrarian aristocracy and isolation. The neo-Confederacy so far is a literary movement, but it attempts a
cultural revival. Briefly it declares for a reversion to the civilization of the
Old South, mildly deplores the passing of slavery, and calls for a wall around
the South that it may keep out industrial invaders . . . , as an effective sentiment around which to rally a genuine regional movement, this nostalgia for
the old South may prove of more avail than the formulae of technicians and
the undifferentiated aspirations of southern liberals. . . . From any technical
or economic angle the traditionalists have the least to present to regional
planning; from the viewpoint of cultural ideals their gifts are greater.[58]

Strange to have poets and essayists rather than technicians and statisti-
cians tackling social problems at the outset of the 1930s, but hadn't poets
Byron and Shelley, Blake and Wordsworth done the same for their culture
under the sharp whip of the French, Greek, and Industrial revolutions?
The intention was to refortify the older southern isolation, strengthening
it through a revival of inner cultural and historical, aspirational and in-
tuitional means, rather than through single-tracked industrial or commercial methods from outside. There was a need to insert a leadership
group or class into the southern matrix, but it should not be a technological or managerial one, but rather a literate and intellectual one, with
some warmer feeling for sociability and the other arts of leisure. For the
TVA and the Agrarians alike, the initial thought was for a larger scope
and a rising quality in valley life that might become middle class.

   Davidson wanted, above all else, to counter the condescension he felt
emanating from northerners: "The residue left by the Dayton [Scopes,
"Monkey," or Evolution Trial of 1925] and the Scottsboro case was the
picture of valley residents as ignorant, bigoted, backward, moronic, de-
generate caricatures of humanity, who might in their better moments
be quaint or picturesque but could not be thought of as counting for
much."[59] The Agrarians henceforth set their faces against H.L. Mencken
and the Atlanta liberals. The superficialities of those journalists appear
to have most bothered the Agrarians. The liberals in the Tennessee Valley
would attempt to compensate for being "pariahs" by calling for more
progressive colleges and schools, more unions and industries; while the
conservative intellectuals, among whom Davidson counted himself, would

recall the deeds and customs of their southern ancestors echoing along the corridors of time and "begin to resent the entire modern regime."[60] Modern "progress" was seen as soulless. The Agrarians wanted to turn the clock back, while everyone else wished it to leap forward by means of manufacturing and the TVA. This divergence created a schism, although anger, isolation, frustration, bewilderment, and bitterness were feelings all Americans shared at the time of the Depression.

Would these local groups be virtually powerless to bring about a true southern cultural renaissance in the face of the federal government's good, if clumsy, intentions?[61] At the end of his essay on the Tennessee River, Davidson seems to be saying that if southerners were including too much post–Civil War emotion in their arguments, it would be equally permissible to presume that the TVA was depending too completely on a cold, austere, one-service technology. "The Tennessee Valley Authority, riding on the wave of the economic uplift, promised through economics to wipe all sins away. Behind its mechanical provisions for taming the Tennessee River was the open hint that social goodness would be substituted for supposed social depravity through the distribution of government-controlled electric power."[62] The ethical imperative was being overly engaged through mechanical devices and electrical engineering. Economic and moral exigencies were being wired up in a most unpromising and possibly disastrous way, according to Davidson. The New Electricity was as unreal and mythological as the Lost Confederacy.

The Agrarians had their own internal quandary, too. They could not find good enough images to demonstrate their case and revitalize their cause. The one physical image that Davidson approved of in the TVA was Norris Village. It seemed to him "to tell the visitor that modern America on the whole has done rather poorly with its resources, and it reminded him what might have been, if only . . . if only! It was a lesson in something or other that Americans ought to learn."[63] Agrarian writing took a great deal of its power from nostalgia and vagueness. The "lesson in something or other," as with Henry Ford, would come from some modified, more intricate pattern of human settlement and ecology, a physical rearrangement that would bring about a better norm of human association. This quest also had a long history in the North, especially with the Progressives who fed ideas into the various New York regional plans, and with Daniel Burnham's 1909 regional plan for Chicago. While the Agrarians' word pictures became more colorful and feeling, exactly what they wanted in the way of a framework of architecture and planning in the "innocence" of the valley landscape became vaguer. They knew what they were against, but what were they "for," except for the rusticity of Norris Village? In "The Southern Agrarians and the Tennessee Valley Authority," Edward Shapiro indicated what they were against in terms of originating places and ideas. The Agrarians "pointed to the city's flashy and cosmopolitan artists as proof of its estrangement from traditional American culture. Truly American art, they proclaimed, could be produced only by artists rooted in a provincial and conservative society, such as the South, and not by deracinated and bohemian urbanites who prefer 'sophistication over wisdom; experiment over tradition; technique over style; emancipation over morals.' New York, . . . which attracted 'all the celebrities and semi-celebrities of Europe into its orbit,' exemplified the metropolis' alienation from the American hinterland."[64] The Agrarians wanted to feature local tradition before national experiment, feeling before logic, and style ahead of technique. The question remaining today is whether it would have been better for the TVA to have introduced several more local-appearing congeries like Norris Village, in order to allay the native suspicion and fear of technology. Arthur Morgan and members of the early TVA staff had the instinct for that kind of amelioration. The Agrarians' only distinct statement in that realm appears to have been that their ideal villages should rest in the same kind of landscape as that of the rich farm country around Nashville.[65] The danger in romanticism could be that the allusions would be too faint and too remote. Davidson never seems to have known why he liked Norris Village. Ransom admitted much later that the Agrarians had failed adequately to establish their landscape preference;[66] and Davidson, with his permanent inclination to think large and tall, as he declared in his poem, "The Tall Men," yearned only "to flee to the remote refuge, the Eden-like valley." No specific substitute scene was offered. Ultimately, the aspiration of the Agrarians was to create a new generation of literate, enlightened farmer personalities, a transported honest-broker class of Jeffersons, Madisons, and Monroes from Virginia, who knew well how to deal with the land in more appreciative fashion.

The leverage of the TVA in this situation came from the fact that its monuments were so visible, clear, and prominent. They had skill, decisiveness, and determination marked all over them. There was little or no native reticence. This was why the monuments would take quicker preeminence over any softer imagery the Agrarians might elicit. For the

TVA, any futurism was never enough; for the Agrarians, it was already too much. Robert Penn Warren, as an Agrarian, outlined still another, similarly unresolved picture of the idyllic village, this time in unsublimated confrontation with the TVA. Of all the Agrarians, he had the strongest sense of the human condition, the Tennessee countryside, and the southern past. Out of this sense, he wrote *The Flood: A Romance of Our Time*, about an imaginary town, Fiddlersburg, soon to be innundated by a great new lake. In actuality Fiddlersburg was Loyston, soon to be under Norris Lake. The urgency of that event impacted on entrenched local tradition. The suddenness and comprehensiveness of the pending flood, as it might overwhelm people and individual lives, put memory in jeopardy. The principle of "dignity" looked different to the two parties, TVA and those who were to be removed. On the last "official day" of the town's existence, the fear was expressed that the inhabitants would become wandering shadows lacking proper identity, since coding for the latter would be concealed beneath the floodwaters, as also at Unitia, Tennessee.(26) It was the old southern problem: where did the individual, faced with uncontrollable change such as those brought on by the Civil War or the advent of the TVA, next belong in time and space? Wherever the inhabitants might turn up next in body, their spirits would have to be left behind, submerged, in Fiddlersburg. In the book, it is the size and scale of the TVA operation that most frightens the native southerners, accustomed as they are to slow and tiny cultural changes, in surroundings that are constantly reassuring since they are so familiar. Yet another question also had to be asked, the ultimate question of the southern conscience: whether the town *ought*, because of its sins, to be washed away in one great evangelical purge. Maybe that was the only way to shrug off the overwhelming weight of the time — all in an instant, through one great catastrophic upheaval? One of the difficulties, of course, had been that the people had been so turned in upon themselves that they found it impossible to treat with others. As Robert Penn Warren put it:

> Hell, the South is the country where a man gets drunk just so he can feel even lonesomer and then comes to town and picks a fight for companionship. The Confederate States were founded on lonesomeness. They were all so lonesome they built a pen around themselves so they could be lonesome together. The only reason the Confederate army held together as long as it did against overwhelming odds was that everybody felt that it would be just too dammed lonesome to go home and be lonesome by yourself.[67]

The Depression exacerbated that southern sense of loneliness and isolation, represented by the box and log houses out of which the early families were removed, together with their lonesome churches (19). By the
time of removal, the feeling had become so acute that the Agrarians
wanted to fence it in rather than out. To increase the dichotomy and
distance, the TVA engineers seemed to the Agrarians to lack seasoning
and maturity, to be superficial as the newspaper people, to display no
depth of feeling in their eagerness to get on with technology.

> The dam was going to be great, the young engineer said. Going to be near
> a hundred square miles under water, going to back up the water for twenty-
> five miles, he had said, gesturing south, up-river. Most of the land not much
> but swamp or second growth. And what good land there was—hell, they
> didn't know how to farm it anyway. But with power and cheap transporta-
> tion it would all be different. A real skyline on the river, plant after plant.
> Getting shoes on the swamp rats too, teaching 'em to read and write and
> punch a time clock, and pull a switch. It was going to be a big industrial
> complex, he said. He liked the phrase, industrial complex.[68]

The anxiety was that the frontier yeomanry had slipped further down
into a peasantlike condition, unacceptable in an America with once-
high hopes.

The notion most passionately rejected by the Agrarians was the factory system. It was likely to threaten the "civilization of the small farmer."
The TVA factory system of the future was represented in Penn Warren's
phrases as "a real skyline on the river, plant after plant." He saw that
the TVA engineer, no less than the Agrarian, was dedicated to fetishes,
but the former's idols were so bumptious. Fading—the vanishing of villages (26), houses (23–25), ideals, and interests—was a phenomenon of
the South, but were not these engineers making it happen faster then
it should? Warren has his central character try again and again to locate
in Fiddlersburg the grave of an old tailor who had befriended him long
ago in his youth. Finally, he gives it up as an unnecessary task, because
"there is no country but the heart."[69] It was the same conclusion that
the Asheville novelist Thomas Wolfe had reached, that "you can't go
home again," a realization that made the heartache more painful each
time the realization again swept over the petitioner. The "agrarians,"
"distributists," and "decentralists" of the South included "Southern Agrarians, historian and journalist Herbert Agar [of the Louisville (Ky.) *Courier-
Journal*], contributors to *Who Owns America? A Declaration of Inde-*

pendence and Free America [the second manifesto of the Agrarians], and members of the Catholic rural life movement."[70] They "argued in behalf of a peaceful, middle-class revolution leading to the widespread distribution of property, the decentralization of economic and political authority, and the decentralization of the city,"[71] but, amid this fragmentation of concerns, most distressing to them was rural poverty and dispossession.[72] People would leave the cities to decentralize, but where else in the South would they resettle, where end up? The Agrarians, while encouraging people to seek appropriate places elsewhere, never quite said which and where those places were. It was a curiously reversed attitude for Americans — the idea that self-effacement and dissolution might lead to a retrenchment and a greater sense of self-worth, but also typically American in that the imagery depended more on words and less on a clear visualization.

How The Strip Mining Crisis Affected People

Constant loneliness and uncertainty about locations and identities haunted the 1930s. Air pollution and further displacement haunted the TVA in the 1950s. This newer crisis peaked between 1954 and 1955, ostensibly coming from the building of large steam plants beginning in 1949 but actually tracing back to the Watts Bar Plant of World War II.[73] By the early 1970s, much of the steam power generation had been converted to nuclear.[74] Freeing power for the Atomic Energy Commission (AEC) for Oak Ridge and Paducah was the underlying reason for the Dixon-Yates dispute of 1954–59, during the Eisenhower administration, over the steam plant projected to be built across the Mississippi River in Arkansas by private companies in order to supply Memphis, which then was supposed to release some of its TVA electricity to the Atomic Energy Commission for uranium enrichment.[75] Just previously, in 1951–53, had come TVA's race to finish the Shawnee Steam Plant before the Electrical Energy Incorporated's Joppa Steam Plant across the Ohio River. Both plants were created to supply the AEC's gaseous diffusion installation at Paducah, Kentucky. The Joppa plant came in later and, as a private undertaking, cost more than the public Shawnee plant. No longer could demands for energy be satisfied by hydroelectric power alone during the 1950s, so the interest in coal, which once had been passed over in favor of the "cleaner" water power, reawakened. Seams of coal ran under much of the Tennessee Valley, from Virginia to eastern Kentucky and Tennessee, down into central and southern Tennessee around Chattanooga, and ended in northern Alabama, near Birmingham. Altogether, these areas incorporated the southern Appalachian and Cumberland mountain regions, and were called the Eastern Mines. Contrasted to that group were the mines of southern Illinois and western Kentucky, called the Western Mines.

By 1955, the AEC was consuming half the electricity that the TVA was producing, the requirement for federal installations having risen from 2.2 billion kilowatt hours (kwh) in 1951 to 21.8 billion kwh in 1955. By 1955, the TVA was consuming 12 million tons of coal per year, while in 1951 it had needed only 1.2 million tons; at the start of the TVA, it had needed absolutely none.[76] Local supply and demand also altered rapidly, since the TVA came to buy half of the coal mined locally, at the same time that national demand was declining.[77] That fact eventually gave the TVA much more consumer influence than it had bargained for—far more, indeed, than it really had wanted. The TVA may have once been a backwater experiment, but by 1955 it had become the chief source of federal electricity and the largest coal customer in the nation.[78]

This "favorable" position made for desirable, or undesirable, side effects, the evaluation depending upon one's angle of vision. For instance, TVA, as a result of a new process of pulverization, discovered how to use ungraded and unwashed coal.[79] This was regarded by the TVA as a great step forward in productivity, but it also led to the reopening of "dog hole" mines in the mountains, bringing on "spot" contracts for delivery in four weeks or less. Such procedures undercut the bigger coal-mine operators, who usually received contracts for one to fifteen years and had much larger overheads in machinery and labor costs to carry, especially if they had labor contracts with the United Mine Workers of America, with the union's higher wage scales and better fringe benefits. Doghole owners backed a truck up to an improvised mine entrance, quickly shoveled ungraded coal into the truck, and then delivered the coal to the nearest TVA stockpile. This impromptu, handcrafted method of acquiring coal was described by no less a person than John L. Lewis as emerging out of "the pushcart policy" of the TVA.[80]

The debate was intensified when sixteen small operators in the Whitwell-Palmer coalfields near Chattanooga instituted an antitrust suit against the TVA; the Louisville and Nashville Railroad; the United Mine Workers; and the Western Kentucky Mining Company, which had Cyrus Eaton as chairman.[81] In the long run, only Eaton's mining company was left in the suit. The UMWA had lent the Western Company 25 million dollars to mechanize its operation. The Louisville and Nashville Railroad was assumed to be awarding the western Kentucky coal-mine operators lower rates, and the TVA was accused of discriminating against the Eastern Mines by purchasing its coal mostly in the west. The claim was that three hundred thousand miners had been put out of work elsewhere by such actions. The TVA's culpability was said to have increased when it let out such long-term contracts to the western Kentucky companies.[82] The plaintiffs in the suit asserted that slowly they were being forced out of business.

The official rebuttal was cast in terms of the necessity for the federal government — i.e., the TVA — to purchase through competitive bidding. Adding insult to injury was the observation that the disaffected miners should look toward more efficient machinery, better management, and greater capitalization in order to reach a higher plane of productivity — an exhortation that was interpreted in some quarters as gratuitous advice. In 1961, mine owners of southeastern Tennessee volunteered to find a pilot mine for the TVA to manage, so that TVA could illustrate concretely to the mine managers exactly what it meant.[83] Indeed, in that same year, TVA began to acquire its own reserves of coal for future mining, building up a backlog of 382 million tons by 1975.[84] The TVA attitude and its initiative with coal were epitomized by the Paradise steam plant on the Green River in western Kentucky. The plant was built on top of a large coal mine, cutting coal transport expense. To illustrate how productivity might be stepped up elsewhere, a huge, electrically-powered shovel (27), in which the operator reached his cab via a five-story elevator, was constructed between 1959 and 1962, costing millions. The scoop gulped 173 tons of coal with one lift. The average miner in the eastern Tennessee fields could dig only 7.55 tons a day. The regular output per miner for the more efficient western fields was 40 tons a day. Symbolically, as poetic, resisting southerners always had predicted it would, giant technology now bid to overwhelm the southern landscape. The most highly mechanized strip mines turned out 100 tons per man day in the early 1960s.[85]

Around the Paradise plant was a six-thousand-acre tract of land that

27. Peabody Coal Company Shovel, Paradise, Kentucky, Steam Plant, 1959–62. The Tennessee Valley ideal of "efficiency" had gown to this enormous size. Strip mining like this produced profound gulches. In this one, human figures can be located beside the thick ground cable bringing electricity to power the shovel. They appear surprisingly small, due partly to the fact that the shovel is such a departure from the size usual for earth shovels. At the time, this was the biggest self-propelled vehicle in the world. Destined to take out coal for seventeen years, it was made by the Bucyrus Company which also manufactured the steam shovel in which Theodore Roosevelt was photographed on the Panama Canal in 1906 (8). Peabody photo courtesy of Tennessee Valley Authority Archives.

had been stripped for the TVA under the name of the Sinclair Mine of the Peabody Coal Company.[86] In the early 1960s, TVA set out to provide a modest demonstration at the site, reclaiming three hundred acres through fertilization and by planting ten species of trees. By 1969, thirty-four species of wildlife were using the area.[87] The original hope had been to recover the whole six thousand acres of the Sinclair Mine. The Paradise Mine area and steam plant site then became the prime demonstration area, a first sign of response to strip mining.

The environmentalists, however, coming into their own about 1965, felt that this reclamation effort was too little, too late, even though in 1965 the TVA further expressed its changing attitudes, when the organization added to its contracts statements recognizing the necessity for reclamation.[88] A significant barrier to reform was the TVA's realization that stripping yielded two and a half times more coal than underground mining. As early as 1958, TVA spokespeople had been quoted as asserting that "bigger, faster stripping shovels, more powerful bulldozers, improved explosives and bigger haulage trucks have all helped to step up productivity."[89] Achievement seemed to be measured solely in terms of quantitative results coming directly out of mechanisms. This singleminded thrust toward greater haste and efficiency had been advancing in the valley since World War II. The district had started to run faster and faster with its technology, and now could not stop stumbling forward, while groping ahead with its engineering yardstick to try to regain its balance. However, native people interpreted this latest sortie into productivity and efficiency in the same way that they had interpreted the establishment of the dams and hydroelectric power in the Depression — as an insidious assault by outside agents upon their own individuality and special habits of living. Similarly, strip mining by TVA was interpreted by some as an assault on the independence of the small mine owner, just as in the 1930s the effort to stop erosion had been interpreted as an assault upon the independence of the agricultural small-holder, who, like the miner, clung desperately to the side of an eroded hill. As one observer of strip mining in the early 1970s put it, "Instead of trees and ground cover, ugly brown scars mar the mountainsides."[90] The landscape had never been treated very sympathetically by the second and third generations of TVA leaders, anyhow. Only the first generation had cared, and not all of them. The Conservationist mindset that, in the Far West, had resulted in national forests and parks, could not take hold firmly enough in this valley, because in the early 1930s "enterprise" and technological change had been the official watchwords for it. Such stances on the part of the TVA were not so much evasions, as they were seen as being in the 1960s and 1970s, as they were further expressions of genetic codes established at TVA's conception. Electrical production had been seen as the first cause of a rising standard of living, and by now the power program commanded 90 percent of TVA assets,[91] representing for many a self-fulfilling prophecy. By 1975 the TVA was producing 6 percent of the nation's electricity and consuming about a quarter of the valley's stripmined coal.[92] TVA had become a huge utility company, with a voracious appetite for coal: "gradually delivery of cheap power became the only goal of the agency." Other criteria of amenity and progress were being crowded out by this untoward exclusivity: "It was simple to measure the increases in power, far more difficult to calculate the results of promoting agrarian reform or creating new community institutions. . . . It was an altogether natural progression from the vague dreams of radical change to the clear measurement of kilowatt hours."[93]

The humanity and generality of the overall effort grew ever more obsolete. The authority was seen as no longer caring as much for its people or the land into which they had subsided so long ago. Thus, two real disadvantages that could not be ignored arose from the latest TVA attitudes. These disadvantages were so indubitably visible — completely manifest — that they could not be ignored, at least not by everyone. No matter how often the TVA management proclaimed that its purpose was to keep electrical rates as low as possible, and that it was for this reason that it devoted itself to "least cost" coal, those rents and slashes on the eastern hills remained permanent and conspicuous. They were manifestations as negative in character as the great white walls of the first dams, stretching across the valleys from hillside to hillside, had been positive. And, almost like a Biblical first cause, hanging on forever, was the renewed presence of an impoverished people. The plight of the individual miners from the Eastern Mines and their families was even more disturbing to many than the scars from strip mining. In the frigid winter of 1961 there was an earnest hope that the TVA would award a contract for ten thousand tons of coal a week to the hard-hit counties of Marion, Grundy, and Sequatchie in southeastern Tennessee. An unemployed miner observed, "the beginning of all this misery is TVA's trying to match the prices for our coal — hand dug — against the prices of the big shovels in West Kentucky."[94] The seams of coal in West Kentucky ostensibly were thicker. Many children in southeastern Tennessee could not go to school that winter for lack of clothing. It was an old, old story in the vicinity. Many children were undernourished—some starving. These three counties were deeply involved in union problems as well, "growing out of the struggle for the giant coal market created by the vast expansion of the Tennessee Valley Authority steam-powered electricity of its generating plants. There have been a number of outright murders, several cases of arson, dynamiting of homes, of industrial and commercial buildings, railway facilities; firing into trucks and automobiles and general unrest throughout the coal fields."[95] In summer 1975 at the University of Tennessee, a TVA official told students from the Third World countries looking for settlement models that the "TVA had not always understood the human situation,"[96] failing in just the aspect in which its original sponsors most hoped it would succeed, as a final triumph of the Progressive spirit of the 1880s and 1890s. The means perhaps had become the end.

The protest against strip mining had begun when, "in the late 1960s the public had 'caught-on' to the ecology movement and more attention was focused on governmental policy. But until that time, citizen concern with TVA policy in this area was almost nonexistent. The pressure exerted on TVA, . . . was primarily from within, from other governmental agencies and from the [Louisville] *Courier-Journal*."[97] Mass opinion—at least mass intellectual opinion—began to be aroused with the publication in the April 1962 issue of the *Atlantic* of an article by Harry Caudill, Kentucky legislator, titled, "The Rape of the Appalachians." His book of the next year, *Night Comes to the Cumberland*, was printed in Boston by the Atlantic Monthly Press. Caudill's main point was that, through strip mining, the larger TVA region had benefited at the expense of a marginal part, eastern Kentucky and Tennessee, the true Appalachia.[98] TVA eventually responded (through director Frank Smith, first in the mid-1960s) by calling for state laws, interstate compacts, and a federal law to bring strip mining under control, but this became a plodding process because of the TVA preference that the states first take up the legal slack whenever possible. Kentucky, under the reponsibile prodding of owner Barry Bingham of the *Courier-Journal*, had passed such a law, but it was poorly administered. Tennessee passed a weak one in 1968. The TVA itself had not begun to include mention of reclamation requirements in its bid invitations until 1965. Interestingly, coal costs did not rise as a result of that preliminary correction, which was supposed to be maintained until the states had passed adequate laws.[99] Notable was the absence of any fundamental aesthetic considerations to accompany the TVA requirements. There seemed to be a subliminal wish not any longer to perceive the manifest landscape or the condition of its people, but rather to deal more directly with the abstract principles of power production and least-cost contracts: "The mere size of the power program overwhelmed all the other TVA activities, and no division wanted a confrontation with Power."[100] This reluctance to view at large and along multilinear channels had been at least temporarily overridden in the more desperate and more innovative days of the Depression, when the TVA was up to the task of setting visible examples in what all concerned hoped would be a more balanced and lyrical landscape. Whereas the American impulse to muddy and to ignore any careful analysis of landscape or architecture was evidently dominant both before and after the 1930s, a notable exception had been established during that decade by the TVA.

Tellico Dam

The abrupt introduction of steam plants that devoured quantities of coal from the immediate substratum of Tennessee and Kentucky earth, became one of the targets of a new breed of environmentalists. Many of these younger, better-educated, and more articulate individuals who were appearing now in the Upper South, especially at such federal government installations as Oak Ridge, Tullahoma, and Huntsville, held Ph.D.s. There was no admonition to them to make local custom their own, as there had been to the original personnel of the TVA. The newer criticism was apt to be bare-knuckled, the critics armed with images more difficult to push aside. This latest initiative came not from Big Business, Big Power, or Big Engineering, but rather from Big, Boisterous Environmentalism. Protesters began to wonder whether the TVA had not been held captive too long by Big Engineering, while the agency choked the "natural" flow of the rivers. The "new" objections of the 1960s, succeeding the "old" objections of the 1950s against the TVA's methods of purchasing coal stocks, centered on the still older, single-solution, "an almost psychotic fixation on dams."[101] For these revisionists, the latest example of the dam fixation was the proposal for Tellico Dam (1967–75), with its accompanying town of Timberlake. That dam was to be located twenty-five miles southwest of Knoxville and was projected at first to cost 42.5 million dollars, a sum later increased to 69 million.

This mighty project was held up by a tiny fish, the snail-darter, assumed to be endangered by the plans of the new dam and the closing of its sluice gates, because the darter's food supply of snails soon would disappear. This tiny cause in its turn almost destroyed the Endangered Species Act of 1973, because to some the fish appeared an overly trivial issue.[102] From another point of view, that of the sportsman, "the Little Tennessee provide[d] a last opportunity to preserve a unique stretch of free-flowing stream for its recreational and other values."[103] The new Wild Rivers program of the U.S. Department of the Interior had made the large, mirrorlike lakes, created by the TVA out of the slow-moving Tennessee River in just thirty years, appear somehow old-fashioned and suddenly obsolete. Fully stocked and "undammed, the final 33 miles of the Little Tennessee can become the best trout river in the East, its backers say."[104] Fontana Dam had made the water coming down the Little Tennessee from above cold enough that trout could survive but not breed in it, so it had constantly to be restocked.

The TVA management repeated its old argument — that the elements of dam and town soon would upgrade the standard of living in Monroe, Blount, and Loudon counties. This area had remained a pocket of poverty, with an annual income of only $2,200 per capita, 64 percent of the national figure.[105] On economic grounds, the issue would be how much of a mighty tug would be needed to pull the supposedly deprived people up to national standard. From that standpoint, sport fishing for game trout, a pastime rising out of the affluence of the postwar, would be a lower priority.

TVA also reasoned that, because Knoxville had not been flat enough to facilitate the construction of riverside dockage, such as had made Chattanooga, Muscle Shoals, and Huntsville more prosperous,[106] nearby sites were desirable. Five thousand acres would be set aside on Tellico Lake for commercial and industrial purposes. Against this argument was leveled the patently unfair charge that everything the TVA by now undertook came out of a bygone revolution and so was bound to be ineffectual. A canoe parade was organized as a "modern day Boston Tea

Party"[107] to protest the closing of the Little Tennessee. A very old objection, as old as the building of Norris Dam, was resurrected in the claim that the TVA callously evicted people from their ancestral farms. The governor, Winfield Dunn, a Republican, voiced his doubts concerning the benefits to be gained from another dam.[108] The old Primitive Progressive view of the valley was retrieved: that, after all, it was really intended to be a vast national forest or park, without artifice. Because of the proximity of the Great Smoky Mountains National Park, the TVA too fostered the image. "The Tellico Reservoir and developments on its shoreline, in a sense, will expand the national park and provide a wider range of recreation opportunity to those who visit the park area."[109] But the more severe critics were not swayed. Conservation writers such as Michael Frome, as a wild-river "fan," put forth the counterclaim: "Having run out of other things to do, TVA must make work for itself with technological overkill of natural resources." Indeed, it was the ascendancy of technology over nature in the valley that had puzzled or repelled observers since the founding of the TVA, and the value of that ascendancy remained an open question. After TVA's attempt "to browbeat local opposition to Tellico into acquiescence," said Frome, "one should never forget the violent havoc caused by strip mining in the mountain coalfields — largely the direct result of TVA purchase practices and failure to include land restoration requirements until vast areas were devastated." The chasm between "good" and "bad" technology was exploited by Frome as a chink in the TVA armor. The old strip mining complaint was tied to a newer one, but the fact was that the TVA's credibility as a reforming force for good was on the wane. "The relationship between TVA and sound conservation practices is becoming more and more of an illusion."[110] The conviction derived from the various forms of evidence ultimately depended upon visual manisfestations. Frome felt that the visibility of strip mining provided the proof that no set of press releases could offset.

In the beginning, the TVA *had* wanted its morality play in the marketplace to have colorful characters; scenery; and gratuitously contributed, but controlled, engineering interventions. In the Tellico crisis, the pageantry turned into an external playlet, mime, or pasquinade that robbed the TVA of its dramatic initiative. Folk singer Johnny Cash, half Cherokee Indian,[111] and Supreme Court Justice William O. Douglas, a participant in the New Deal and, true to his roots in the American Northwest, still committed to nature as wilderness, were enticed into the fray. Cash was in an entertainment occupation that had considerable status in Tennessee. The ancient capital of the Cherokee's, Choto, a site threatened by the proposed lake, was "rediscovered." The chief of the Cherokees, Jarrett Blythe, was supposed to return from exile in North Carolina to meet with Douglas on the bank of the Little Tennessee. But, due to his age, 76, Blythe sent his regrets and a substitute, Richard Crowe.[112] Douglas, in his own eagerness and advanced age, appears not to have known that he never saw Blythe, for he reported the latter's "actual" words and, in an article four years later, named him as having been present to greet him.[113] Douglas, the old frontier warrior from the Northwest, felt that the TVA by then was being "manned by political hacks" and engineers with "a compulsion for make-work projects for the purpose of self-perpetuation."[114] The *Knoxville News-Sentinel*, in its headline described the event as a "staged affair."[115] Having staged "a happening," the participants withdrew, feeling they had reidentified a blessed homeland, as well as revealed a ravaged soil. A Cherokee act of less nobility, which neither side appears to have wanted to recall just then, was that Fort Loudoun, that gave its name to the dam adjoining the projected Tellico, had been surrendered by the British to the Cherokees in 1760, and "after surrender, the [British] garrison was ambushed and slain on the homeward march."[116]

The retribution by the colonists that this event brought on led to the final opening of the Tennessee territory. The appurtenances of technology, such as the big dams, were held by the modern protesters to be incompatible with the proper appreciation and cultivation of nature in this more passive and "unspoiled" setting. Ironically, this was, in effect, the same concern that the first engineer, Arthur Morgan, who had championed saving the Everglades in Florida, had expressed three decades before and tried to implement at Norris Dam with local parks and nature reserves like Island F. However, Morgan had not demonstrated in person, in the manner of these later dissenters, but rather had written voluminously late in his career about ill-effects deriving from the construction of dams.[117]

In 1978, the Supreme Court held that the Tellico Dam could not be sealed, because the four- to five-inch fish was on the endangered species list of the Department of the Interior. However, in September 1979 President Carter signed, "with regret," the law that made it possible to complete the dam by that November.[118] Some of the snail darters were transported to the Hiwassee River, where they were able to increase, and

David Etnier, the University of Tennessee aquatic biologist who had discovered the fish in the original locality along the Little Tennessee in 1973, in fall 1979 found further specimens in South Chickamauga Creek near Chattanooga, and in March 1980 more in the nearby Sequatchie River.[119]

The Hope of Fulfillment in Recreation Instead: Land Between the Lakes

The golden dream of a renewed country within an old country removed to its farthest point in outflow (in this case to be near Paducah, Kentucky), was revived through the hunger not for economic salvation, but for recreation, that built rapidly right after World War II. In May 1965, it was reported that the recreational use of the TVA had increased sevenfold since 1947,[120] the year that really had signified the end of the war. Actually, the trend had been anticipated during the war, when the National Resources Planning Board (NRPB) had put out a "New Bill of Rights" for the postwar period. "The right to rest, recreation, and adventure, the opportunity to enjoy life and take part in the advancing civilization" was number 9 on this list. Charles E. Merriam of that board considered these postwar "rights" to constitute the "underlying philosophy" of the NRPB's future endeavors,[121] but the list never would be exercised officially, largely because the premature assertion of such "rights" aroused the social anxiety of conservative Congressmen during the war, with the effect that the board was abolished in 1943.[122] However, by the postwar period, the general tide of public choice was running in the direction of Merriam and his cohorts, for by 1961 "TVA recommended to President John F. Kennedy that a large tract of thinly-populated land between Kentucky Lake and Lake Barkley in western Kentucky and Tennessee be developed as an outdoor recreation area [(28)]. These two big man-made lakes run parallel for a distance of about 40 miles with a narrow, 170,000-acre wooded peninsula sandwiched between." By 1964 all this was underway, and for the first time in the evolution of the TVA, the major element to be modeled was an isthmus of land rather than a towering dam of concrete. It was a magic island with spread, an Everyman's eighteenth-century Cythera all over again. The avowed aim was "to show how an area drained of most natural resources can be restored to serve a wide range of national recreational needs and provide an economic asset to the surrounding region."[123] It was again an alternate image or concept, deftly offered. Even those who regarded it as only a token effort recognized its cleverness, contrasting with what they by then thought of as TVA's gradually acquired but habitual clumsiness: "TVA apparently saw the Land Between the Lakes project as an opportunity for the agency to polish its tarnished image. The praise it had earned in earlier years for its enlightened pursuit of the public interest had been eroded by its complicity in strip mining, in water and air pollution, in land speculation. Here was a chance to champion the cause of environmental protection and conservation."[124] The old worries about too great a scale of technical or political commitment to too great a land mass would be assuaged by concentrating on a smaller project, lying visibly open, unexploited, and isolated in the still quiet waters (4).

At Land Between the Lakes, the TVA actually began to resemble the national park that the earlier Progressives had envisioned as their best model, just as, in the next decade, the Tellico Lake district would be charged with doing. All that hampered final realization of the model was the ambiguity inherent in the geographical takeover. At Land Between the Lakes, TVA was seen to be helping something to be "like" a national park, by not quite becoming one. The TVA "had every landscape device but a national park. That it didn't need, because it *was* a national park, not of canyons, flumes, or domes, but of planning techniques, technical advances, and even some social experiment."[125] In one respect Land Between the Lakes did conform to the basic requirements of the western national park: it would contain no permanent inhabitants. That lack was criticized severely by those who were altogether against the undertaking. They didn't like the absolute remoteness being resorted to, the final break with the aspiration to reform populations still in need of assistance. "The Land Between the Lakes is billed as 'one of the newest and most exciting outdoor recreation areas in America.' The words reverberate with irony: It's a nice place to visit, but you wouldn't want to live there. And now, at last, nobody does. Under the trees is dark and night in the Land Between the Rivers."[126]

Barkley Lodge (1966–70), built on Barkley Lake by the Commonwealth of Kentucky, was a latter-day remembrance of a western national park or forest lodge, like those at Yellowstone, Yosemite, Crater Lake, or Timberline. (29). In fact, Barkley was more sumptuous and cathedral-like than they, and also than anything ever seen in the Depression locale

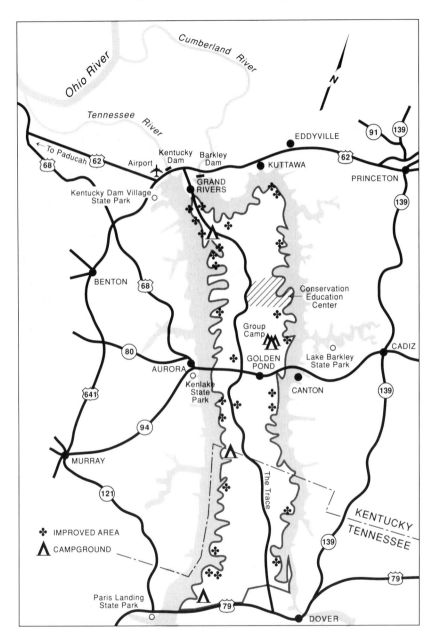

28. Land Between the Lakes. With their river-like shape, the pair of lakes offered a rich, extended coastal experience for any who wished to cruise them (4). Map courtesy of Tennessee Valley Authority Archives, redrawn by University of Tennessee Cartographic Services, 1990.

29. Barkley Lodge, Lake Barkley State Park; by Edward Durrell Stone,
1966–70. The state lodge signified the nation's emergence from the harshness
of Depression and war. Postponed gratification no longer would be
necessary. More and more, the Tennessee Valley Authority tried to get the
seven states to take responsibility for on-shore recreation. The deep, two-
deck porch of the X mansion (23) has been revived. Stone's model blends
austere Modern architecture with the memory of the southern plantation
portico and rustic image of the faraway western national park or forest
lodge. Photo courtesy of the Kentucky Department of Public Information.

of western Kentucky. Affluence finally had come to those who merely stood and waited. Made possible by World War II, the lodge, through a kind of elegant mannerism, embodied the relief and euphoria felt by the common man. Under Wank's design influence, the dams for the common man too had been elegant. An elaborate siting plan (30) gave the lodge viability and a sense of conviction as a tidy, geometrically-cut focal point in the rougher natural setting, much as the first dams had been presented. In this monumentality, the lodge followed a midwestern, postwar assertion that landmarks should show their symbolic importance and be oriented better, as with the postwar Saint Louis Gateway to the West, designed by Eero Saarinen; the Federal Reserve Bank of Minneapolis by his friend, Gunnar Birkerts; and the John Deere Headquarters near Moline, Illinois, by Saarinen and his architectural heirs, Roche and Dinkeloo. The group of pyramidal office buildings of College Life Insurance in Indianapolis, by Roche and Dinkeloo, carried that trend farthest, because the units were obviously monumental and offered spectators a sequential view from different angles as they sped along the city's outer highway belt. All these projects represented an attempt to help mid-America recapture its confidence, a mission, forecast by the earliest TVA buildings, with their similarly exquisite, self-conscious, composed, monumentality.

More of the same virtuoso handling of the midwestern scene turned up in the landscaping of Land Between the Lakes. Roderick Nash, a Californian, noted in his introduction to Frank E. Smith's *Land Between the Lakes* that the treatment of the site had to be aggressive, in order to offset its environmental blandness.[127] An "artificial" naturalism had to be added, in the form of liberally sprinkled picnic areas, campgrounds, trails, ecological education centers, bird sanctuaries, Boy Scout adventure camps, buffaloes, iron furnaces, and a restored 1850s settlement, the Homeplace. Because of its flatter terrain and quieter waters, mid-America was an area lacking opportunities for western-style backpacking, mountain climbing, skiing, rapids-shooting, and surfing. Moreover, there is none of the East Coast's sense of sudden, springlike release from the clutch of urban density and endless string development, with resorts often caught between major conurbations. Instead, the midwesterner and the resident of the Upper South are more likely to go toward their space-endowed recreation at a leisurely pace and in a stabler and less changed mood (31), often without discerning when they have left the realm of work and entered the precinct of play. Much unfurls; little or nothing closes in. The rivers, or the artificial lakes made out of them, appear nearly endless under the open sky. (4). It is an internal kingdom. Human activity, by and of itself, matters less in this larger setting. At Land Between the Lakes, the TVA ended where it once was supposed to have begun—quite happily, within a peaceable, idyllic kingdom.

The original arcadian dream of landscape innocence and purity, set out in the generously illustrated 1938 TVA report, *Scenic Resources of the TVA*, had been resuscitated with Land Between the Lakes in general and with the founding of the Homeplace in 1978 in particular. The latter was a cluster of sixteen log cabins and outbuildings on 250 acres that had been brought together from the immediate vicinity to Route 49 just south of where Kentucky turns into Tennessee. The compound was used for displays of 1850s folkways—square dancing and fiddling, harvesting crops, and handling farm animals. In the 1930s this amount of reverence for the past never could have been summoned, and log cabins of this vintage usually were abruptly torn down. The study of vernacular architecture and folk customs in the last few decades has brought new prestige to such structures and restored the opporunity to catch a glimpse of meaningful precedents, although it has also to be kept in mind that even in 1850 warnings were being issued about overuse of the land and consequent erosion. With the Homeplace construct, the TVA at least could be seen better as the outcome of a series of laminations in time.

About 1953–54, a decade before Land Between the Lakes itself was started, increased attention was addressed to boating as an embodiment of the new leisure (32). *Fortune Magazine* noticed in 1954 that "the ability of Americans to spend far more for fun than for clothing or shelter of course stems from the spectacular increase in the consumer income, up from the $160-billion to $170-billion neighborhood just after the war to nearly $250 billion last year (though about half the increase was due to price rises)."[128] Awaiting this new and unaccustomed affluence were artificial bodies of TVA water. "Such clientele has helped the boat business grow robustly in areas with large natural bodies of water. But between the Alleghenies and the Rockies its growth, helped by several score artificial lakes and reservoirs, has been explosive."[129] Chris Craft boats took up an old textile mill in Chattanooga as a factory. "The public has swarmed to these lakes like bees around a hive. This year the TVA lakes alone should account for some 25 million 'visitor-days' (officialese mean-

30. Barkley Lodge from the Air. This structure exemplifies a major characteristic of the Tennessee Valley Authority and the Midwest during the postwar period – the search for large architectural symbols. Geometry was used to focus attention on the singular complex. By contrast, a romantic mist covers the distance beyond, as it did in Bingham's mid-nineteenth-century painting of the trappers descending the Missouri (5). In a curious allusion to TVA's New Deal past, the shape of the complex resembles that of a WPA amphitheater in the woods (83), with its cabins perched all around the central lodge, as around a 1930s campfire. Photo courtesy of Tennessee Valley Authority Archives.

ing a visit by one person for part or all of one day)."[130] The popular *Saturday Evening Post* had noticed that explosion the year before: "Despite the political and economic controversies that raged about TVA, before and after its inception, one inescapable fact stands out quite clearly after its first twenty years: its dam construction created what is possibly the outstanding recreational development in America, radically altering the lives of millions of people living within the drainage system of seven Southern states, and affecting many more millions who visit the region."[131] The result was lucrative, even though not greatly influential: "The tourist dollar is now the third largest source of income in the Valley."[132] The communal opportunity was plainly there and would grow greater "in the next five or six years," following on "the creation of new facilities."[133] The unplanned opportunity for a bonanza existed, but there were built-in inhibitions, too, deriving from the past characteristics of the local people and the fact that recreational facilities had not existed in the vicinity. The *Saturday Evening Post* remarked:

> You have to remember that whatever you see most likely didn't exist two decades ago, and that most of the development dates from less than ten years ago. The people, maintaining the independence for which the region has always been noted, have become innkeepers, but they don't kowtow to tourists. The change in attitude from the native who preferred pellagra to tourists is reflected in a remark attributed to a West Tennessean who profited from tourist trade: "Each tourist is worth a bale of cotton — and a lot easier to pick."[134]

There was no tradition of hotel keeping. By the 1980s, the area around the lakes centered on Murray, Kentucky, and had become one of the most popular and inexpensive retirement foci in the country.

The annual reports on reservoir properties in the beginning-to-boom 1950s remark that "regular requests" were now coming in for vacation space "from once considered 'far-off' states of Michigan, Kansas, Louisiana and Ohio."[135] Kentucky Lake and Dam were the biggest in the system, so "Kentucky Lake, which has about one-fourth of all recreational developments on the TVA system as measured by the value of facilities and equipment, enjoys heavy patronage from such midwestern cities as St. Louis, Chicago, Detroit, Louisville, and Evansville."[136] Pools of great recreational and economic worth had appeared suddenly, as a dividend to the original program of the TVA. Would their unexpected advantages be allowed to increase the general armamentum of the TVA to carry on its work?

31. Camper in Land Between the Lakes Park. The long Cadillac, the aluminum fishing boat on top of the car, and the extended camper-trailer attached, indicate that the interstate highways of the 1950s have been completed. Everything is big. Such mom-and-pop sorties were impossible during the 1930s and 1940s. That kind of contrasted compartmentalization of the experience of decades led to misunderstandings between generations, the "generation gap." The grove of scrub pine has a southern look. The folding aluminum camp chairs and the shelter signal the fact that the armaments factories had been relieved of their military obligations. It was time for celebration and relaxation. These are still plain folks, but not necessarily deprived ones. One more war has produced vast material rewards for the visitors. Photo courtesy of Tennessee Valley Authority Archives.

They would not. All through the annals of that decade can be discerned a reluctance to capitalize upon the windfall, and an eagerness to redistribute its benefits as rapidly as possible to state, county, and local authorities, ostensibly as a gesture of goodwill. This meant that the TVA would look like an empty shell in contrast to what it might have

32. Boat Launching, Kentucky Lake. In the early 1950s, a large undertaking of recreational boating began on the TVA lakes. The outboard motors became larger than the prewar ones, the boats bigger, along with the wagons that pulled them over the new interstate highways. Photo courtesy of Tennessee Valley Authority Archives.

been, more and more like a federal entity proceeding in the guise of state government. The shores were not to be shaped by the agency that had created their value. Perhaps it all happened so fast that there wasn't time to think? "The first long-term lease granted by TVA for operation of a boat dock and related tourist facilities such as restaurant, lodge, and vacation cabins was issued just before the close of fiscal year 1946. The site was at Jonathan Creek, on Kentucky Lake."[137] That was the first instance of a private shore compound within the series of TVA lakes. The process accelerated very rapidly after that. "By the end of the fiscal year 1948, on Kentucky Reservoir there were 34 commercial, public, semi-public, or private recreation developments on TVA reservoir properties made available for the purpose of lease, license, or sale at public auction."[138] That extraordinary increase in an activity initially made possible by the TVA occurred within two years, at the same time that TVA's best thought apparently was being directed toward how to obtain more electricity from steam, and later from nuclear power. By 1954, "Kentucky Dam and the nearby State Park entertained more visitors than the Great Smoky Mountains National Park, the most popular of all national parks,"[139] and "Kentucky Reservoir alone received an estimated 4,500,000 person-day visits, more than used the Blue Ridge Parkway, the nation's most heavily used recreation area under federal administration."[140] Something untoward was definitely occurring within this region, for Great Smoky Park had appeared only shortly before World War II. Kentucky Lake now had surpassed with its own popularity (29) even that most popular of national parks!

In Kentucky Lake, a great national asset suddenly had been built, but the TVA reports of those years make amply evident the agency's obvious wish (or compulsion?) to off-load it, to divest. Previously, there had been a similar eagerness to lease or sell off every one of the TVA workers' villages. Fontana Village had been leased away as early as 1946, and Norris Village had been sold in 1948. Big Ridge Park on Norris Lake had gone to the State of Tennessee in 1949. Gilbertsville, the village for Kentucky Dam, had been let go in 1948, and Pickwick Village in 1949. The Hiwassee Dam Village had been given up in 1948, and Wheeler Dam and Watts Bar villages in 1949 and 1950, Watts Bar becoming a particularly viable private resort. The cities of Guntersville and Decatur in Alabama were assisted by the TVA to develop "municipal park-boat harbors." Properties first leased to those municipalities for that purpose were then transferred by the TVA directly to them in 1955. By 1953, all along the valley an interest was manifested in family accommodations to augment those previously furnished for single fishermen, another unanticipated demand that was greeted with a degree of disbelief similar to that which met the immediate popularity of the "park-boat harbors."[141] The new facilities were to be "for families in the lower and middle income brackets," perhaps signifying by omission that the well-to-do need not apply for rooms. The affluent—and there were to be so many of them after World War II—thereafter would not be considered a viable section of participatory democracy. The consciousness of recreational responsibility was enlarging but not far enough yet to escape the older, prewar conviction that the TVA ought always to be a charitable and pragmatic organization, as far as its direct contact with people was concerned. The creation of Oak Ridge had been an epic incident in the TVA story, one that could not be revealed or shown to the American public at the time, although it affected the whole world. Kentucky Lake, with its environs, as an alternative mode of American existence that could be displayed to all the rest of the world, likewise constituted an epic incident that had implications far beyond the Tennessee Valley, in the whole South and Midwest. But the attention of TVA's top managers appears to have been centered elsewhere during the 1950s.

What else would give rise to anxieties that would curb TVA's commitment to the landscape, as distinguished from the lakes themselves? One eternal question was which group deserved the unearned increment deriving from the engineering improvements. Theodore Roosevelt already had expressed a firm opinion on that score in his 1902 veto of a bill to have a private firm build a dam at Muscle Shoals: "Wherever the Government constructs a dam and lock for the purpose of navigation there is a waterfall of great value. It does not seem right or just that this element of local value should be given away to private individuals of the vicinage, and at the same time the people of the whole country should be taxed for local improvement."[142] How would it be more justified now to hand out such benefits to small, rather than large, entrepreneurs all along the new lakeshores? Private power companies might abuse water rights in order to produce and distribute electricity, but in what way was it less abusive to hand out too localized perogatives for the newly burgeoning leisure-time activities, ultimately to result in poorly-planned, poorly-designed, cheap recreational villages, second-home and retirement subdivisions, and cluttered marinas, too conspicuous against the shore, all reached by too-familiar stretches of roadside strip development? In the setting of a world-class environmental experiment (many foreign leaders, intellectuals, and architects followed a straight line to the TVA district as soon as they arrived in the United States, to see what Americans *actually* could accomplish), local color and quick profit perhaps had been emphasized and assisted overmuch, while at the same time the critical, colossal presence of the TVA itself had been very much downplayed. Had the dissembling occurred because of a longstanding dread of federal ownership or direction, resembling the "topophobia" of conservative Dutch patroons that kept them from developing the Hudson River Valley, in which Franklin Roosevelt grew up?[143] The TVA scenario for Kentucky Lake recreation appears to have reflected a similar fear of federal ownership and mastery, arising out of what seemed to be, but wasn't, an endless expansion of the territory to be handled. Now TVA turned in the direction of divestiture and fragmentation. Maybe the geography had not been sufficiently readied for such a holistic operation on such a large canvas, after all? Perhaps sadder even, the beauty and caring that once had been evident in the imagery of the TVA, which consequently had made TVA a rallying point for so many architects, engineers, political liberals, and refugees, now lost their power to convince. The presence of beauty was still advocated by the TVA, but the ability to create and install it was much less in evidence. Was this outcome to be interpreted as the failure of just one more scheme in this "vicinage" that during the nineteenth century had spawned so many utopian visions?

Getting rid of shoreland was, to be sure, a possible way of reducing one kind of responsibility in order to take up another, such as increased power production. The drift toward early divestiture could be discerned already in 1950, the first year of the "bad" decade. TVA's *Reservoir Properties Report* of that year hinted as much:

> The original impetus to development of docks and resorts on reservoir properties was given some years ago when the TVA made such sites available by cancellable licenses. Later, additional impetus was given by superseding the licenses with long-term leases when circumstances warranted. Sale of properties where establishments meet specified criteria is expected to supply even more incentive to private operators to expand their services and facilities for public enjoyment of the reservoirs. There was a minimum of activity in leasing sites for new cottages in established TVA cabin site areas, and no new areas were opened for lease, inasmuch as plans are being made to recommend the sale of lots now made available by lease.[144]

A few years before, in 1947, it had been reported:

> On Kentucky Lake, for example, approximately 50 tracts of privately owned land have been subdivided into nearly 600 lots for vacation cabins, and about 125 cabins have already been erected in these subdivisions. Some of the structures are substandard, measured by the most inexacting requirements of recreation technicians, and are on lots smaller than is the rule in TVA cabin site areas. . . . The development of privately owned recreation land has been so rapid, however, even in the face of [postwar] shortages of building materials and labor, that regional and local planning agencies have been unable to keep pace, especially since the role of some of the agencies is largely educational

and advisory; moreover, that role can be enacted in most cases only when the private landowner or builder initiates a request for technical assistance.[145]

In 1949 the same jarring anxieties resurfaced:

> Such rapid growth of the recreation industry as has occurred at Kentucky [Lake] points up the need for technical direction to guide sound development. Thus far, efforts in guidance have not been able to keep up with the Jack-and-the-Beanstalk growth. TVA seeks to strengthen state, area, and local agencies and groups interested in various kinds of recreation development, but more intensive direction than TVA is authorized to provide is needed.[146]

From 1947 to 1950 it could be seen that matters were getting out of hand, but nothing was done.

Norris Dam and Village had been built in Depression-era haste. Oak Ridge had been built in wartime haste. Now the shores of the lakes were being modeled in postwar haste. However, this third phase was seeing much less application of expert design skill than the earlier two. In 1955, the problem of awkwardness still persisted in an organization that had once prided itself on smooth coordination:

> Several poorly planned private subdivisions during the year point up the need for more effective subdivision control by counties. Henry County, Tennessee, appointed a planning commission this year but only after almost all of its comparatively small reservoir shoreline had been subdivided in 2,500 lots, some as small as 40' by 90'. Such developments behind TVA-owned shorelines present management problems for TVA.[147]

Land use and structural size and appropriateness at least were alluded to, but nothing was said about such matters as the necessity of service functions and police and fire protection; schools; utility access; drainage, very important in suddenly rainy Kentucky and Tennessee; road use and condition; style and appearance; landscaping; or the general availability of capital. What made fulfilling potential even more elusive was the fact that western Tennessee and Kentucky were even farther than the eastern portions of the two states from a faint memory of the tasteful materials and proportions of the Georgian-Federal tradition. In the West, the new "Georgian" houses built by people lately removed by TVA from Land Between the Lakes were likely to be too long, too wide, or too short, with porches poorly placed and Classical detail either misappropriated or entirely absent. At one point, the architecture of this area had looked too old and rundown. Now, it looked too ignorant and new. The

1930s and the late 1940s, once the war was over, had stimulated interest in new and cheaper synthetic materials such as plywood; pecky cypress; cast stone; bent iron; aluminum doors, windows, and screens; dry wall; and asphalt shingles and tiles. Plastic leather and wall-to-wall nylon carpeting appeared in the houses and resorts built around Kentucky Lake. Postponed, pent-up expectations caused everything to be seen in the flicker of blue-white fluorescent light (just newly marketed), on these "luxurious" materials, bidding fair to make every man or woman a king or queen at last. Domestic taste, of course, the TVA was not permitted to inculcate or pass judgment on. Some of the more arresting names of the new resorts on Kentucky Lake were Hester's Spot in the Sun, Early American Motel, Big Bear Resort, Cozy Cove, Ken-Bar Resort, Southern Komfort Resort and Marina, and Paradise Resort. Paradise, Kentucky, nearby, paradoxically was the site where the huge Peabody Coal Company shovel (27) had been built in 1959–62 and the hundred or so inhabitants removed. The countryside was overflowing with unleashed enterprise, as once it had been filled with painful memory and despair.

The upgrading that did occur came from pressure put on the states by the TVA to accept and develop new parks. This effort produced creditable results: "According to 1952 figures compiled by the National Park Service, Kentucky ranked fourth and Tennessee seventh, among the 46 states reporting, in total expenditures for state parks. Kentucky was one of four states reporting total expenditures exceeding two million dollars. The others are, in order, New York, California and Ohio."[148] In one leap, Kentucky and Tennessee had joined the company of the big industrial states. That same year, 797 acres on nine reservoirs were transferred by the TVA to the State of Tennessee. By 1954 Kentucky was second in state park receipts, after New York.[149] By 1955 the Commonwealth of Kentucky had twenty-four state parks, of which two, Kentucky Dam Village and Kentucky Lake State Park, were "in many respects, the backbone of the entire Kentucky state park system."[150] In the 1950s, not only did the state of Kentucky build the slightly upscale, slightly flamboyant Barkley Lodge (29–30) on Barkley Lake, but also other state lodges of superior architecture, sitting securely within the landscape, began to appear from west to east, including those at Lake Cumberland; Carter Caves; Jenney Wiley Park; Buckhorn Lake; Greenbo Lake, made from materials gathered from the site; and Pine Mountain.

At Kentucky Lake and along the shores of the other TVA lakes, a major opporunity was missed. TVA simply seems not to have been ready to create an optimum environment in the valley, whereas, three decades before, it had claimed that it was ready. In actuality, TVA had created a recreational magnet greater than that of any national park or forest. Implementation of the original goal had suffered, however, because the more conventionally-minded TVA officials could not credit that fact. The area could have been seen as a novel "national park" in another, more accessible place in the middle of the country. The outcome should have been recognized as the fitting culmination, on a larger scale, of what had been begun with parks and community planning around Norris Lake, at the eastern end of the system. Its seriousness and worth should not have been denigrated because it had no smokestacks. Amid the rigidities of American thought would linger the conviction — perhaps most frequently reinforced at county seats, in state capitals, and in the national capital — that the TVA always had been some kind of "socialistic" land grab, rather than what it was at the outset, a sincere effort to experiment with, and bring forth, new ways of forming the total environment through the arts of engineering, architecture, land planning, and public administration, that had measurably improved during the first decades of the twentieth century. Could a twentieth-century democracy be wholly effective without a laboratory for environmental research and development? Since no monument was ever built to either Roosevelt in the national capital, ought the TVA to be considered in part a visible commemoration of their assumption that the federal government should be involved with the actual landscape of the hinterlands? The blame for the later inadequacies of the TVA can, in part, be laid on the disruptions of World War II and its interruption of the continuity of the cultural memory. Idealism then went wandering off in search of a new theater of operations. The energy and prosperity of the immediate postwar furnished a new clientele, and that made the TVA mission as a supporting or splinting element for the local population that much less critical. And what had caused it to be so fascinating in the first place as the greatest magic trick of the greatest prestidigitator of them all, Franklin Roosevelt, attracted the lightning of political disapproval after the war, paying the price for its visibility. The American government's habit of concealing as much of its intention and action as possible, in order not to attract that lightning of disapproval to itself, was more easily and often manifest by the less conspicuous devices of building, banking, and environmental controls from indirect, subsidizing methods, such as mortgage guarantees and tax writeoffs, that influenced the fabric and density of what appeared as suburbs, beginning in the late 1940s; the redevelopment of inner cities, in the 1950s; and even the preservation and rehabilitation of older buildings, beginning in the late 1960s. That kind of *sub rosa* influence affected the expression on the visage of America just as surely as more conspicuous compositions such as the TVA did, but the former method had the strong advantage of convincing people that they were in a "pluralist," "grass roots" democracy, where *they* made all the final decisions. "I" decided to build this house or this skyscraper, save that old building with its ornament twinkling in the sun and shade. However, the disadvantage of using that mitigating approach too exclusively, over and above the approach of the original TVA, is that so much of the quality of public architecture thereby is neglected, does not reflect the best of the design potential of the time, makes the term "public sector" too often synonymous with indifference and "shoddiness." The common public environment and all the coordination that it requires — not only because more and more utilities and infrastructure are entailed, but also in order to furnish an ambiance fit for the moral, aesthetic, and spiritual growth of its citizens — are often ignored too. This lack of attention to coordinating features registers as a greater detriment when it is realized that the United States, in its cultural memory, historically has been more dependent than most other countries on nature and landscape as a romantic expression, because of the time when the country began and the dimension of its discovery and westward expansion. Today we too often forget that the American architectural tradition did not begin, like the European, with the palace, castle, or church, or even with the city hall or factory of the late nineteenth century. Instead it began with the humble house type of all its centuries. The area around Land Between the Lakes, which possessed a lot of nature and a lot of humble houses, always the beginning ingredients in America, under the tutelage of the TVA became neither a Paradise Regained nor a Paradise Lost, principally because the dream was permitted only to half-happen. Time was not allowed fully to improve and ripen the TVA enterprise there, so it turned out largely to be an Eden evaded.

4 | The Imagery of Structures Triumphant

The Sudden Call for a Fresh Imagery

America always was a "never ready" country. Its architectural imagery was supplied on the spur of the moment, as if the nation had never possessed an authentic visual tradition or any disciplined interpretation of its visual heritage. The TVA now needed to demonstrate how architecture could make life more complete and better rounded in a particular valley, but no-one in Roosevelt's New Deal appears to have been completely sure of what that demonstration would consist of visually. Earle Draper began to build up a style for the authority by hiring Roland Wank as its first architect, selecting Wank because he felt that Wank was entirely original and not "wedded to any *style* or *tradition*."[1] In the 1930s, there was tremendous eagerness to obtain a fresh start and indeed little time to do otherwise; little wonder the International Style of architecture was received so readily all whole when it was imported from Europe during the decade. In retrospect, it is more obvious that Wank's ultimate inspiration was an entertaining French Art Deco, preceded by the solid Viennese Moderne, a combined mode distinguished by its emphasis on smoothly launched and sumptuously displayed power. This perspective derived partially out of the awe that had been generated by World War I's revelation of the power of the machine — an awe particularly marked in France and Italy at first, then handled more subtly in Germany and Britain during the later 1920s. In France, especially, the turning wheel and whirling propeller, the penetration of the porthole, the rise of stair and smokestack, and the stretch of the searchlight and wireless on the ocean liner often could be seen in the decorative arts. A sheen was to be had from all the resilient, silky surfaces. Big, simple, dynamic forms were wanted. The order and cleanliness of a new display seemed to promise a waiting future that might be initiated by any eager young group, such as the 1920s aviators, for example. A TVA lawyer, in retirement in 1969, recalled the spirit of jejeune daring and buoyancy among TVA personnel at the time:

> I think that the very fact that the TVA started at the bottom of the depression enabled them to obtain people that they could not have obtained at any other time. They had people there, particularly in the engineering area, who were unquestionably among the greatest engineers anywhere in the world. . . . What was true of engineers was true certainly (maybe to a lesser degree), but was certainly true with respect to all the rest.[2]

The order anticipated in much of the eager display was intended to stand in distinct contrast to what was then seen as the disordered and sometimes drab regional (and national) scene surrounding it. The desire for a bright new order grew intense. Franklin Roosevelt and the New Deal personnel favored the quick, hard-hitting distribution of attention-getting devices, incipient trademarks, to counter the impression that the South had been abandoned in the hour of the country's greatest economic agony. The TVA would offer an unparalleled spectacle of technological skill and abundance, fitted right over abject proverty. In the FDR manner, the agency was to be presided over by an unsurpassed pool of the brilliant and talented, while reveling in the turmoil produced by the abrupt juxtaposition of two value systems. The dice had to be well shaken in order to be sure that new numbers would fall out!

What would most strongly represent a clash of value systems and the energy issuing therefrom would be the appearance, all of a sudden, of outsized, three-dimensional, but peculiarly refined architectural entities distributed over hundreds of miles in a valley. The terms *spread* and *synchronization* captured the very essence of the ideals of the TVA. There had been a great many "imaginables" on the list for twentieth-century America, but until the advent of the Great Depression, few had been implemented. The difference in the 1930s was that the client, American society at large, at last was ready to apply itself, to rouse itself sufficiently, to transform some of its previous hopes into actual, concrete, operating realities. In this sense, the TVA was a pursuit of "imaginables" already envisioned for four decades or more. At the same time, the abruptness of the alteration in the economic and cultural situation of the society meant that the surface motifs sometimes would take on an artificial, ahistorical, improvisational character. The TVA was supposedly built for all the ages, but the actual designs often were done under the extreme stress of emergency *ad hoc* conditions during the Depression and then during World War II and the Korean War.

The architectural arts had not been readied sufficiently for a preemptory call by such a large public body. During the 1920s, the United States was as short of gifted designers (as well as mature enough historians and critics of architecture and engineering) with an affinity for public causes as it would be short of research scientists for public causes during the wartime 1940s. In previous decades, the artists' and scientists' roles had been judged not to be central to the interests of a burgeoning, compromising, populist democracy — one that, as in Tennessee, only shortly before had been overwhelmingly frontierlike and agrarian. Therefore those professionals who took seriously the more advanced expressions of the visual arts or the research sciences almost inevitably had to be Europeans imported into America — the Gropiuses and Mieses; the Einsteins and Fermis; the Saarinens, Neutras, and Wanks. Thus, as far as the awareness of the average citizen was concerned, the pall of indifference to the more advanced stages of the arts and sciences only occasionally lifted. The greatest merit of the novelty in engineering and the arts, now showing up in the TVA, was its adaptability. Leo Marx explained the unique American situation in his seminal book, *The Machine in the Garden*, referring to literature and not to the TVA: "The sudden appearance of the machine in the garden is an arresting, endlessly evocative image. It causes the instantaneous clash of opposed states of mind: a strong urge to believe in the rural myth along with an awareness of industrialization as counterforce to the myth."[3] The excitement inherent in TVA's confrontation of cutting-edge technology with a recessive meadow-and-forest, folk-art-and-frontier society therefore was not to be unprecedented. Such a confrontation had gone on in the whole long literary history of the country, too.

The closest the TVA ever came to totally assimilating its technological constructs into the landscape was with four dams of rolled earth and rock fill, built during and just after World War II. South Holston (1947–51) (33) and nearby Watauga (1946–49), at 285 and 318 feet, respectively, reached heights comparable to the wartime concrete-gravity, high-head dams of the TVA. Hiwassee, for instance, was no more than 307 feet

33. South Holston Dam, 1947–51. Such rock-filled earth dams, constructed after World War II, represented a more rudimentary and elemental earth form, as they couldn't have spillways, elevator or control towers, or traveling gantries on them. Photo courtesy of Tennessee Valley Authority Archives.

high, and Douglas only 202 feet. The stepped terraces of the rock-fill earthern dams gave them a striking resemblance to the Mayan pyramids of Central America. These dams seemed to relate farther back in time, as well as closer to nature. On the Modern side, they were characteristic of the TVA insofar as they were current and feasible but did not attempt to go beyond the present state of the art. Concrete dams had come into their own around 1900–1910, but it was not until the 1930s that earth dams were investigated systematically for seepage and stability. By

1940 the theory of such rock-fill earthern dams was complete in all its essentials. South Holston and Watauga were first proposed in 1942. The new possibility was that they could be built so high, as high as 538 feet at Trinity Dam in California in 1958. But, of course, earthern dams had been built for centuries. Arthur Morgan had employed five, between 65 and 120 feet high, in his 1913 Miami River conservation project. The gates in them were closed only in floodtime, to protect Dayton. Always dealing with water in different ways, on those dams Morgan also made the hydraulic jump practical as a stilling device at the end of spillways.

If, as many believed, irreconcilables could be matched, it followed that if anything fresh and vital were to appear, it would have to come from the private sector. That belief arose out of the great mechanical progress that had been achieved since the Civil War through private capital investment, energy, and initiative. In the case of the TVA, such a view was furthered by the prescience, in the early 1920s, of Henry Ford's social ideals and planning images for the area around Muscle Shoals, Alabama. Intimations of subsequent TVA greatness came from this archetypal American mechanical genius and private capitalist; from his private-enterprise architect with a very large corporate firm, Albert Kahn; and from the private artist whom the Fords hired at the end of the 1920s and the beginning of the 1930s to depict their River Rouge Plant near Detroit, designed by Albert Kahn. Somewhat surreptitiously, the TVA emerged from antecedents created by the biggest capitalist, and, almost as deceptively and inconclusively, subsided into the huge and secret war works of the U.S. Army at Oak Ridge. So its architecture and engineering, land planning and river planning, were its own and someone else's at the same time. After 1938 there was a persistent trend of the TVA to take on the protective coloration of a state, county, or municipal agency. Always it had to deal with a series of oppositions in authorization and visualization.

Originally, mass production had been intended as a people's payoff for the indefinable erosion, since the Civil War, of public ability to generate civic beauty and integrity, at least in the North. That was reason enough for the southerners, with their recollections of handsome and charming cities like Savannah, Charleston, and New Orleans, and of the pillared dignity of the old plantations, to see themselves as a more virtuous breed, even amid presently threadbare and ruined surroundings. Much depended on which century or decade a person felt entitled to live in conceptually. In that context the TVA was another step by the northerners beyond the Chicago City Beautiful Movement of 1890–1910, intended to offset the southern intuition that the South alone, if given the chance, could provide architectural commodity and delight for its citizens. Material resources were to be made visible on TVA premises, in order that an important federal promissory note could be issued to the southerners. The note held in it one or two special clauses. It was not due now, exactly. It was intended for the future, counteracting a regrettable past. Chief Architect Roland Wank assured visitors to the first power house of the first TVA dam at Norris that the sound of electricity and the sight of splendid materials all about them were but harbingers of what the housewife could expect inside her home, once the Depression was over. Postponement of gratification was one of the chief themes of the Depression — universities taught it to people who earned their way through, banks proposed it to those seeking mortgages or car loans, preachers preached it as a promise of the next world, and later marriages and fewer children testified to it. During Depression and war, many a young person was kept going in a faraway jungle, battlefield, or factory, by a memory of her or his mother in the kitchen, over a stove, ladling out oatmeal, basting a turkey, or slicing a pie. Through the lonely times, one "kidded oneself along" with such imagery, whether small or large, domestic or monumental, old or new, maudlin or just plain sentimental. So the rise in the standard of living anticipated through electricity would outwardly have to do with the recovery of civic or architectonic pride, but inwardly more to do with the promise to domesticity. It was Wank's opinion that "the design should reflect not merely the austerity of the present but the greater affluence of the years to come."[4]

The language could be tentative to begin with, as with Wank's hints connecting future kitchens to present power houses, but in the end the meaning could not be mistaken. Thus Gordon Clapp, a latter-day TVA chairman, said in 1955, "And some have even hailed [the TVA] as the nation's belated compensation for the humiliations imposed upon the South during the grim years of the Reconstruction."[5] A feast of postponed reason had been activated in 1933, at the lowest point of the Depression, because the baggage of guilt from the Civil War grew heavier then. What had now been evinced in the TVA, Clapp reported, was a "creative federalism." Unlike Britain, Germany, Austria, Finland, or Sweden, the United States had made no direct effort to reconstitute or standardize housing through public incentives during the 1920s. In a manner of speaking, Progressivism had replaced its emphasis on housing with something larger, such as power houses, or something smaller, such as the interior of a happy kitchen. As Roy Lubove explained the situation,

> As the planning profession's concern for housing and related issues diminished towards the 1920s, the housing movement itself floundered. World War I sharply inflated housing costs, producing severe shortages throughout the nation. The speculative developer withdrew from the low-income market. In a period of scarcity and rising costs, restrictive legislation became not only ineffectual, but also irrelevant. It was not surprising that the vigorous housing reform movement of the Progressive era declined. . . . Restrictive legislation could not build houses, but Americans were not yet prepared to concede the necessity for government financial [or design] participation.[6]

The greatest emphasis, with regard to the private single-family house, would be on acquiring mechanical devices in the core of the house, such as lighting, heating, plumbing, and cooking equipment, as can be well seen in the utility cores of the houses at Norris Village, near Norris Dam.

The British accomplished a considerable amount in semidetached housing in the "Homes for Heroes" movement between the world wars, partly because, guided by Garden City tenets, they integrated their housing designs with nature and with the site layout. This progress gave Britain a strong lead in both the theory and the practice of housing from 1930 to 1960, culminating in New Town construction right after World War II. In that frame of reference, TVA was another form of compensation for the previous lack of attention to domestic accommodation, regardless of the variety of environmental responsibilities that it took up at a regional scale. Despite the claim that the British New Towns were units unto themselves, they tended to be placed as satellite cities around metropolises, and nothing like that was attempted with the TVA villages, whose plans also were influenced by Garden City thought. What was attempted at Norris Village aimed rather toward the "efficiency" kitchen and some reminiscence of vernacular Tennessee cabins. But in America as yet there had been no investigation into regional folk architecture of the 1890s, including the crafts, comparable to that undertaken in Finland, Austria, or Britain, although Tennessee and Kentucky were the places where such research should have taken place, had the will been there. The English were aware of Kentucky-Tennessee crafts. But, even though Henry Ford, Arthur Morgan, Mrs. Morgan, and Mrs. Roosevelt had an abiding interest in vernacular architecture and mountain crafts, it was difficult to incorporate that factor in TVA expression. David Lilienthal never wanted to, and later, when he was trying to evict Arthur Morgan, he was frank to say so. In the 1930s there was a great eagerness among TVA's administrative "doers" to close their efforts off from the past entirely. One great disadvantage of that stance was that it precluded their keeping up with the unfolding theory of Modernism too. So the leaders sometimes appeared to be flying double-blind aesthetically. Life was going by so fast that it was difficult effectively to take notice of the whole of it, so in the TVA some decisions were bound to be made impulsively and with increasing diviseness

The Art Deco or Moderne styles in furniture, crafts, and architecture arrived in New York City as a shock wave from the Paris Exposition of 1925. Stimulating, frothy, and jointly Austrophile and Francophile in derivation, within the TVA they were regarded as ready remedies for the acute slump in national morale. An indigenous or native style hardly could have been tailored out of whole cloth on the spur of the moment, because inside the country there had been no regular habit of amassing data and wisdom for such an artistic endeavor. Chronic lack of the requisite kind of accumulated visual and aesthetic information formed part of the roster of national deficiencies. Instead, a style or styles from across a wide ocean would be instantaneously transferred and applied. Much of its acceptance in the U.S. derived from unformulated or unexpressed, invisible intellectual and artistic longings. Perspicacious intellectuals were already sensitive to the possibility that the world might come closer together in harmony and unity, that Wendell Wilkie's visionary "One World" might be achieved, *if* sophisticated thinkers could persuade the totalitarian authorities to relent. "Internationalistic" imagery would be useful in heralding that eventuality, tastemakers knew. Art Deco, which had been current in France for over a decade when the TVA was initiated in a remote part of the United States, had been billed in France as an "Internationale, Decoratif, and Industriel" style. Such adjectives exactly captured how TVA proposed to use Art Deco as a stylistic suffusion. When it finally reached the valley, the style had the welcome and unwelcome freshness of wet paint. Its color, plasticity, and liveliness held definite emotional advantages, contrasting with the gray despair of economic deprivation and the prospect of war. But Art Deco tended to be weaker than it ought to have been for use in a democratic setting. Compared with the less ambiguously determined, simplistic, synthesized symbolism of fascism, from the same continent, Art Deco could look pretty and tepid, irrelevant and diffuse. Nevertheless, it projected an undeniable confidence, much as Roosevelt himself beamed with confidence and optimism — while "Rome" got ready to burn itself down!

A More Definitive Attitude toward the Landscape

TVA had two approaches, one too broad, the other much too confined. With TVA's architecture, difficulty was caused by the American receptivity to so many proliferating inspirations, to eclecticism and catholicity. With the landscape, the hindrance may have come from the opposite postulate, from sources that had been too narrowly defined and procedurally specialized. Progressivism and Conservationism had led to a preoccupation with specific forest tracts and particular riverways, with less interest in how those features might be joined with other opportunities for improving the total environment, especially in terms of achieving a comprehensive improvement in mood. The methodology for handling forest resources, which Gifford Pinchot, for so many years, had prepared under the aegis of Progressivism and Conservationism, was apt to be exclusively absorbed in the productive aspects of forests, grazing lands, and water rights, and hardly ever with questions of the control or enhancement of broader emotional or scenic values. Pinchot "at every opportunity . . . stressed the utilitarian value of the forests."[7] After a very promising start in *The Scenic Resources of the Tennessee Valley*, surveys of scenic and naturalistic assets of the Tennessee Valley were very rare.[8] In the TVA, landscape innovation and amelioration, when it leaned at all, leaned more in the direction of Pinchot's national forest model than toward the national park model of John Muir and Theodore Roosevelt. If the original creators of the TVA had been oriented more toward the national park model, the treatment of the geography of the valley might have been more cohesive, complementing the synthesis of hydraulic control that was accomplished through the regulation of the dams. Of course, the national park system itself never had been much interested in the imagery of human occupation or settlement. Under the national park model, some of the architecture would have been more rustic, adhesive, blended, and agrarian as well. That more rustic TVA alternative today can only be sensed in such structures as the newer Fontana Dam Lodge in North Carolina or the equally pleasant and impressive Barkley Lodge on Lake Barkley, belonging to the Commonwealth of Kentucky. Agrarianism was imaged only as a counterpoint at Norris Village, and then only on a very small scale.

Pinchot's reference to practicality and payout became more of a pledge as time passed: "The apostles of the gospel of efficiency subordinated the aesthetic to the utilitarian. Preservation of natural scenery and historic sites, in their scheme of things, remained subordinate to increasing industrial productivity. Forestry, Pinchot argued, did not concern planting roadside trees, parks or gardens, but involved scientific, sustained-yield management."[9] This "scientific" approach meant that when TVA officials like Howard Menhinick or Robert Howes advocated historic building preservation in the cause of providing more cultural memory or identity for the valley, the directors would largely ignore them. The cachet for history and tradition was attainable only through closely associated facilities for recreation, as, at last, at the Homeplace on Land Between the Lakes. Recollective villages like Norris, in woods distributed along the land's flowing contours, were not to be repeated.

Quite strangely, in view of how much he had propelled the earlier actions and ideas of the TVA, Gifford Pinchot, the founder of the National Forest Service, seems to have appeared at TVA only once. He arrived unannounced and uninvited in fall 1938. He had been stirred to visit, he said, by recent reports in congressional hearings about the TVA's poor forestry practices. He received a two-hour briefing by Harcourt Morgan, made chairman of the TVA in 1938 following Franklin Roosevelt's removal of Arthur Morgan a few months earlier (the first cause of the congressional hearings referred to). In the briefing, Pinchot ran head on into H.A. Morgan's rather limited notion that, in the end, forestry should be another, but only a supporting, branch of agriculture. If the farmer could be persuaded to bring his row crops down onto bottomland, his hillside then could be turned into pasture. But, Morgan noted, "If it was so badly gullied and just so far gone that that was out of the question, why then we planted trees on it."[10] This type of small-scale arboreal afterthought would be entirely contrary to what Pinchot had had in mind throughout a lifetime for mile after mile of federal forest.

During the mid-1930s, a well-regulated watershed was a great ideal, sometimes even played up in films. The best-known such picture was *The*

River (1937), directed by Pare Lorenz. On 31 January 1938, the head of the TVA Forestry Division had resigned. He had been most indignant over what was *not* happening in the way of "an official, acknowledged policy" for conservation. In a July 1938 article explaining his resignation, he said, "The eroding hillsides, the burning forests, the valley's streams flowing red with mud, wait for a clear-cut policy decision on watershed protection by the Board of Directors of the TVA."[11] It was not merely a matter of passive indifference and total inertia on the part of the directors, he thought. Those so devoted to the production of electricity (Lilienthal?) merely tolerated forest and wildlife management insofar as it helped with that production. Others (H.A. Morgan and John McAmis, with the blessing of Lilienthal?) appeared to want to make the TVA over into

> an adjunct to local agencies, such as the land-grant colleges and extension services of the seven valley states. . . . Knowing little or nothing of either forestry or wildlife, this group holds firmly to early pioneer ideas of land ownership, land use, and conservation. It thinks of forests as small scattered woodland tracts. . . . As the farm woods is, in their minds, only a minor part of the farm, so the forester or the wildlife expert is only a minor technician, quite unfit for large responsibility in land management and policy determination.[12]

The message "that the TVA is an unique effort on the part of the people of the United States, through the federal government, to solve the major problems of a great watershed," simply was not getting through. In August 1937, responsibility for Norris Lake Forest had been taken away from what was now called the Department of Forestry Relations, so that "the existence of the forest as a single entity was broken up. Untrained people were placed in charge of the various parts of the work, the name of Norris Lake Forest was dropped, and many of the carefully selected and trained personnel were let go."[13] Since Norris Lake Forest had been the only "constructive use of any of the rural lands of the TVA," this was the specific decision that caused the forestry head to resign.

The farm size Harcourt Morgan and his associates regularly championed was half an acre to ten acres. Pinchot, by now a somewhat toothless old lion, went away mollified by the courtesies extended by the TVA, telling the *Knoxville News-Sentinel* only that he was still disturbed that the TVA was not turning to the Civilian Conservation Corps (CCC) camps for personnel to fight forest fires.[14] This latter inadequacy was part of another conscious TVA policy, formulated to encourage the states to take up the duties of fire fighting and tree growing in anticipation of the day when the CCC program might be phased out of the valley. This end came, of course, with the onset of World War II, four years after Pinchot's visit. Shortly after the war, the sentiment for divestiture was revived under the cognomen of "The Historical Roots of the TVA": "Except for forest lands along the margins of its reservoirs, TVA controls no public woodlands; its major aim in carrying out its program, in cooperation with states, counties, forest agencies, and landowners, is to encourage improved management on the four-fifths of forest land in the Valley which is privately owned."[15] By 1967, TVA's two tree nurseries, at Clinton in East Tennessee and at Muscle Shoals in Alabama, had been abolished.

Attitudes toward Agriculture

Within TVA, what were considered "correct" attitudes toward agricultural interests further qualified attention to forestry. This attitude toward agriculture in turn depended upon the cause-and-effect relation that the TVA had set up between its fertilizer-producing facilities and the farmers of the region who appeared to need and want it. To obtain access to cheap fertilizer had not been easy. When they came to TVA in 1933, Harcourt Morgan and John C. McAmis had journeyed to Muscle Shoals to examine which World War I facilities might be available for manufacturing fertilizer. They found Wilson Dam and its spillway intact, but the nitrate plants were inoperative, and there were no electrical generators. In any event, Harcourt Morgan did not want the nitrate itself; rather, he was anxious to convert the production to phosphate, because he thought that would be better for the Tennessee Valley soil. "A cheaper phosphate is needed to make the legumes grow to fix the nitrogen. But we must have electric power to make our phosphate." So eager was he to obtain the phosphate that it became difficult to discern what comes first in the agricultural cycle. Thus, "when we check erosion we are in a position to balance live stock with the production of grains or cotton or tobacco or some other great commercial crop. This helps restore the natural cycle and the rest can be accomplished through the production of phosphates at a price within the reach of every farmer." But, on the other hand, "we check soil erosion through phosphate and our cycle is complete."[16] Phosphate magically supported all gestures toward the good. The upshot was that, in 1934, a pilot plant producing phosphate was built at Muscle Shoals, and the conversion from nitrogen to phosphate began.

H.A. Morgan had been born in Canada, but John C. McAmis had been brought up in a large family on a small farm in East Tennessee. Therefore he was inclined to put his faith in human rather than in chemical resources. His enduring pride was in the fact that he "understood" the small-holders' plight. In 1914, in the wake of the Smith Lever Act, he was asked by the University of Tennessee to become an extension specialist in agronomy. In 1933 he was appointed director of TVA's Department of Agricultural Relations. His first big opportunity to commit to his faith came with the eviction of the families from the bottomland that was to become Norris Lake. As a result of the Depression, each house seemed to be full of people. About a third of these people came from Union County. As that county did not have a county agent, one was provided, with funds that came equally from the University of Tennessee, the TVA, and Union County. So began TVA's practice of supplementing its own efforts with "extra" extension agents hired to communicate information about scientific farming and forestry practices directly to the clients. This task the universities were incapable of handling alone, according to McAmis,[17] because their structures were too specialized. McAmis's favorite monetary pipeline apparently was from the federal government through the land-grant universities and university-operated agricultural extension services, and finally into farmers' cooperatives. McAmis did not like to see broken up what he thought of as extended family units, each with barter systems, mutual borrowings, and a house with grounds around it, which acted as a refuge for family members down on their luck. Many had come back from Detroit and Chicago, or from the California to which so many "Okies" and "Arkies" lately had fled, because there were no jobs at all to be had. McAmis did not like either the usurpation of rich bottomland by the flooding of the great dams.

The Interpretation of the Architecture

The TVA raised so many issues and inspired so many different purposes that analysis by the agency was done neither well nor completely, least of all in terms of its architectural symbolism. In the TVA story, this analytical insufficiency was highlighted in the spring 1941 exhibition of

TVA architecture at the Museum of Modern Art (MOMA) in New York. No doubt the curators felt that they were carrying on the mission that Thomas Bender says New York City editor Herbert Croly articulated in his 1909 book *The Promise of American Life*. The exhibit's originators would define a cultural course for the United States that would "avoid choosing between an irrelevant European tradition and an inchoate indigenous culture by committing themselves to a creed, a democratic promise to be defined by intellectuals."[18] Henceforth, there ought to be a new license for such intellectual curators to explore "the larger cultural and political issues."

But in this instance, that aim was not to be fully satisfied. At one moment it went to the left of the conceptual target, at another to the right. Bender felt that, with the previous Modern Architecture show at the Museum of Modern Art in 1932, there had been a break from that procedure of looking toward "the larger cultural and political issues." Curators Philip Johnson and Henry-Russell Hitchcock, in their accompanying catalog-book *The International Style* (1932), had devoted themselves more to the meaning of architecture per se, but in the course of that exercise had removed themselves from the discourse on the whole of society and its democracy. The 1932 book had occupied itself so much with style definition that it no longer grappled with the implications of everyday occupancy of architecture by the society at large or, on the other hand, with the deeper symbolic and philosophical considerations. The MOMA exhibit and book on Internationalism showed, "on the one hand, intellectuals unable to engage architecture as a vehicle for general intellectual discourse, and, on the other, an architectural criticism that cannot get beyond itself to engage and illuminate more general political and cultural issues. . . . It made great claims for architecture, even while drastically reducing the content and meaning of the architecture it celebrated."[19] Together with Siegfried Giedion's *Space, Time and Architecture* (published in 1941, the year of the TVA exhibit and of America's entry into World War II), *The International Style* "tightened the discourse of architecture, making it essentially self-referential, a quality that persists in the field to the present."

A sharper image of TVA architecture might have been gained from Johnson and Hitchcock's more detached, more rarefied methods of analysis, but those authors failed to produce a second book on the rules and stylistic trends of the TVA. Perhaps through their silence they reinforced the common impression that TVA architecture was a further evolution of the International Style, which it surely was not. Bender thinks of Lewis Mumford as being more comprehensive and hence more praise-worthy in his outlook, but, apart from a few generalities, we do not learn a great deal about TVA architecture from that source either. In his cri-tique of the exhibit, Mumford did not discuss style. He described Hiwas-see Dam as "striking" and thought the controversial (as far as the TVA itself was concerned) generator building design at Guntersville was "ex-tremely good." But we do not learn whether those buildings were in any way Internationalistic or reflected any other style or precedent. His big-gest regret was that the museum had not allocated enough wall space to the photo essay, so it had had to be confined "to the main structures," making it "impossible sufficiently to indicate the architectonic treatment of the whole landscape."[20] Hence the reader gained little additional knowledge about either the expression of the "whole landscape" or the character of the buildings in it. The analysis was well intentioned, but without much penetration. Mumford did try, as Lilienthal would not have, to link the appearance of the TVA buildings with what he con-sidered a substantive American vernacular tradition by noticing their resemblance to "grain elevators and storage warehouses."

The approach of World War II was adding still another filter to com-mentators' visions, and maybe putting patriotic stars in their eyes. The exhibit introduced a spring into Mumford's step, for reasons less of de-sign than of collective security and a public affirmation in anticipation of upcoming military hostilities. He asked, "Aren't we entitled [after hav-ing looked at the exhibit] to a little collective strutting and crowing?" But where had one strutted from, and where was one supposed to strut to, and what was one supposed to crow about? One method of aesthetic analysis, that of Johnson and Hitchcock, may have been too narrow and esoteric; while the other, that of Mumford, did take in social concerns but may have been too broad to deal adequately with actual architecture.

When he opened the TVA exhibit at MOMA in April 1941, Lilienthal was equally vague but even more exhilarated, probably because the pros-pect of war produced an adrenalin rush in all speakers. In having been daily in charge of the TVA, Lilienthal had an obvious advantage over Mumford. However, the only reference to a literal object in his talk was to a flagpole. The erstwhile New York listener was provided with no specific descriptive or interpretive images of TVA buildings beneath that flag:

> Millions of Americans, we told ourselves, will see these structures. They will see in them a kind of token of the virility and vigor of democracy. . . . When people see these dams that they own and were built for them we wanted their hearts to be moved with pride. We wanted them to look upon the flag flying over these structures and feel a renewed love of their country and faith in the future. In this impulse we were right, as it turns out; for today at this breathless moment in the history of this nation we need greatly to declare again and again the faith we have in the future of America. And it is just this faith that the TVA project presents in a form that the eye can see and the imagination of men dwell upon.[21]

It was possible to write or speak of art, democracy, or technology, as these experts did, but otherwise difficult for them to align and connect those subjects, just as Thomas Bender had asserted.

Wank Becomes Chief Architect of TVA Because of the Cincinnati Railroad Station

When Earle S. Draper, as head planner of the TVA, invited Roland Wank to become its chief architect, it was because Wank gave more promise of "spark" and imagination than any of the native American interviewees.[22] What was needed was a sign, a token, revealing a symbol, made manifest in the public arena. The job offer was conveyed to Wank as he and Draper rode a speeding train, recalling the meeting in which Roosevelt and Norris had founded the TVA; the latter, of course, had taken place speeding south to Muscle Shoals, while Draper and Wank were headed for New York. The Cincinnati Railroad Station (1929–33) that Wank, as chief designer, had just finished for the New York firm of Fellheimer and Wagner, appeared to fit the requirement (34). There was a certain native or local character to the manifestation — Draper had only to traverse Kentucky from Tennessee to examine the station in Cincinnati, on the Ohio River. This consolidated station stood at the center of a revamped track and bridge system, such as Frederic Delano had dreamed of for Chicago at the turn of the century. The building was big and monumental; shining forth at the end of a long mall, it symbolized steam power produced from coal, much as the TVA structures would be intended to represent electrical power produced from water. Of the TVA buildings exhibited at MOMA in 1941, Lewis Mumford and David Lilienthal were to remark that the most notable feature was their "Americanism."

34. Union Terminal, Cincinnati, by Fellheimer and Wagner (Roland Wank, Designer), 1929–33. For onlookers of the time, this structure seems to have represented the next step beyond the civic architecture of the 1893 Chicago Columbian Exposition. Courtesy Cincinnati Historical Society.

The location and engineering of the Cincinnati train station were very much in and of the regional heartland, but the architectural vision, the art for the TVA to conjure with, was definitely overseas and exotic in source and mood. Americans did not easily compose civic symbols at the time.

What made the rotunda of the Cincinnati Station, 106 feet high and 180 feet in diameter, prophetic of all the future turbine halls of the TVA was its removed, echoing, cavernous feeling (35–36). However, while the turbines would "function" in splendid isolation, the Cincinnati space was meant as a concourse. Specialized shops and restaurants, and even a small theater for those waiting for trains, were grouped around the

35. Rotunda, Cincinnati Terminal. The sense of enclosure within one great space was carried on in the TVA power houses (36). Courtesy Cincinnati Historical Society.

rotunda. The thick, stable walls were decorated with huge, prominent, glass mosaics by Winold Reiss. They depicted Cincinnatians in striking poses at work in local industries, a refreshing novelty in times of dire unemployment and forced idleness. With this urban landmark amid a railroad network, Wank showed that his imagery remained closest to his conceptual birthplace, the imperial city of old Vienna, as envisioned by its most recent architect, Otto Wagner. The Moderne Monumental character, along with the superior transport engineering encased in lavish materials, owed much to Wagner's work. A sheen, reinforced by color, emanated from all surfaces. Wagner had often anchored his veneers of tile or marble with aluminum bolts. Aluminum banding inside the TVA dam houses and the checkerboarding outside on the concrete likewise drew inspiration from the Viennese School. The TVA's power house en-

36. Guntersville Dam Power House Vestibule. Here we see the same curvilinear embrace within a large hall as in the Cincinnati Terminal (35). The wall-surface of the Guntersville vestibule has a similar sheen, although the overlay of mosaic murals of workers, a popular subject during the 1930s, is left off the TVA building, maybe because of the absence of industrialism in the Tennessee Valley. The big wide door under the stairs is like the one at Cincinnati, while the large scaleless window above is typical of 1930s Moderne style. The proportions of the vestibule with its grand staircase are those of the metropolitan picture palaces of the day (another "palace of the people"), while details such as the glass bricks, indirect lighting fixtures and aluminum handrails are carried out most carefully. Photo courtesy of Tennessee Valley Authority Archives.

sembles, and that of Norris in particular, related directly to Wagner's control buildings for the locks on the Danube Canal, such as at Nussdorf (1894–98), Kaiserbad (1904–1908), and Donaustrasse (1906-1907) (37). The Viennese structures were clad in white marble slabs and cobalt blue tiles, with intricate machines inside them to apportion the waters.[23]

At Cincinnati, the building recalled Wagner's work in an interest in a simple white volume; the odd rhythm of the fenestration; slightly mannered proportions contrasting big and little; an emphasis on symmetry, crowned with a dormer skylight (34) — a motif Wagner had used at Donaustrasse; and an emphasis on a civic presence. Upon his arrival in the Tennessee Valley, Arthur Morgan had drawn lessons from the rivers of Europe, most particularly the Danube.[24] Before the Kiwanis Club in Knoxville in 1934, he had discussed the future strategy for the TVA in light of flood control on the Blue Danube, beginning in 1055 A.D.! Morgan even had been to Vienna to look the Danube over.

Roland Wank came to New York from Austro-Hungary in 1924. Winold Reiss, the mosaicist of the Cincinnati Station, had arrived in the United States from Vienna in 1913. Joseph Urban, later known for his Ziegfeld Theater in New York, as well as his movie and Metropolitan Opera sets, had come to New York from Vienna in 1912. He was drawn into domestic design after a southern planter saw a mansion by him in a film, *Zander the Great*, and demanded a real-life replica for Dallas, Texas. This first experiment on Urban's part resulted in construction of a number of fanciful resort homes in Palm Beach, Florida, between 1924 and 1926. Urban died while supervising the Viennese-American color scheme of the Chicago Fair of 1933.[25] That "Century of Progress" exhibit had considerable influence on the interiors of the TVA buildings that were begun the same year. The need in both cases was for a tinting touched with grace.

The political and technological rules altered rapidly in the first half of the twentieth century, making it more desirable (but also more difficult) to formulate and sustain a consistent visual system complete with its own aesthetic coordination. The question was: could an ordered art ever come out of pronounced civil disorder? Wank often was described as being calm and urbane, but he came from the Balkans, where World War I had started and where severe political repercussions in the wake of it had included, between 1920 and 1925, food shortages, riots, and anti-Semitic demonstrations. The political and ethnic unrest in the Austro-Hungarian Empire at the end of the nineteenth century had contrasted with the brilliance of its artistic and intellectual production, giving the lie to the assertion that art could not flourish except in harmonious times. What excitement Wank must have felt when he was invited to direct the artistic aspects of an enterprise that (he must have thought at

37. Lock Control Pavilion, Danube Canal, Donaustrasse, Vienna; by Otto Wagner, 1906–1907. Wagner's water-control pavilions undoubtedly influenced Roland Wank's design of the Cincinnati Terminal. An imperial style became a democratic style. The dormer skylight on the Viennese example turns into the main entrance motif at Cincinnati (34). Two metal wreaths as accents become a round-faced center clock in Ohio. The freestanding columns and beaded and striped trims on the railroad station are Viennese in derivation. A close but earlier parallel was Eliel Saarinen's Helsinki, Finland, railroad station (1905–14), with its great arch entrance; that building also was inspired by Viennese designs, particularly those of Joseph Olbrich. Photo courtesy of Georg Riha.

the time) would be strongly backed by a judicious and far-seeing American government! His utterances at that moment and later after his retirement reflected his persistent optimism about the TVA. He described the current inadequacies of American society as "cracks in our inherited framework," for which he proposed "to hammer out a new integration'[26] (or, as it would have been called in Vienna, a *gesamtkunstwerk*)[27]". Throwing off triviality, TVA's would be an all-embracing type of art, of the sort that chronically had been missing from the American consciousness, except perhaps in the case of the premonition from Burnham's City Beautiful Movement.

The lack of total perception was not the only "crack" in the "inherited framework," however. Even more disrupting, from the Depression through World War II, when most of the TVA construction took shape, was the unprecedented alteration in the basic usages of machinery. A second technology often overtook and rendered obsolete a previous one that had involved heavy capital investment. The Cincinnati Railroad Station — in its technical, as distinguished from its expressive, manifestation — was totally unsuitable as a precedent for the TVA, since the passenger train and its coal-burning locomotive so soon were to fall from favor. By the same token, the TVA was predicated upon hydroelectric power rather than on steam, but ultimately it returned to steam through coal a generation or less later and then went on to nuclear power, due to the technological dislocations of World War II. During the destruction of the Cincinnati Station in 1974, critics made it plain that the structure had been built too large, too late, and too lavish to survive the postwar period. Planned to handle 216 trains every twenty-four hours, through an intricate web of reorganized track and signal stations, by 1972 only 4 trains ran daily. Altering power sources were likely to invalidate the basic structural and visual ideology.

Norris Dam (1934–36)

A dam is a unique, impressive object, "a hunk of concrete," as one TVA architect described it. As an artifact and a symbol, Norris Dam was as distinctive in its own way as Roland Wank was as an architect or Arthur Morgan as an engineer. Juxtaposing unusual and unlikely buildings and settings, people and places was Roosevelt's great treat. This kind of unlikely juxtaposition was epitomized at Norris in the way the first TVA dam contrasted with the first TVA village. The dam, together with the comprehensive development of its site, set an example of coordinated expression that was never again equaled. The initiation of the Norris Dam design was auspicious. "Chairman Morgan brought out the sketches of the Bureau of Reclamation, tossed them across the table, and asked for [Wank's] judgment. Mr. Wank requested a few days to look them over and returned later with sketches of his own which were accepted by the Board. The Bureau plans, for example, had no provisions for visitors. The TVA plans did."[28] The directive to build Norris Dam had emerged so abruptly that the TVA did not have the staff ready to design it and so had to fall back on the Bureau of Reclamation. This recurred with Wheeler Dam. But by the time of the third dam, Pickwick, the TVA was prepared to take over.[29] The TVA staff eventually was able to carry out every detail of constructing a dam and shaping its environs, right down to contracting. This unified implementation kept TVA's crews and their temporary buildings moving along the valley in a process called by "forced account," meaning without taking public bids on the projects. Perhaps it was Morgan's wish to be independent of the Bureau of Reclamation and the Army Engineers that caused him to be so receptive to Wank's visions. At the same moment that Morgan withdrew from the Classical precedent established in the Wilson Dam (38) at Muscle Shoals and followed by the Army Engineers in their 1927 study for a dam (39) on the Norris site, he turned toward Detroit factory examples. After Wank had shown his alternate proposals for Norris, Morgan picked up the phone to call Albert Kahn, the long-established Detroit factory designer often employed by Henry Ford. After the call, Morgan told Wank to take his proposal to Detroit for checking. Kahn at once approved of it.[30] This early meeting was to produce further results. In 1937, when a difference of opinion arose between the architects and the engineers over the surface treatment of the Guntersville Dam power station, the architects again won their case and were allowed to sheath the upper part in gray buff rather than yellowish brick, upon the approval of Kahn, who then took on the function of a permanent consultant to the TVA.[31] Guntersville turned out to be quite a striking station (36). After this dispute, there was a joint committee on each project, under the control of a chief designer

38. Wilson Dam, Muscle Shoals, by Army Engineers, 1918–25. Thought not progressive by Roland Wank, Arthur Morgan, Earle Draper, and many later critics, Wilson Dam nevertheless fostered some continuities, such as the Beaux Arts style that came out of the Chicago Columbian Exposition of 1893. As an expression of civic responsibility, this style had an effect on Frederic Delano and others. The dam was built of concrete rather than Renaissance marble, but its repetitive Neo-Renaissance concrete bays nevertheless echoed the regular rhythm of the bays of the dam. Photo courtesy of Tennessee Valley Authority Archives.

PLATE 76

CORPS OF ENGINEERS, U. S. ARMY

DOWNSTREAM ELEVATION

EAST ELEVATION

For General Plan, see Drawing No. 1
For Location and West Military, see Dwg No. 4
For Floor Plans, see Drawings No. 5, 6 and 7
For General Section Thru Power House see Dwg No. 2

TENNESSEE RIVER SURVEY
COVE CREEK DAM

POWER HOUSE
DOWNSTREAM AND EAST ELEVATIONS

U.S. Engineer Office, Chattanooga, Tenn.   Scale: ¼"=1'-0"   Dec. 1, 1927

Submitted

Drawn By   Traced By

39. Cove Creek (Norris) Dam Proposal by the Army Engineers, 1927. This elevation came from an earlier flood control study of the Tennessee River by the Army Engineers, that contributed preliminary data for the TVA. The drawing follows the Beaux Arts style of Wilson Dam (38), which Roland Wank, Arthur Morgan, and Earle Draper were trying to break with. One bizarre, non-Classical touch was the electrical switching equipment right on the roofs. Drawing reproduced from Maj. Lewis H. Watkins, *The Tennessee River and Tributaries* (Washington, D.C.: USGPO, 1930).

whose responsibility [according to Harry Tour] was to sit in the conference
with the engineers and the site planners and others and get their reaction to
what kind of a project we were going to have, how big it was going to be,
and what its physical size was going to be and all of that. Then he was the
one that would sit down and make the initial sketches leading up to the adop-
tion of the final design and work with the engineers in the evolution of this
design. . . . It was Rudolph Mock for a while. It was Roth for a while. (? Fellow of the AIA, first name Frederic, works for Vincent Kling). Then it was
Joe Passoneau for a while. Different ones would stay for a few years and they
were probably the most important members of the staff.[32]

The position of the architect as the main conceptualizer thus was affirmed
by the incident at Guntersville. Following it, there was a slight time lag,
for it was not until the design of the first TVA steam plant in 1942 at
Watts Bar, mostly done by Wank and his aide Mario Bianculli, that the
Kahn influence became fully visible. In 1944, upon leaving TVA, Wank
became chief designer at Albert Kahn Associates in Detroit, where he
was responsible for a number of industrial plant designs, including the
Philadelphia Inquirer; the Bond Factory in Rochester, New York; and
eight major automobile factories for Ford and General Motors, of which
the Ford Motor plant in Saint Louis perhaps was most typical.

The phenomenon of Norris Dam's final appearance (40) instigated by
Wank and supported by Morgan and Kahn, had two new aspects that
particularly commended it to idealistic Americans who thought that, in
visual terms, something better could be achieved with engineering. First,
Norris Dam made manifest a very seldom seen *monumentality* on the
land, and, second, it put emphasis on the *prospects* obtainable from the
top of the dam road and the visitors' plazas on both banks. There was
a desire, on the part of persons who could think at large, to bring greater
dimension and perspective to large-scaled architecture, engineering, and
now site planning; indeed, such a wish had been the genesis of the TVA
legislation as well. Frederick Gutheim, who, through Senator Norris, had
been a major influence on the formulation of TVA law, was also one of
the very few able to comment intelligently on its building designs. In
his person was encompassed a rare awareness of both TVA's legal and
its artistic expression. In 1940, just before the war began, he remarked,
"In the Tennessee Valley we have shown (at a time when many had come
to doubt it) that a public architecture can be a great architecture. We
have demonstrated that the functionalist dogma is not broad enough,
wide enough, or deep enough to support the esthetic even of functional

buildings." The missing elements had been a public monumentality and contextualism, a demonstrable proof of a greater permanence. "Skin and bones" architecture was not to be enough. The visible structures themselves, as an ensemble, would outshine even the invisible electrical medium. "These are demonstrations which I believe transcend and exceed anything the Tennessee Valley Authority understands it has attemped or claims it has achieved: they hold the seed of a future brighter than even electricity has promised," Gutheim predicted.[33] This novel theme — sheer size and weight leading to a well-tempered public symbolism —

40. Bird's-eye View, Norris Dam, Tennessee Valley Authority; by the Bureau of Reclamation and Roland Wank, 1934–36. The two visitors' outlooks and the connecting road over the dam, Wank's ideas, can be seen. The many involutions of the shore mark the lake as Norris. Photo courtesy of Tennessee Valley Authority Archives.

was brought out again by Gutheim in his obituary for Wank. Wank "saw to it that [the dams] were approached as one would the Acropolis, seen suddenly as one came around the wooded flank of a mountainside and [was] presented with a heightened sense of scale."[34] The symbolic mountain (in this case the closeup face [41] of Norris Dam) was an apparition much admired in nineteenth-century America, as the Athenian Acropolis was admired by those who studied Classical forms in the early academies.[35]

Paul Zucker, the distinguished historian of Baroque city squares, did not view the dam as if it were a Classical acropolis, nor as a freestanding pre-Columbian or Egyptian pyramid, as a few others did, but rather as a form that was so carefully synchronized with its surroundings as to blend with nature. Zucker suggested the metaphor of a Gothic cathedral: "Regardless of whether the contour of the dam is curved or straight, it is its grandeur and clarity of form and its kinship with nature that strikes us. The deep gorges below create space sometimes comparable to the interior of a cathedral: the walls may be natural rock, the climbing slope of the dam and the falling masses of water, and the roof is the sky."[36] Such a vision was like a Teutonic philosophical act in encompassing all recognized phenomena, the imagery not too different from that of the early German romantic painter, Caspar David Friedrich, with his Gothic cathedral ruins in the woods. The monumental buildings in civic spaces that America ordinarily lacked should be integrated with nature, Zucker felt; that would be an American distinction. Zucker, whose roots also were Viennese, was oriented differently from Wank, who, although he stated that the dams "should blend with the natural surroundings," also held that they and their appendages ought to be designed so as to include "an urban atmosphere should a city [someday] encompass the reservoir."[37] Cities like his native Budapest, Vienna, and New York someday might spring up in the Tennessee Valley? Arthur Morgan, showing the American reformer's zeal for getting on with it and for moving nature along new paths, also spoke of his preference for retaining unspoiled natural spaces around his major engineering projects: "As a direct result of projects of which I was chief engineer, or otherwise concerned, there are now preserved in several states more than a dozen areas of unspoiled nature, some embracing several square miles. This, with some exceptions, is not the case with the Corps of Engineers."[38] In connection with Norris Lake, he had Island F cleared of people so that the local ecology could be studied more carefully over decades.

41. Elevator Tower and Entrance to the Power House, Norris Dam. Ideal urban imagery is transferred to the countryside. Wank's Viennese checkerboard concrete can be seen. The corners are slightly beveled for a softer effect. The aesthetic aims of the treatment were to break down scale, give better proportion, mitigate staining, and dissolve light reflection. The tower above resembles the train control tower of the Cincinnati Terminal, although this example also reflects Wank's interest in the German Expressionistic architecture of the time. New York skyscrapers are not entirely out of mind either. Photo courtesy of Tennessee Valley Authority Archives.

By the early 1930s there was an urgent need to replenish the quiver of American public aspirations. The Civil War, World War I, and the Depression had brought varying amounts of popular disillusionment. As war again approached at the end of the 1930s, the necessity increased for showing that democratic experiments could express dignity and staying power. Much as Hitler boasted of his projects lasting for a thousand years at rallies in German public squares, carefully arranged and orchestrated by his own architect, Albert Speer, Lilienthal, at the 1941 Museum of Modern Art exhibit of TVA architecture, declared that his TVA would also endure for a thousand years.[39] Talbot Hamlin, the erudite and open-minded architectural historian from Columbia University in New York City, likewise explained the TVA in the light of rising militancy overseas: "It has long been the boast of totalitarian thinkers that only in their system is efficient large-scale planning possible. In the achievements of the TVA the United States has proved the contrary, and for that the TVA projects gain an importance which may even transcend the entire practical performance."[40] Was the TVA program, then, no longer solely a funded form of Civil War reparations to the South? Hamlin, in the pivotal year of 1939, concluded that the European dictators had only laid down a smokescreen of "false efficiency," and that large-scale plans of a constructive nature could be conceived and executed in a democracy, too; the TVA had proven its peculiar worth by exhibiting so much "humanity and charm" in its buildings.[41] Americans often had looked for a gentleness and receptivity in nature that they could not find in their factories or cities. It was precisely on that basis that Yosemite Valley, with its soaring granite dolmens, had been saved as a national park, at first under the auspices of Abraham Lincoln and finally under Theodore Roosevelt. Customarily, a democracy did not wish its governing machinery to be too visible to those governed, but, in the stress of the times, the TVA was a frank, clear, outward expression of responsible awareness and coordinated counteraction.

Had a sufficient number of mutually well-understood symbols from the American democratic past been accumulated to be carried over into the TVA? For the town of workers at Wilson Dam at Muscle Shoals during the First World War, there was a plan literally laid out in the shape of a Liberty Bell from Philadelphia — uncracked![42] What survived from the past could appear specious and brittle because the country was really "never ready." Repertoire hadn't built up because there was so much dependence on improvisation. In 1934 sculptor Jo Davidson modeled a bust of Franklin Roosevelt that was considered both "fine" and "serious" by its admirers.[43] Being in the company of the president already, he also modeled an Atlas figure that he suggested should be placed against the spillway of Norris Dam, holding the waters back.[44] Wank vetoed such a nonfunctional concept at once, while the TVA engineers objected that it could interfere with the free fall of the waters,[45] although until 1969 the spillway on Norris Dam was only used four times. There was a heightened awareness of the physical presence of tremendous power that nevertheless had to be manipulated gently to create a contrast with the brutal posturings of dictators. This awareness was reflected in the pictorial sections of the Sunday *New York Times*, in newsreels in every theater across America, and soon in the superb photos in *Life* magazine and in the interiors of the TVA power houses. Hamlin described the interiors as "bright and almost joyous,"[46] and they did indeed exhibit an opulence in their veneers of aluminum, chromium, marble, travertine, terrazzo, and ceramic tiles, with lots of plain aluminum lettering proclaiming the one conviction that the surrounding structures were proudly dedicated to "The People of the United States of America." However, the social and political message of the TVA found strength by not becoming too literal in its symbolism, particularly within the power houses. There an impression of dynamism was conveyed by the vibration felt through the feet when one was on the balcony, or through the hum of the generators, a special high-speed hum like that associated in that era with radio beams in the cockpits of planes trying to find their way home. As one moved about the huge room, the hum altered in frequency. Such vibrations and hums could be taken in two ways. One was on the level of the people themselves, the "demos." For them "God was in the details," as Mies van der Rohe liked to explain about his highly refined skyscrapers and industrial buildings in Chicago. This was the approach that Gutheim had reservations about, as inviting overdesign. Marian Moffett and Lawrence Wodehouse have reported, based on their investigation of the TVA drawings, that "utmost care was taken to study qualities of proportion, scale, and texture in the design of the major projects," even into the "overall effect of a wider pier or a taller ballustrade."[47] Harry Tour, the TVA architectural engineer, perhaps grasped the intent best: "They hear the hum of the generators and they are impressed by the size and importance of it but they don't understand it. It is completely foreign to their knowledge of things in the past, but the things that they do understand are handrailings on stairs [36], doorknobs, lighting fixtures and floors."[48] These were the immediate, tactile gratifications.

A more visionary level was that of the American intellectuals, ready to recognize larger conceptual issues and changes. Among them was Gutheim, who recalled Henry Adams' prophecy that someday the dynamo would be worshipped as fervently as the Virgin Mary had been in the Middle Ages.[49] Stuart Chase, the economist, also took up the spiritual, rather than the materialistic, theme. In 1936 he said,

> Compromise or no, to see the authority in operation is a spiritually refreshing experience. To look at the clean, strong walls of Norris Dam between the hills of pine [40]; to feel the will to achievement, the deep integrity of a thousand young-minded men and women, schooled in the disciplines of science, free from the dreary business of chiseling competitors and advertising soap; to realize that resources are building rather than declining and that the continent is being refreshed; to know that, over the whole great valley from the Smokies to the Ohio, men's faces turn to a common purpose and a common goal — intoxicates the imagination. Here, struggling in embryo, is perhaps the promise of what all America will someday be.[50]

For so much in the 1930s, realization had to be postponed, but the TVA would not have to be!

These intellectuals were looking for an opening that might be spiritually charged and shaped with an entirely new mission. The impulse to look down a valley in order to gain a new prospect, an aperture into the future, was as old as the opening, following on the American Revolution, of new awareness of the national landscape. In 1783, Jefferson had looked from Jefferson Rock, above Harper's Ferry, down the Potomac Valley in order to conjure up the higher hope of a future America by detecting "a small catch of smooth blue horizon, at an infinite distance in the plain country, inviting you, as it were, from the riot and tumult roaring around, to pass through the breach and participate of the calm below." In 1774, he had privately bought the Natural Bridge near Lexington, Virginia, so that from its top posterity might look down Cedar Creek Gorge and take vicarious pleasure from the visual wonders of this new country.[51] Gutheim had spoken of the downstream vista from the top of the dam at Norris, of "the magnificent natural view downstream, the thickly wooded hills rising above the trickle of water in the river, the rugged, still untamed nature."[52] So one had to ascend in order to realize the nature and extent of the valley below. Gutheim had involved an expanded convention of sight that by now — following on World War I and coming up to the Depression, when the building of skyscrapers ceased for awhile — had developed for the American city, rather more than for nature. The glimpse into the magic canyons (42) of the future American city was encouraged by imaginary drawings of the 1920s by Hugh Ferriss, Harvey Corbett, Ralph Walker, or Raymond Hood. Hood's Rockefeller Center was in some ways an in-town TVA, and Hood was very concerned to enable viewers to look down from the height of a skyscraper on roof gardens in the New York project. This sense of wandering through canyons amidst enchanted towers was brought out as well in the Americanized sets of the film *Metropolis*, directed by the German, Fritz Lang, and released in 1926. The 1929 scheme for the 1933 Chicago Century of Progress layout, initiated by Corbett and Hood of New York with Hugh Ferriss as consultant, included a series of pseudo-skyscrapers from which the movement of crowds across suspended bridges and roof and through passageways, three stories up, regularly could be observed and assessed.[53] The urge to emulate in Tennessee the dynamic spatial and emotional relationships of a futuristic New York or Chicago — two cities then unique in terms of height and density of population — also can be noticed in TVA's Pioneer Freeway, modeled on the Bronx River Parkway (1909) and especially the Westchester County Parkway (1922) of the New York metropolitan road system.[54] This cultural longing to look down in a masterful way on rapid and intense movement was reincarnated in 1967 in Atlanta by architect John Portman, whose lobby of the Hyatt Regency Hotel (43) became the prototype of numerous other caravansaries across the country. Portman's design was a new-old idea, based on seeing great distances from odd angles, particularly from a considerable height. Most important because it struck a responsive chord of memory, the same sort of upscale, polished imagery appeared in much of the TVA photography of the 1930s, often taken from an angle above or below the actual subject (41). The Atlanta Hyatt lobby also was American in that, although its purpose was commercial and mercantile, its anachronistic message (which turned out to be so popular) was expressed in open, expansive, idealistic, and liberating ways. Its enclosed focal feature, like that of TVA's power house halls with their generators centering attention, was a mechanical device, the rapidly rising elevator.

42. Crowning Towers; by Hugh Ferriss, 1929. Looking down into a canyon
was a time-honored American indulgence, dating from as early as 1774,
when Thomas Jefferson used his own funds to purchase the Natural Bridge
in Virginia so that posterity might enjoy the vista below, symbolizing the
nation waiting to be developed. A viewing terrace appears at the lower
right. The prospect was urbanized and aggrandized during the 1920s when
skyscrapers multiplied, and movies, such as *Safety Last* (1923) in which
Harold Lloyd hung over a Los Angeles street canyon, became popular.
Sketch reproduced from Ferriss, *Metropolis of Tomorrow.*

43. Lobby, Hyatt Regency Hotel, Atlanta; by John Portman, 1967. The Hyatt Regency capitalized upon the older American penchant for looking down into a canyon of concrete and steel (41–42) to inaugurate a new kind of hotel interior that became very popular and was replicated across the country. The views from above, like those from the top of a TVA dam or a visitors' balcony of a power house, could take in several angles and focus on machinery; in the hotel an elevator cab. It premised an ideal 1920s–30s interior city, sufficiently extravagant to make up for any fallibility about the future. Courtesy John Portman & Associates, Architects and Engineers.

Where Did the Norris Paradigms Lead?

The Norris Dam compound set a high standard for coordination, difficult for any later dam to equal. Its outstanding achievement was in establishing a bold beginning for future cooperation among architecture, engineering, and landscape architecture. The dam materialized itself, represented democracy in a noble way, and engaged the attention of many of the best thinkers of the time, including many from abroad. While the TVA perhaps did not produce any industrial designs entirely up to the calibre of those by Albert Kahn in Detroit, nor provide anything with quite the dramatic impact of a Le Corbusier postwar concrete monolith (he did come to see the TVA) or the highly imaginative and poetic compositions of Frank Lloyd Wright (who must have been influenced by it, especially at Falling Water), Norris Dam nevertheless held out its own challenging public promises. It showed that, despite Wright's often-voiced doubts, teams could bring forth commendable architecture, and that all the worthy architecture in this country did not have to be privately created. The TVA drew on another model of organization, not often enough recognized as being unique to American architectural practice — the big firm, of Chicago, Boston, New York, or Detroit. The transfer of big-firm expertise and consensus to the TVA was made possible by Albert Kahn. As Grant Hildebrand points out in his definitive biography of Kahn, his firm, having hundreds of architects and engineers, was set up to get things done in a hurry. "More than any other architect of his time, he put this approach [of using parallel experts in "related fields" important for the decision-making process] into actual practice, showing that it was capable of processing large amounts of data toward a large-scale solution of demonstrable usefulness, which could be carried out rapidly."[55] The main advantage of such large-scale, coordinated office management was that it enabled the firm to respond very quickly and with considerable "know-how," as the TVA office did in the Depression; and as Kahn's firm did in World War II, in record time furnishing shelter for Chrysler's tank production and Glenn Martin's bomber manufacturing buildings.

Fontana Dam (1942–45) as a Stopgap

Fontana Dam (44) represented a natural climax to the TVA dam-building sequence, but, because of the exigencies of war, it was not permitted fully to express that position in a visual manner. Fontana was begun less than a month after the attack on Pearl Harbor. The tallest of the TVA dams, it rose to 480 feet. Because of its height, it was said at the time that one Norris Dam was being built on another. But the height and other dimensions could not be exploited fully in visual terms, nor could the humanizing opportunities of the site be developed fully. After Norris and the other early dams, the need to respond to sudden and unanticipated war alarms grew, keeping such potentialities — at least from the Depression-oriented perspective — from crystallizing.

The first frustration arose from the fact that, although Fontana Lake would be on the edge of the new Great Smoky Mountains National Park, which had been developing for a decade and had been dedicated in 1940 by Franklin Roosevelt, there was no way that this remarkable new amalgam, which included several beautiful nearby national forests of ancient hardwoods, could be used for leisure activity. At this time there was no leisure. The second major frustration derived from the fact that wartime shortages inhibited the use of the TVA design innovations and accentuations. Without them, the inordinate size and spread of the dam (45) created difficulties with both perceptual scale and proportion — difficulties that later would interfere with the steam plants' relating comfortably to their physical surroundings.

There had been thought of building the power house half a mile downstream in order to gain additional "head" for the generators. This proposal was given up because not enough steel was available and because this site would have required extra security to protect against sabotage. Instead, the power house was built flat against the dam and was made of concrete. It was given a sufficient number of windows to make it look like a conventional Modern hospital, school, or factory of the time. Although the structure works fairly well when the viewer is on either riverbank and can see the dam as a frame behind it, from further downstream, where the blank, uninterrupted expanse of the dam face looms so huge behind it (45), the power house counts only as a small box, an empty orange or egg crate. What are really missing for the power house

44. Bird's-eye View of Fontana Dam. The Great Smoky Mountains dwarf the highest of the TVA dams. The road up to the dam takes the sudden revealing turn toward its face that architect Wank advocated. Photo courtesy of Tennessee Valley Authority Archives.

45. Fontana Dam, Tennessee Valley Authority, 1942–45. Fontana was kept plain and simple in contrast to Norris Dam (40–41) because of war emergency. It therefore had no comparable spillway, elevator tower, or gantry crane to provide the syntax of accents and scale. The visitors' outlook shelter could not be constructed until after the war. The uncertainty of scale, that makes it look like a blank white sheet hung across the valley, can also be seen in the odd relationship between the height of the power house and that of the two diversionary tunnel outlets at the foot of the bluff on the right, that were put offside because no spillway had been provided. They are about four-stories tall. The tunnels can shoot water 150 feet in the air and 400 feet downstream from molded bases. Photo courtesy of Tennessee Valley Authority Archives.

dimensions to relate to are the details of the dam that had to be omitted because of time pressure. There is no traveling gantry, no elevator or control tower silhouetted against the sky and setting the scale. Dams in the Far West of this size, with such high bankings, often would be built on the curve, creating dramatic shadows, but here there was no time for such subtlety. Instead of an elevator shaft, an inclined railway was planned to run in the open air from the visitors' outlook down to the power house, but none of it or the outlook could be built until after the war. There was no spillway with training walls to offer textural and shadow relief, and no break in the long horizontal line of the top of the dam. It is impossible visually to detect the relative size of any feature. Instead of a normal spillway, two diversionary tunnels (45) were drilled to the right of the dam, as one looked upstream toward it. In a photograph, they look no larger than two openings of a farmer's drain tile in a field. The tunnels appear to have no dimensional relation to the power house. They take on adequate phenomenological stature only when the looker is informed that each is 34 feet in diameter and discharges water hundreds of feet in the air and 400 feet downstream, guided by free-form discharge buckets.

Because work had to go on night and day (46) and security considerations were involved, a searchlight system was set up (47). It illuminated everything, but it also flattened the scene, so that whatever was three-dimensional no longer looked it. The power house appeared even more flat and miniscule (47). Outside in daylight, the building seemed to have six stories or more, but inside it was mainly one great generator hall (48), 63 feet high. Function was misleading the observer of form? The official report on Fontana mentioned that inside, there was "a cathedral-like quality to this room that is both impressive and inspiring."[56] The point is well taken, similar to the suggestion that, at least in painter Charles Sheeler's eyes, the River Rouge automobile factory was a "Classic Landscape." Medieval church interiors also employed an indeterminate space as a revolt against the Classical devotion to human proportions. Without a figure up against the window wall of the TVA hall, it would be extremely difficult to gauge the height of the window panes (9 feet). To awe and entice visitors through unregistered heights was indeed a goal shared by the Gothic and the TVA architect. However, the differences between the Fontana power house and the nave of a cathedral may convey even more than their similarities about the motivations, beliefs, and blind spots of the Modern Age, and may suggest caution in drawing too-ready analogies to ancient buildings, with their greater complexities and refinements. The Fontana interior differs from a Gothic nave in not filtering light in through its structure; rather, it floods it! No stained glass intervenes. Consequently, while the space may be enigmatic in the medieval way, the illumination is not. Instead, it is very candid and revealing of the structural parts. The colonettes and vault ribs of a Gothic nave would be constantly uprising and slender, but they would not flaunt the presence of stone in the way that the power house frankly exhibits its concrete piers. Medieval piers would be more modeled in light and shade. The window bars of Fontana are said to be "massive mullions," and the "concrete surfaces were left untreated and exposed." Moreover, when the piers attain 36.5 feet from the floor, they are interrupted by a massive and emphatic crane girder that projects 9 feet into the room in order to carry the 300-ton-capacity moving crane that no cathedral ever would have to countenance. On that interrupting lintel, brushed aluminum capital letters (48), impossible to gauge as to size, would read: "1942 – Built for the People of the United States of America – 1945." The spatial experience of the Fontana generator room may be to a degree scaleless and unfathomable, as in a Gothic nave, but the structural members are secularized and insisted upon, and not dematerialized. After more extended aesthetic analysis, the power house remains in the twentieth century. Shortages of time, materials, and willingness to invest meant that the nuances of the Middle Ages could not be recaptured, despite the welcome malleability of concrete. A Gothic cathedral normally took hundreds of years to build, a TVA power house, as the aluminum sign affirms, two to three years.

The horizontal grace that the TVA can accomplish is demonstrated better at the Fontana site with the bridge on the Tennessee River about a mile below the dam (49). The bridge won an American Institute of Steel Construction award for 1945. It lacks the bravura of the bridges of Robert Maillart in Switzerland, or even of the high-arched concrete bridges of the California Division of Highways of the 1930s, but this bridge was built in wartime. In fact, it was made of leftover pieces of the dam construction (46) trestle.[57] This inheritance determined the length of 120 feet for each of the three main spans.[58] The bridge is supported by four slab piers of concrete, a hallmark motif of the TVA bridges, with a light fret on the girders of vertical stiffeners. Tying the wilderness banks together, it could have appeared in a Charles Sheeler painting.

Never unpleasantly oversophisticated, the TVA bridges displayed a refreshing simplicity and directness.

Yet in this most remote of spots, far from any city and surrounded by mountains rather than hills, the most sophisticated of wartime technology prevailed. TVA asked the Aluminum Company of America (ALCOA) to give up its water rights for building Fontana Dam. Its manufacture of aluminum took place in nearby Alcoa, Tennessee. A lot of electricity was required to make two pounds of alumina into one pound of alumi-

46. Working on Fontana Dam by Day, 1944. The construction is more uniform than on the usual dam, due to the need for wartime speed, but it still breaks down into modular parts. Short supply steel from the high trestle furnished reusable spans for the downstream bridge (49). The wooden slip form for the pour of concrete is at the top of each bay rise. The trestle carried the trains bringing concrete. Photo courtesy of Tennessee Valley Authority Archives.

num. By the time the TVA was founded in 1933, the aluminum company had already built three hydroelectric dams in the vicinity: Cheoah in 1919, and Calderwood in 1930, both on the Little Tennessee River below the Fontana site; and Santeetlah in 1928, on the Cheoah River.[59] Glenville and Nantahala, also high-head dams, had been finished above the Fontana site in 1941 and 1942 respectively. However, the need for planes

47. Working on Fontana Dam by Night. Because it was wartime and there were sabotage possibilities, the dam was intensively illuminated. The workmen appear as dissolving flecks. Photo courtesy of the Tennessee Valley Authority Archives.

became ever more urgent, so Roosevelt called for more and more alumi-num production. From the whole Tennessee Valley, 395,000 more kilowatts were demanded. Fontana was designed for three generators of 67,500 each. The first went into operation on 20 January 1945, the second on 24 March. ALCOA had been trying to hold up the building of a new

48. Interior, Power House, Fontana Dam. Outside, the dam appeared flat and the power house looked like a six-story building. Inside, the power house was one great story high and emphatically three-dimensional. Scale and proportion were enigmatic. Photo courtesy of Tennessee Valley Authority Archives.

dam at this location in order to gain advantage over the much smaller Reynolds Metals Company, which, at its Muscle Shoals plant, would have used a share of electricity output from Fontana. However, none of the disputants had taken into account the character of Harry S Truman, chairman of the Senate Committee on War Production, which was in charge of investigating defense material tieups. Exasperated, Truman declared in June 1941, "I want aluminum. I don't care whether it comes from the

49. Bridge Below Fontana Dam, Leading to Fontana Village, 1944–45. This bridge represents Tennessee Valley Authority team or committee architecture at its best. The deck has a subtle rise and curve not visible from this distance. It is not ahead of bridge design in the rest of America and the world at the time, but it displays a reassuring capability. The TVA bridges exhibited regular variations but remained pleasantly true to type. One of that type's trademarks was the slab support pier. Photo courtesy of Tennessee Valley Authority Archives.

Aluminum Company [ALCOA], the Reynolds Metal Company, or Al Capone!" This expostulation was credited by two later observers with bringing on "a landmark agreement in government-industry relationship."[60]

The dam was rushed to completion in a steep-walled gorge with an annual watershed rainfall of 75 to 80 inches, a great deal of water to route. The average in the whole Tennessee Valley was 52 inches. In the Ohio River Valley, the average was 35 to 37 inches. Work hours would not be reduced from eight to five and a half, as had occurred at Norris Dam in an era of high unemployment and need for adult education. Nor would there be an effort to build a people's park around Fontana as an area of unmatched recreational potential for the East and South. Dam construction went on day and night, in three shifts seven days a week. This stiff regimen led to employee absenteeism and turnover. The government was anxious not to appear to be regimenting or drafting workers for forced labor camps, as was being done in Europe, so workers remained free to take jobs elsewhere.

Long before, Arthur Morgan had proposed that Fontana be the key dam in a mutually supportive electrical system. But he had been rebuffed. As he described the sequence:

> This was by far the most important single project on which I came into conflict with the other members of the board. . . . On May 19, 1936, the day after he was reappointed to the TVA board [over the strenuous objection of Morgan], David Lilienthal brought Dr. Harcourt Morgan to my office and told me that we would have a business meeting. He then announced that the purpose of the meeting was to take construction of Fontana Dam out of my hands and put it in his.[61]

First the commission actually to design Fontana was withdrawn from Arthur Morgan, then the project itself was cancelled.

> On June 2, 1936, Mr. Lilienthal and Dr. [H.A.] Morgan held a meeting without my presence and without giving me notice, though I would have been available in two days. At this meeting they voted to suspend negotiations by the TVA to obtain the Fontana damsite but left open the possibility of negotiations with Alcoa relating to the interchange of energy. I have discussed at length how crucial TVA control of the Fontana site and its water was to the Authority's integrated management of the Tennessee River and its tributaries [a unified system was paramount—W.C.]. In suspending negotiations by TVA to obtain Fontana, the decision of the board majority drastically compromised the position of the TVA as it related to the public interest in Fontana, and constituted what might be a serious violation of the most important physical aspect of the TVA concept.[62]

This move, of course, made Fontana unavailable for urgent defense need five years later.

Because it would have had insufficient time to cure out within an emergency schedule of twenty-one months, the concrete at Fontana was quick-cooled in fifty-foot-wide and one-hundred-foot-high sections, with the contraction joints between, later to be grouted. This method had been used first at Boulder Dam. Coolant pipes of one-inch diameter and five-hundred-mile length were installed in the horizontal joints.[63] The heat of curing thus could be dissipated more rapidly and cracks avoided.

In the end, the first power from Fontana did not go to ALCOA and the manufacture of more bomber wings, but rather to Oak Ridge, Tennessee, where an even more deadly weapon was being prepared.[64]

The Watts Bar Steam Plant (1940–42)
Without a Consistent Representation

The first expression is one of democratic amenity. If Fontana Dam represented a reversion to a single-minded purpose in the face of an upcoming war, then the complex at Watts Bar would represent its splitting or fraying in the face of the same demands. The site was prepared for a dam and shortly thereafter had a steam plant imposed upon it. The kernel of the latter's design plainly lay in the optimistic, upbeat prewar mood geared toward overcoming the Depression, without a hint of war visible in it. The turbine room of the steam plant (50) completely reflected the "humanity and charm," the "bright and almost joyous" manner, that Talbot Hamlin had described as typical of TVA interiors. Even the official account captures a bit of the euphoria first associated with the aims and purposes of Roland Wank and his assistant, Mario Bianculli:

> Within the building the center of interest is the turbine room, and the layout and decorative treatment were studied with a view toward making this room an example of architectural excellence . . . ; walls are faced with light-blue terra cotta; and the floor is surfaced with gray ceramic tile. Nothing was allowed to detract from interest in the huge turbo-generators. Rather, the enclosure is subordinated to focus attention on these imposing yellow and tan machines.[65]

It was as if these architects wanted one more fling, one more hypothetical encounter with a brighter past, before war broke out in earnest. As if, while a bit of yesterday was retained, it was already tomorrow, and the war was over. The time frame was very large, but the cold, hard fact

50. Visitor's Gallery, Turbine Room of the Steam Plant at Watts Bar; by Wank and Bianculli, 1940–42. The generator room and visitors' gallery were designed as if no war were in the immediate offing, despite the fact that by 1940 the Watts Bar compound had been made off-limits for civilians. The room looks like one for a hydraulic system attached to a dam, but the viewer knows immediately that it is run by steam because the generators are on their sides rather than upright. Watts Bar was the first steam plant of the Tennessee Valley Authority. Photo courtesy of Tennessee Valley Authority Archives.

51. Generator Room, Steam Plant, Oak Ridge; by the Kellex Corporation. This was
the biggest generating plant in the world. Watts Bar (50) had yellow and tan generators with Viennese chrome trim; the walls were Viennese blue tile. Here, at Oak
Ridge, the tonality is monochrome, the electrical apparatus is crowded in, the pipes
are exposed, and spare parts lie about. Austerity and minimalism have been reasserted.
Photo by J.E. Westcott, reproduced courtesy of U.S. Army Corps of Engineers.

was that, in November 1940, before war had been declared, the whole Watts Bar compound had been designated a restricted area, off-limits to civilians, and troops had been sent in to guard it. The description of the steam plant ignores all that:

> Without an easily accessible vantage point from which to view this room, some of the effect might have been lost. Therefore, a spacious overlook balcony was built across the east end of the turbine room from which employees working in the adjacent offices and visitors passing through the entrance lobby can see the entire room without interfering with the operation of the plant. . . . The public lobby is an impressive room 20 feet wide, 30 feet long and 26 feet high. Both ends, one in the exterior wall and the other between the lobby and turbine room, are entirely of glass. Each side wall is faced with marble from the floor to the ceiling.

What a contrast to the intensely crowded, drab gray wartime generator room (51) of Oak Ridge, so tightly and blindly chained to its short-range purpose of reducing uranium! The Watts Bar interior even echoed the foyer of a New York film palace of the 1920s, as did any number of TVA buildings, including the foyer of the Guntersville power house (36). Because the Watts Bar power house was eclectic in style, it was also extended in stylistic time, running all the way from 1920 to an intimation of the 1950s. All that was missing from the fanciful inventory was moviepalace crystal chandeliers or colorful murals of Morocco or Monterey.

> The balcony and stair railings are steel with plate-glass panels. From the reddish brown terrazzo floor through the golden-buff marble walls into the blue ceiling, the effect is light and colorful, appealing and interesting. The large combination lighting and ventilating fixtures in the ceiling, flanked by outer rows of recessed lights, give the upper part of the room an attractive sparkle.[66]

The Viennese term might be "sparkle," the American term "spark," as per Earle Draper. Apparently there was a desire to achieve an animation, a vigorous revitalization, through being in a TVA Garden of Allah surrounded by shiny machines, bright lights, and warm ceramics — a happening in a remote place that could not be reached even after a laborious journey because it had been closed by the military! In this landscape, the mood-change from peace to war was profound.

The second expression at Watts Bar was one of consolidated efficiency. The outside of the Watts Bar steam plant (52) presented quite a different picture, suggesting cool logic and rationality, veering toward the distinct, frozen moment and the big, independent artifact. The subunits were kept independent, but they were connected by conveyor belts, covered over, just as the generators were encased in the turbine room. The working parts were identified but not revealed. This silent, abstract order surely emulated the Detroit auto plants of Albert Kahn. Kahn had, of course, come into the TVA arena in 1937 as an adjudicator in the Guntersville power house design dispute.[67] His general approach was encapsulated in Charles Sheeler's famous paintings (53) of Ford's River Rouge factory (1916–32)by Kahn, the most complete industrial assemblage in the world at that time. An ore boat could come in at one end of it and a finished car out at the other end, in an elaboration of assembly line production. At that time, both Watts Bar and the River Rouge plant were seen in the context of decayed and obsolete factories all over America, closed down since the 1929 market crash. That spectacle, which would disappear entirely with the advent of the war and stepped-up production, in the prewar days was even more demoralizing than the dilapidated cities, which would receive greater attention after the war, in the 1950s. Clarity of concept and its visualization were appreciated all the more because they were so little in evidence during that era. Between 1927 and 1932, at the behest of the Fords, Sheeler took photographs and painted pictures of River Rouge. River Rouge was an industrial estate more outspread and more internally contrasting than it appeared in Sheeler's *American Landscape* (53). The painting was called a "landscape," but it had no valley, no trees, no hill. Elements were subtracted, estimated, or moved about in order to achieve a more perfect union, just as in the contemporary International Style of architecture. There was a compulsion to differentiate and label each part within the more ordered industrial precinct, to stamp it with the seal of abstract functionalist approval. The Watts Bar Steam Plant (52) was, to a degree, a further distillation of the painting. Emphasis on the parts made the plant appear more "real." "In order that each element be well-defined, the structures are covered with 3/8-inch asbestos-cement board screwed to the supporting framework. Flush metal copings and well-spaced windows impart a sort of blocky character and tie the component parts together into a harmonious assembly,"[68] the official account said. A further display of almost painful geometry, achieved by placing specialized structures at measured intervals, was made in Sheeler's *Classic Landscape*, owned by Edsel Ford, depicting the River Rouge plant in 1931. The com-

52. Outside of Watts Bar Steam Plant, 1940–42. Watts Bar was TVA's first experiment with coal-burning in place of water generating, in response to the call for war readiness. Therefore it was essentially a model or pilot plant – simple but extremely orderly in its emphasis on the unit parts and their dynamic relationship. It was the first major evidence of the influence from Albert Kahn's Detroit auto factories (53). In the coal hopper shed at the right there is a coal car that can be turned entirely over and emptied by the cradle holding it. Photo courtesy of Tennessee Valley Authority Archives.

53. *Amerian Landscape*, Charles Sheeler, 1930. Sheeler's painting of the Ford River Rouge Plant by Albert Kahn presents it as a cleaner, more orderly, and more compact enclave than it was in reality. The visual statement, like that of the Watts Bar Steam Plant (52), is mostly about how each part ought to have its own identity yet fit into the main scheme. Painting reproduced courtesy of Museum of Modern Art, New York; gift of Abby Aldrich Rockefeller to the museum.

plicated but eventually precise relationships among the factory components encouraged their association with Classical Greek or Roman forms in an age better equipped mechanically and scientifically but not as well ordered. The actuality of the Watts Bar Steam Plant more mimicked the orderly painting than approximated the huge spread of the River Rouge works. With only three simple units, separated by decent intervals of space, and lacking human scale, the steam plant was more an ideogram or pilot plant than a full-scale adaptation of the original River Rouge model.

This sort of TVA essence of engineered precision on a great scale in a partial vacuum was epitomized again in Charles Sheeler's 1939 painting *Suspended Power*. It showed a monumental propeller (54) about to be lowered into its scroll case at the TVA Loudoun Dam. Again we see a stop-action, trying-for-timelessness effect, executed in an era that has had little permanence. The personal concentration of the two figures below the propeller, while it hovers above them like a giant bird, drains the whole surrounding room of air or atmosphere. At the actual installation, there would have been present many workers, who would have pushed and pulled at the object from below (55) but not have given it a great deal of observation and contemplative thought, since in the TVA that type of operation always was carefully preplanned. The propeller would whirl around in the hydraulic turbine only after it had been seated in its proper place. Its staticism in the picture constitutes a rare and serene moment in the disturbed culture of the 1930s.

Soon a more militant, detached attitude will appear. By 1942, when the Watts Bar installation was being finished, the weight of anxiety over a war that had gone on since 1939, and had often seen the Allies losing, was beginning to have its affect. The dam's control building was on a high bluff (56) above the dam. Against the bluff, the control building looked to be a defensive entity unto itself — like a German concrete bunker, a solid part of the Maginot or Siegfried Line, a defense of the Normandy Coast against Allied invasion. There were many similar radar and lookout towers along the U.S. Atlantic Coast. The building was detached from the body of the dam below and from the landscape across the lake. Only a single high-tension tower signals back to this shore (57). This was already a historically sad landscape because many tragic and costly Civil War battles had been fought along these shores down to Chattanooga. The landscape looked gentle and nostalgic, while the visitors' balcony

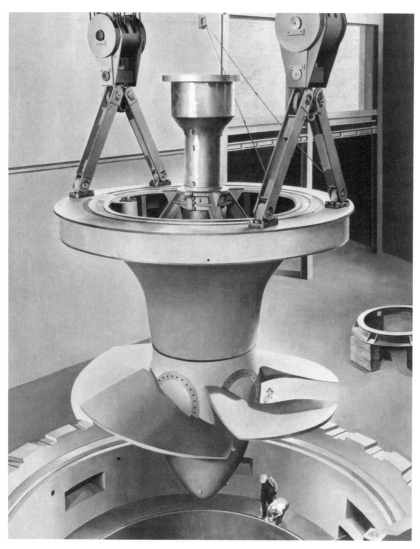

54. *Suspended Power*, Charles Sheeler, 1939. The painting features a large, poised, single object at Fort Loudoun Dam, in a way the 1930s distinctly relished. Perhaps because the rest of life seemed so petty during the Depression, it became that much more satisfying to focus upon a monumental entity. Painting reproduced courtesy of the Dallas Museum of Art; gift of Edmund J. Kahn.

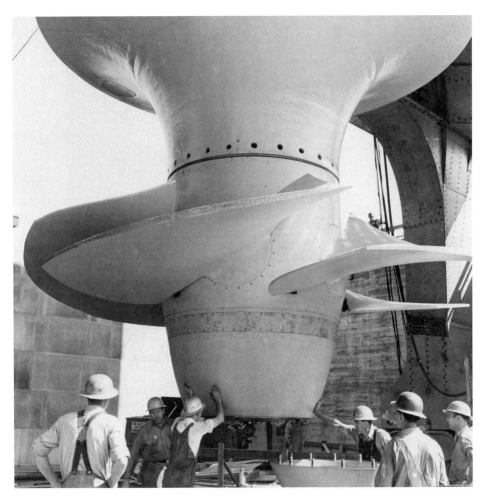

55. Seating a Turbine Propeller, Fort Loudoun Power House. The chief difference between the reality of this photograph and the legendary Sheeler painting of the same subject (54) is that the action of several more workmen is kept continuous. It is a variable pitch propeller. Photo by Charles Krutch; reproduced courtesy of the Tennessee Valley Authority Archives.

that gazes out upon the view from too great a distance appeared grim, austere, and too plainly built, with none of the older TVA lyricism, because of the mood and the war shortages. The prewar lustre of public places, so evident in the observation balcony over the generator room (50) inside the Watts Bar Steam Plant, here in the outdoors had entirely evaporated. Watts Bar architecture represented three time periods, not entirely successfully joined.

The Slightly Earlier (1937–40) Hiwassee Dam Setting Illustrated the More Organized Peacetime Regimen of the Early TVA

The earlier visitors' buildings, such as those at Norris or Chickamauga, were square or rectangular multiforms. At Hiwassee (58), as at Watts Bar, this building's shape was a more curvilinear monoform (57). The full circle roof at Hiwassee undoubtedly owed something to the image of C.A.P Turner's 1905 mushroom slab capital invention or Maillart's Swiss version of 1908. Even more likely as an immediate inspiration was Frank Lloyd Wright's Morning Glory mushroom column of spun concrete, used in the Johnson Wax office in Racine (1936). The TVA architects, however, took the slightly more office-approved, conservative stance with their mushroom columns, as with their bridges, not flinging caution to the winds like Wright and Maillart. The Hiwassee mushroom roof has steel stanchions holding up its outer rim. But it is daring and compatible in the way that it shares the void beneath its roof with the lake and mountains beyond (59), in a gesture much more congenial than that of the Watts Bar outlook toward its landscape. One could look right through this pavilion. And the engineering of the dam and the contours of the parking area complement its daring with rhythms of their own, rhythms of arabesque pattern that bring the whole scene together. A road passes over the felicitious arrangement of the downstream dam face and through the traveling gantry on top, as under a Roman triumphal arch. Here is a premonition of a broader landscape composition synchronized with the devices placed in it. With this project, something splendid is about to take place — the dam site next after this one might truly deserve the title "organic." But of course that development could not occur, because the war, with all its haste and expediency, too soon intervened. We get only a hint of how things might have been ordered for the better in the way that the tops of the terraces of rock on dams like

South Holston (60) run on into the benching of the river banks there, and how the texture of its rock fill finds some response in the groups of hills and trees beyond. Those dams seemed to begin to belong.

The Watts Bar Site Was Not Coordinated

The Watts Bar steam power house was essentially a pilot plant (61). From the air it looked like a child's model steam engine on the green baize cover of a card table. The only landscape treatment of the yard was a well-placed drainage ditch. There was no other coordination of the site with the adjoining dam or with nature. The plant had been placed at Watts Bar because an electrical switchyard and transmission lines were already in place for the dam. When the war threatened, another personality from Detroit, William S. Knudsen, arrived in the valley to call for power and more power.[69] He was now the government's manufacturing czar, but he had been president of General Motors. Eventually, six more steam plants were built, beginning in 1949 with Johnsonville (62). This facility was full-sized, nearly three times the size of Watts Bar. Johnsonville could turn out more electricity than Norris, Fontana, Hiwassee, Douglas, and Cherokee dams combined, so one can see why the authority so readily turned away from hydraulic power sources.[70] By 1970 no more than 20 percent and by 1975 no more than 12 percent of the electricity available in the valley was coming from water power. The rest derived from steam, superheated by either coal or atomic power. 1950–51 were years of increase in coal-fired steam, following the outbreak of the Korean War in 1950. Small wonder, then, that there was less and less interest in site attachment or reconciliation. The machine was present in bigger and bigger increments, but there was no longer any serious care for the garden. In form, the Johnsonville plant was a linear extension of the Watts Bar main unit, but it also began a new format for TVA, in that the subunits, such as the water treatment plant, the control center, and the utility building, stood independently in the yard. There was an attempt to differentiate materials such as steel, brick, glass, and aluminum siding so as to better identify subunits and break down scale, an effect that was carried further by contrasts in paint colors, particularly dark maroon that would not show the coal dust.[71] The compound had its own coal receiving harbor. The coal storage yard (63)

56. Control and Visitors' Outlook Building, Watts Bar Dam, 1942. Detached and mounted on a rocky bluff because there was no building space available below at the dam, this prominent structure demonstrates that TVA support buildings had lost the warmth and confidence they had previously evidenced. More than that, this building seemed even to communicate an embattled anxiety. By now the war was going badly. Photo courtesy of Tennessee Valley Authority Archives.

was so large and received the fuel from so many directions, including from trains, that the precise on-line delivery of the pulverized fuel that had obtained at Watts Bar (52) did not occur at Johnsonville. In this yard, conveyors seemed to come from several directions, even from underground. The monster scale and more incidental arrangement in the big coalyard created a slightly primeval or sinister impression, an atmosphere never sensed before within the TVA, as if prehistoric animals were grazing and feeding there, with the stocking-out conveyors poised like praying mantises.

Small wonder that the final integration of technology with the landscape became, from the design point of view, less and less of a likelihood. World War II had elevated the demand for electricity from ALCOA and Oak Ridge far above previous requirements. In the immediate postwar period, a federal installation for advanced rocket research was founded at Huntsville, Alabama, and staffed by ex-Nazi technicians, while the Arnold Engineering Station, with its giant wind tunnels, was established at Tullahoma, Tennessee. In the early 1950s, the newer gaseous diffusion plant of the Atomic Energy Commission (AEC) was located at Paducah, Kentucky, as a second Oak Ridge. The TVA built its Shawnee Steam Plant there, but still was unable to give the AEC as much electricity as it wanted. At the time, Oak Ridge and Paducah together consumed twice as much power as New York City. Evidently, after World War II the federal government had come to believe that the TVA was the traditional source of the electrical "free lunch," although it became more and more evident that the supply would never be entirely "free" and that producing it would have more effect on the environment than had been bargained for. During the war, William S. Knudsen had insisted that Douglas Dam be built, and "Dr. [Harcourt] Morgan fought a losing battle against the building of Douglas Dam, because the lake would cover some of the best farm land in the Valley." This objection would surface again in the late 1960s, in the environmental debate over building Tellico Dam. As long as crisis followed crisis in the nation at large, however, there could be no ultimate reckoning of resource damage or dissipation. The Valley Authority was kept on constant alert as to its projects, never left in peace by outside agencies. That was well demonstrated by the rapid overriding of H.A. Morgan's caveat on the critical occasion of undertaking Douglas Dam. With the outbreak of war, it became doubly difficult to resist the cry for electricity. "Somehow, TVA never got over

that excited demand for electric power. Power was never again incidental."[72] In the original 1933 charter of the TVA, the production of power
had been incidental. Section 23, Item 3, of the original TVA Act had
called only for "the maximum generation of electric power consistent
with flood control and navigation." Since TVA's prewar ideals and support somehow had gotten lost amid later crises, the persistence of the
Cold War after World War II gradually forced the TVA, largely in its
own self-defense, to become an appurtenance of national defense policy.
This cooptation led to another shift into nuclear energy to supply electricity after 1960. As former Assistant Chief Engineer Harry Wiersema
explained from retirement in 1969, the rise of nuclear energy and the
subsequent departure from coal, which in turn had succeeded hydroelectricity, was due basically to defense considerations: "As we studied
the problem we could see that nuclear fuel was something that was a
by-product of the defense program. So for that reason it could probably
be produced cheaper than coal."[73]

The power yards and grounds became larger, and their components
more ungainly, at least to the naked eye. A few of the features appeared
to divorce themselves completely from the surrounding landscape. The
Kingston coal plant (64), the largest such facility in the world when it
was built (1951–55), in 1976 added two completely freestanding stacks
of 1,000 feet high, as tall as a hundred-story building. This addition
reflected a desire to reduce sulfur emissions, which the four earliest stacks
of 250 feet and the next five of 300 feet had not satisfied.[74] There was
an aspiration to become more and more detached from the earth plane
and to rise higher and higher to the clouds in an effort to reduce pollution, but beyond pollution on surrounding fields still lay the question of
far-reaching acid rain. Kingston mostly supplied electricity to Oak Ridge.

How Do the TVA Dams Compare
with Those in the West?

Of what did the genuine distinction of the TVA dams consist? In size
they did not surpass those of the West (65), built by the Bureau of Reclamation. At 480 feet, Fontana was the tallest dam of the TVA series, but
in the 1930s and early 1940s, Boulder Dam on the Colorado, Shasta on
the Sacramento River, and Grand Coulee on the Columbia, all built by

57. Balcony, Visitors' Outlook, Watts Bar Dam Control Building, 1942. Outside the immediate orbit of the steam plant, the Watts Bar Compound has a random and accidental feeling. The steam plant (61) was placed on the premises because the dam switchyard already was available. The landscape had a history of a proud and sad sort in the Civil War battles that had been fought in the vicinity. The Control Building appears not to want to get in touch with that history, however. The building seems more cheap and detached, more solitary and aloof from its landscape, than any other Tennessee Valley Authority example. The lonely gantry of the dam stands apart at the lower right, while a single high tension tower stands across the lake, delivering power to somewhere else. Photo courtesy of Tennessee Valley Authority Archives.

58. Hiwassee Dam Visitors' Outlook, 1937–40. A visitors' outlook slightly earlier than the one at Watts Bar Dam (57) is more indicative of the Tennessee Valley Authority's prewar confidence and optimism, its "humanity and charm." Relatively conservative structurally, like the Tennessee Valley Authority bridges (49), the Hiwassee structure is at the same time open and buoyant, almost like a holiday kiosk, such as the earliest shelter, at Norris (7). Photo courtesy of Tennessee Valley Authority Archives.

the bureau, were considerably higher than Fontana, at 726, 602, and 553 feet respectively. Of the five largest concrete dams in the world at the time, Fontana was only fourth. Since the TVA dams were not larger, were they instead cleverer and more sophisticated in design? The plans (66) of the "World's Five Largest Concrete Dams" show that Fontana does not come at the top in those attributes either. Boulder and Shasta are graceful arch dams in plan, while Fontana is of the straight-across gravity type. Boulder is reassuringly symmetrical, with its four intake towers balanced in pairs and its power house forming a reverse capital U against the dam below. When Boulder (or Hoover) Dam discharges water, it does so symmetrically from both banks in huge streams, whereas the two

59. Visitors' Area, Hiwassee Dam. Although in this picture the site preparation is not finished, the viewer gets more of an impression of synchronization and intimacy with the natural landscape beyond than at Watts Bar (57). There is a free-form flow and sense of ease; the war has not yet begun. Still more organic ensembles might have been achieved if the war had not broken out just then. Photo courtesy of Tennessee Valley Authority Archives.

60. South Holston Rock-Fill Dam, 1947–51. Although the Hiwassee project was on its way to a full unity between site and architecture (59), South Holston Dam promised even more in the way that it fused built features with land abutments, always an awkward visual break, and especially so at Hiwassee (59 and 68). Photo courtesy of Tennessee Valley Authority Archives.

discharge tunnels for Fontana (45) are in an eccentric position on the right. Boulder Dam (1931–36) was begun a few years earlier than any of the TVA dams.

Recently Richard Guy Wilson, in a thorough description and evaluation of Boulder-Hoover Dam, reveals it as a forerunner of Norris and the later TVA dams in terms of aesthetic development. The richness of its materials and the pageantry of its iconography, as organized by sculptor Oskar Hansen and color consultant Allen True, were, if anything, more extravagant than those of any TVA dam. The subject matter of the two

61. Bird's-eye View of the Watts Bar Steam Plant Site. The grounds of this first Tennessee Valley Authority steam plant, as distinguished from the buildings, appear completely unplanned and little treated. The joint breaks of the advent of war and the first use of coal were responsible for this haphazard treatment. A coal stocking yard lies beyond the coal receiving shed. More distant nature gets little acknowledgment either. Photo courtesy of Tennessee Valley Authority Archives.

artists were earthly labors of the time and Southwest Indian motifs. The consulting architect, like Roland Wank, happened not to be native born. He was the English-trained Gordon B. Kaufmann of Los Angeles, who came to the United States in 1913. Did the accident of foreign birth make him and Wank more alert to American possibilities?

At the rim of Boulder-Hoover Dam were four towers designed like miniature Art Deco skyscrapers. The two inner ones were used as elevator lobbies, with dark green and black marble walls. The visitor descends in the elevator from them, 528 feet down to the power house. He

62. Johnsonville Steam Plant, 1949–53. The steam-generating plant built next after Watts Bar for the Korean War is obviously an exponential enlargment of its predecessor (61). The support buildings are more numerous and specialized, and the coalyard is much bigger. Interest in site development and appreciative contact with the landscape has virtually disappeared. Photo courtesy of Tennessee Valley Authority Archives.

then "exits into internal galleries decorated with True's terrazzo designs," Wilson says, "and finally arrives at the great turbine chambers in the powerhouse. The setting overwhelms: immaculate hygienic preciseness, large shiny green and black casings of generators, chrome flashing and pipe railings, and repetitive piers of the enclosing wall, all accompanied by a pervasive loud hum."[75] The refinement of detail, surface tension, color, and spatial and auditory compression may be greater here than in any of the TVA power houses.

If the TVA dams cannot surpass this dam of the Bureau of Reclamation either in size or in splendor of materials and depth of coloration, upon what then is their claim to attention based? Their viability really depends more on their collective existence as many dams in one system (67), synchronized, and in tandem the most complete set in the world in such a river system. Within such an unmatched sequence, design and engineering experience could be cumulative, construction teams could progress from one site to another, and equipment could be reused. Even when machine equipment or design details, such as gantry cranes or checkerboard walls, appeared unique in their invention, they had a way of turning up in slightly different versions several times over at different dams and bridges so that the norm could appear a comfortable product of standardization at the same moment that it provided the refreshment of invention and variation. Through such a complete system, quantities of water could be exchanged at will and electricity easily sent first to one place and then another, as happened particularly during World War II. In the Northeast blackout of 1965, TVA reciprocated the original donation of expertise from New York City by sending a few of its experts up to advise the metropolis on how such deprivations of lighting could be avoided in the future.[76] The dams in the mountains could shut down for snowmelt in the late winter or early spring, or open for drought in the late summer, running up to December, to perpetuate navigation, power production, and water supply. The fall of water was from a river elevation of 1,745 feet at Nottely Dam, the highest point; to 354 feet at Kentucky Dam, the lowest. Chattanooga alone was saved from $2,622,431,000 in flood damage between 1936 and 1984.[77] Perhaps most important, an imaginative spectator could conceptualize and visualize a different, more versatile, more affluent American culture growing up all along the system's shores.

At its best, as at Hiwassee Dam (1937–40), TVA exhibited an excellence arising from four unusual approaches: (1) the implicit revelation in all its works of coordination and interrelation, that tended to animate its every expression; (2) care in the design of both the large and the small object; (3) interest in innovation and in the kind of stewardship that, after integrating the innovation, ensures good order; and (4) honest excitement over what at that time promised to be an era of improved civic, domestic, and internal interest for the region. The downstream face of Hiwassee Dam (68) shows it to be without the seductive curve of Bureau of Reclamation dams such as Shasta (1940–45), that might provide a more interesting and subtle play of light over the dam's surface during day and night. However, Hiwassee compensates by testifying to a better

63. Coalyard of the Johnsonville Steam Plant. The coal storage area has grown so large and is served by so many conveyors that they appear dispersed, coming from too many directions, and the logic of their continuity is much harder to fathom. Photo courtesy of Tennessee Valley Authority Archives.

64. Kingston Steam Plant, 1951–55. The two major smokestacks have become so tall and prominent that they dwarf even the enormous power houses. At 1,000 feet, they are only 350 feet shorter than the twin towers of the New York Trade Center, built on Lower Manhattan by the New York Port Authority. This extraordinary manifestation of the smokestacks, like the gargantuan earth shovel at the Paradise, Kentucky, Steam Plant, has occurred in the name of the old Progressive-TVA watchword, "efficiency," in this case for the dispersal of pollutants like sulphur dioxide. There is really no site landscaping, but the service roads have been formed into a cat's cradle of complexity. Photo courtesy of Tennessee Valley Authority Archives.

order and emphasis. Shasta, by contrast (69), looks very loose and casual. It has no buildup to accent features like the two traveling gantrys of Hiwassee. What captures the eye instead is Shasta's reckless, abrupt projection of five black penstock pipes from the dam surface down to what may be recognized, after a little effort, as the power house removed from, and at an odd angle to, the dam. Hiwassee coordinates its banking and dam face so that the viewer perceives the power complex as a series of four regular terraces down to the river. On the terraces, the switchyard frame and transformers are placed sequentially. The dam face is seen as a continuous surface slope to the river, accented by the guide walls of the spillway and by the equally careful delineation of the knees of the spillway gates above it. The main horizontal feature of the ensemble is a deck, upon which the control building and the 275-ton capacity traveling hoist are placed. A daring pièce de résistance is the 120-ton gantry (70) above on the crest of the dam. In a way the control building, hoist, and gantry are too conspicuous, unique accents, but they fill their role with a dashing self-sufficiency and containment and stand as great improvements over the earlier awkward and unwieldy gantry forms such as that at Wheeler Dam (1933–36), the second TVA dam. The gantry form began to be improved starting with Pickwick (1935–38), the third dam. At Hiwassee, the cynosure of the family of TVA dams, nothing turns out crude or hardedged, despite the tight grouping. This project sustains the prewar impression of "humanity and charm" that Talbot Hamlin ascribed to its phylum. On both the lake and the downstream sides of the dam, the horizontal joints have a three-and-three-quarter-inch wide chamfer for light and shade effects.[78] Because of the view from the higher walks and roadway (70) down on the roof of the control building, a second roof of textured precast concrete slabs standing on flat round pedestal blocks was laid on top.[79] Above on the dam roadway, where the extremely versatile gate gantry presides and travels, similar thought has been invested in visual refinement. The curbs of the walkways are extra high so that vehicles will not jump them and injure viewing pedestrians. Two perennial discomforts occurring to drivers and passengers when crossing bridges are restudied: "The parapet is designed for visibility out over the lake or down the river gorge as seen from a vehicle driving across the dam. High solid parapets have in the past been unsatisfactory in this respect."[80] How often have bridge railings everywhere thwarted the down or upriver views! Second, the light standards are seventeen feet high (70) and kept seventy-five to eighty-five feet apart so as to reduce the flicker in the eyes of the driver at night as

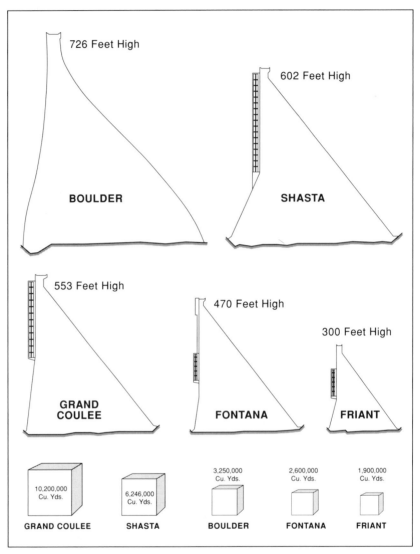

65. The World's Five Largest Concrete Dams. Fontana, the tallest of the TVA dams, does not begin to match the height of Boulder, Shasta, or Grand Coulee Dams. Adapted from Pacific Constructors, *Shasta Dam and Its Builders*, 1945. Drawing by University of Tennessee Cartographic Services, 1990.

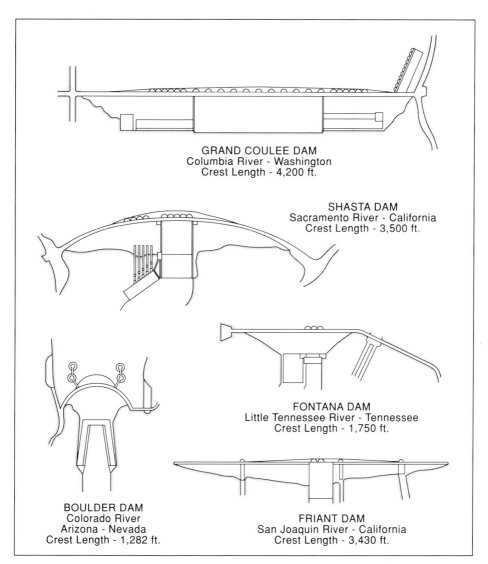

GRAND COULEE DAM
Columbia River - Washington
Crest Length - 4,200 ft.

SHASTA DAM
Sacramento River - California
Crest Length - 3,500 ft.

FONTANA DAM
Little Tennessee River - Tennessee
Crest Length - 1,750 ft.

BOULDER DAM
Colorado River
Arizona - Nevada
Crest Length - 1,282 ft.

FRIANT DAM
San Joaquin River - California
Crest Length - 3,430 ft.

66. Comparative Plans of the World's Five Largest Concrete Dams. Fontana Dam is the least symmetrical of the five and lacks an arch shape. World War II caused it to be lopsided. From so high above, such delicacies of design count, as with a landscape brooch. Adapted from Pacific Constructors, *Shasta Dam and Its Builders*, 1945. Drawing by University of Tennessee Cartographic Services, 1990.

he crosses the dam at vehicular speed. This problem had resulted from low, indirect, experimental lighting on the parapets at Norris Dam.[81]

The capacity of TVA architects and engineers to register dignity and calm, order and interval, éclat and finesse, in the power house can be demonstrated by comparing those structures at Norris (71) and Shasta (72). At Shasta, belonging to the Bureau of Reclamation, the roof was supported by a spindly series of fan web trusses. The ceiling at Norris had more substantial beams. The rhythm of the wall bays at Norris is slower and more stately, the travertine-textured surfaces more splendid.

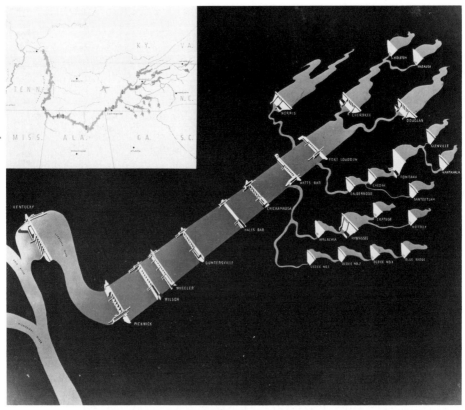

67. Tennessee Valley Authority River System. The remarkably complete and sequential system, with everything following in close proximity, constitutes TVA architecture and engineering's real claim to worldwide distinction. If the coal and nuclear plants (3) were superimposed on this map, the pattern would constitute an indeed great garden of power. Courtesy of Tennessee Valley Authority Archives.

A lintel is a lintel, a post a post, and a tile floor a ceramic tile floor at Norris. The architectural members were made more distinct and even in their rhythms. Stairs and railings were not mere add-ons, dubiously designed, as at Shasta. Most pertinent, the turbine generators at Norris were contained in one housing each and made one major statement, rather than being a series of escalating assertions topped off by a light bulb and supported by a clawfoot standard. The decade of the 1930s, in its more lucid moments, wished for a visual demonstration both that new power technology and its transmission were ushering in a new era, and that people would be able to find surcease from other distractions in encapsulations that were large, simple, self-sufficient, and fully volumetric. The generators on exhibit were proofs of that prowess. In the company of such a masterful object, the spectator felt reassurance. The machine should not be allowed to become one more threat amid so many economic, political, and military threats, rapidly converging at the time. Form needed to follow function in the conceptual, nationalistic, and morally reforming manner, and less along the direct route of impersonal, functional cause and effect.

Where else in America at the time could one find as much elementary poise, and as strong a conviction that events were coming under control and into focus, as in the Norris Dam power house? Wouldn't a person notice at once that he or she could step with full confidence from the visitors' gallery onto the top of the broad generators with safe railings of pleasing design (71) all around them? The tops of the Shasta generators (72) were not as reassuring, and they did not welcome visitors. A person would have to climb the stages. In the presence of the TVA, architecture and engineering might explain that democracy worked, while other less visible, less tangible systems in business and government were breaking down. The aesthetic valuation test had become an eminently spatial, three-dimensional reaffirmation along a whole valley.

The Oak Ridge Installations (1942–45)

The TVA design consensus in the valley began to break down in 1940 at Watts Bar, with the fear of oncoming war. With the retreat into the Oak Ridge installation (73) for the Manhattan secret atomic project, the break became complete. Minimalism, restriction, isolation, impersonality, and concealment became the order of the day. TVA color and "humanity and charm" were deleted from the formula. The public would not be welcome any more. No-one in the TVA, including David Lilien-

68. Downstream Face, Hiwassee Dam, 1937–40. In this mountain area, the valley was so narrow that a platform had to be built out to carry the generator, control house, and moving gantry. The great ability of the Moderne Style to suggest an exciting future (the twenty-first century?) is especially evident on the projected deck with all its contrast and glow. The deck was held up by custom-designed Modern piers and had a slab surface; this latter, and the slab roof on the control building, were designed to be seen from above. The horizontal seams of the dam face are chamfered for visual effect. All comes close to total balance and harmony except the ragged juncture of the earth abutment and the dam face, seen here on the opposite shore — a design feature difficult on any dam. Photo courtesy of Tennessee Valley Authority Archives.

69. Shasta Dam, 1940–45. While the power house is hardly recognizable, it can be located by following the five black penstock pipes that spring so abruptly out of the lower dam face. Shasta Dam does not have the accented unity and coordination of the Tennessee Valley Authority dam types (68). Photo reproduced from Pacific Constructors, *Shasta Dam and Its Builders*, 1945.

thal, was aware of what was afoot in the Oak Ridge Army Reservation when it was sealed off in fall 1942. Stupendous plans were to go forward without any of the TVA administrators, experts, designers, and idealists hearing what and why.[82] Nonbuildings were to be put up in a single-purpose, short-term place to which there would be no access. There were to be no more of the old measured, contemplative, distant sightings from high points and odd angles, that were so congenial to Americans.

70. Road Deck at the Top of Hiwassee Dam, with the Smaller Traveling Gantry. There are several TVA-type visual refinements on this passageway over the dam, such as the lowered parapets for viewing, and raised lamps and sidewalks. Photo courtesy of Tennessee Valley Authority Archives.

The lifestyle of the Upper South and the atmosphere of the Tennessee Valley itself became irrelevant, except for the fact that these Clinton Engineering Works were located there partially because TVA's Norris Dam, with its abundant electricity, lay only sixteen miles to the north-east above Knoxville, and the Watts Bar Dam and Steam Plant were about twice as far away to the southwest. Tennessee was relatively re-mote from prying eyes, and far enough inland from the Atlantic Coast so as not to be easily reachable by enemy air attack.[83] The Army had decreed that major installations had to be two hundred miles from the coast, the Gulf of Mexico, and the Great Lakes. In actual effect, Oak Ridge was to become yet another segment of the TVA that at the time could not be acknowledged officially as such.

The gaseous diffusion plant, K-25, located at the northwest corner of the Army reservation, was the biggest Oak Ridge unit. The building (73) was U-shaped, 2,450 feet long, and an average of 400 feet wide and 60 feet high.[84] It covered 44 acres. It was virtually proportionless and with-out fenestration or facade, except for a small clerestory at the top of the wall. Technology in this structure became blank-faced and imperturb-able, inscrutable. The Oak Ridge factory designs were created not by a leading architect with a record of characterful public or industrial buildings like Wank or Kahn, but rather by the more anonymous Kellex Corporation of New York City. Sheeler's detached, treeless, moundless, figureless, suspended-moment architecture in a landscape had become too real and utilitarian. Instead of establishing a lasting monument in a democracy, the Kellex Corporation had the short-term task of creating something like a vast federal prison or a manned maximum-security gar-rison. The only canon carried over from the TVA to Oak Ridge was cleanliness. It was desirable to keep the interiors absolutely spotless, so that the heavier, more common U-238 isotope of uranium could be changed into the lighter U-235 of plutonium without its becoming tainted. For that, even a thumbprint would be a pollution threat.

The other plant for processing uranium (74), but by electromagnetism rather than the gaseous method, was not a giant U-shaped megastructure, but instead recalled a huge Western frontier town waiting for High Noon. In terms of TVA's ordering and amenity sequences, this plant was even bleaker than the gaseous plant. Pipes and wires moved in every direction. The electromagnetic process allowed for placing buildings in groups, hence the scatteration of 170 different-sized buildings over five hundred acres.[85]

71. Interior of the Norris Dam Power House. The need to create a calm, simple, almost bland interior, to counteract any threat implied by the "machine age," is already in evidence at the first Tennessee Valley Authority power house. Photo courtesy of Tennessee Valley Authority Archives.

72. Interior of the Shasta Dam Power House. This hall looks more cut up and transitory than that of Tennessee Valley Authority's Norris Dam (71). The sum of the parts is not greater than the parts themselves. The designers of Norris worked harder against the adversity of the Depression than the designers of Shasta worked against war. Photo reproduced from Pacific Constructors, *Shasta Dam and Its Builders*, 1945.

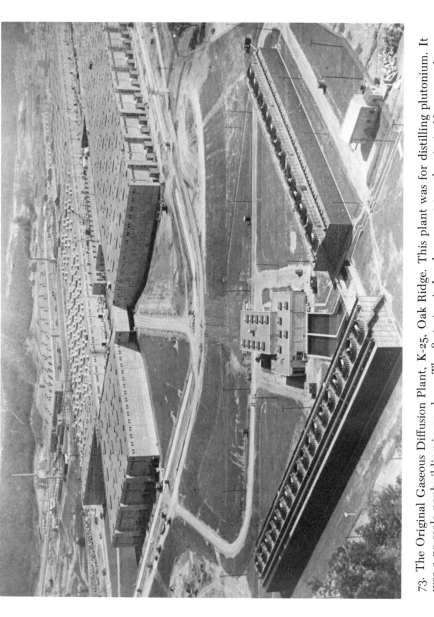

73. The Original Gaseous Diffusion Plant, K-25, Oak Ridge. This plant was for distilling plutonium. It was a no-scale nonbuilding in nowhere. The first atomic bomb was put together in it. No outward expression of inner working parts could be offered to democratic bystanders. U.S. Army. Atomic Energy Commission, J.E. Westcott.

74. Electromagnetic Conversion Plant, Oak Ridge. Although the methodology was not random or disordered, just highly experimental, no outer expression of order or logic was recognizable except to a few highly-trained technicians and scientists. There were no well-proportioned spaces, no visual cues that this was an enterprise of a democracy. In appearance the conversion plant was more like a huge gold-rush or frontier town. The low hills were supposed to protect the works from prying eyes and the town beyond from any possible explosion. Photo by J.E. Westcott, reproduced courtesy of United States Atomic Energy Commission.

Its order could be made comprehensible only in the mind of a physicist or chemist. The idea of sharing engineering knowledge and scientific information with the public was no longer inherent in this federal project, so there was no outward manifestation of symbolic art or taste.

A third increment was the thermal diffusion plant, composed of 160 buildings and covering 150 acres. It was later judged to be impractical and became a laboratory for further nuclear research. A fourth major element was the steam power plant (51), designed by Kellex of New York and Sargent and Lundy of Chicago. At the time, it was the largest steam plant ever constructed and could deliver 238,000 kilowatts on demand.[86] How demanding of electricity the Manhattan Project was is indicated by the fact that, even with this new steam plant, it still drew most of its electricity from the TVA. Instructive in themselves were the differences between the Wank and Bianculli design for the Watts Bar generating room (50), full of good feeling, texture, and color, and that of the Army Engineers by Kellex at Oak Ridge (51). Watts Bar envisaged the long future after the war would be over, while for Oak Ridge "a time limit of three years was set for the development and use of atomic energy as a weapon of war."[87] The generators were crowded into the room at Oak Ridge, and the room failed to adjust to them. There were people present, but, in association with the slightly sinister machines, they did not stand out in the way they always did in a TVA generator room. At Oak Ridge the machine and its auxiliary equipment was paramount over man. The display no longer carried an optimum message, or really any message at all, from the governmental "haves" to the collective "have-nots" of the valley, as had been the first intent at Norris Dam.

Each complex at Oak Ridge was built among hills and ridges (74) to ensure concealment from prying eyes and to protect one complex from the others in case of atomic accident.[88] TVA had been begun out of surplus Wilson Dam, created to manufacture munitions at Muscle Shoals in World War I. Swords had been beaten into plowshares. Now, all that long progression was being reversed, so that a stupendous explosive device could be built at Oak Ridge. A multipurpose mission for the TVA had turned into single purpose one here, but power still bred power. The social danger, of course, was that through sustained habit it might come to be thought that creativity and progress could arise only out of spectacular technological responses to successive cataclysms — economic disruptions or military setbacks, just as some thought only the private sector could be creative. Perhaps Americans could get too used to mounting the barricades, permanently distracting themselves from the steadier pursuit of long-range goals. The original goals of the TVA had been to orient itself more completely to nature and to integrate itself more closely with the valley environment and people. The product of Oak Ridge would be aimed toward destroying foreign cities and their surroundings in an instant and from a great distance. The only apparent carry-over between the two organizations (and could it have been much more than an individualized or token one?) was when, in January 1947, after World War II was over, Gen. Leslie R. Groves of the Army Engineers handed the atomic city of Oak Ridge on to David Lilienthal, then chairman of the newly-formed Atomic Energy Commission.[89] TVA's electricity, originally promoted as a means of improving domestic order and tranquility in a provincial setting, suddenly and inadvertently had become the means of fashioning a weapon to leverage the whole world from that same valley.

The Arcadian Image: A Picture Not Yet Colored In

Admittedly, in the United States it is difficult to maintain the focus and momentum of any broad or long-range plan. Political leaders get voted out of office, procedural minutiae forever get in the way, and physical manifestations are esteemed less highly than fiscal accomplishments. Perfection is never attained. But in the promotion of twentieth-century utopianism, in comparison with the Tennessee vicinity's nineteenth-century utopianism, the tools became more elaborate and drew much more public attention, especially in relation to electricity. At New Harmony, Indiana, the English utopian manufacturing genius Robert Owen, who made his fortune in factories in industrial Manchester, set up no factories or mills and called on no new technology to rescue his people. Lately the technical means may have become the end, the medium the message. The TVA itself, during the second quarter of the twentieth century, was forced by the crisis demands of wars and economic dislocation into an increasingly narrow and feverish pursuit of technology. In fact, this pursuit began even earlier with World War I and the Muscle Shoals Munitions plant.

The alert viewer may take in all those factors and what they mean without ever coming to grips with the more fundamental long-range lesson of the Tennessee Valley Authority and all its structures. The TVA was the first exercise designed to present an alternate image, on a large scale, of what another, rearranged kind of life might be like in America, and to fulfill the initial seventeenth-and eighteenth-century promises to the land. Every country has to set aside places where Nowhere can be a gratifying Somewhere. In Great Britain it may be the Cotswolds, the Scottish Highlands, or the southwest coast. In France it is Normandy or the Riviera. In America at present it may be California, Arizona, or Florida. What is different about Tennessee in this context is that, through the TVA, it suggested some presentiments of what an improved *everyday* American life could be, not tailored exclusively for the holiday escape. Confusion over that point appears to have generated much of the Tellico Dam protest. Technology of the newest kind could be organized in an actual place, so that ordinary (and sometimes far from ordinary) people could be assisted to go about their regular daily tasks in an enhanced environment. People could be encouraged to live in a readapted and improved version of a national park or forest, on a readapted leisure lake.

Actually, there were three basic images of the Tennessee Valley. The first was the electrical utopia, in which a neglected citizenry was salvaged by means of electricity. The second was the bonanza, or curse, of atomic energy. The third, now neglected, was the prospect of the romantic arcady that Americans have never quite found but whose image they don't want to let go of. What could be of concern today, in a more pacific time, is that that third page of the TVA program, the arcadian dream, was so early and so readily torn off the clipboard. What is most disturbing about the Tennessee Valley chronicle is that, even after a number of the most crucial issues facing the nation had been made uniquely visible there, national attention was fixed on the TVA for so short a time. One can observe the short national attention span with particular clarity in the unrealized imaging of nature. Americans have always tended to resort to this type of imagery in times of trouble, as happened also in the Depression with the Back-to-the-Land Movement and with urban decentralization theories. The book by the early TVA planning staff, *The Scenic Resources of the Tennessee Valley*, shows that this group first grasped the recreational and scenic possibilities of the area during one great, grand tour around the valley. The planners looked upon the valley as still unspoiled, although they took note of the coal- and copper-mining districts. Photographs from the beautiful book, such as the view of Lake Lure (75) from Chimney Rock in North Carolina, all attest to that positive evaluation.[90] The Lake Lure vista is a typically

American view of a comprehensive scene — at a distance, from above. It lingers in the mind, an intimation of what the book's authors hoped the Tennessee watershed someday could become. It encompasses an entire geography in a panorama — the softer, older mountains, unlike any in the West; the hardwood trees, more varied here than anywhere else in the world; the lake; and the cloud formations above, softened "with the almost ever-present haze." The inventory that they took also included many picturesque landforms, such as knobs (low hills), bares (meadows on the mountains), and coves (small, secluded valleys, where pioneer families liked to settle). Many caves and waterfalls were illustrated. The woods could be filled with rhododendrons and flowering dogwood in the spring. The view of Lake Lure may induce a reverie in the two girls in the foreground (75) sitting on Chimney Rock, 1,500 feet up. The larger awareness is signaled and encouraged here not by technology but by an art, the art of observing and interpreting. In this way the TVA initially approached its landscape. More subtly, other pages in the book reveal that Lake Lure is artificial, that it is actually a three-mile stretch of the Broad River, dammed in 1927 by the Southern Power Company.[91] Moreover, there are hotels, camps, and roads around this artificial river-lake. The photographer has taken some pains to hide them under a mantle of trees. He is more interested in a visual sleight-of-hand that will permit him (and the TVA) to emphasize the therapeutic powers of nature, universal and unharmed, that constitute the message of the entire book. Pictorial means were used everywhere in the TVA enterprise — in engineering, architecture, and landscape — to reinforce or enhance nature. At the outset, however, unspoiled landscape was not regarded as the least of these. The assumption of the book is that the region has magnificent natural settings still, inhabited by people who should not be allowed to sink any farther into invisibility and ineffectiveness. One way to keep them from doing so was to improve their total environment.

The view of Lake Lure and the book that contained it declared that the ultimate arena would belong to nature assisted by man. To make the new formula work, conscious art would be the catalyst, the reagent, the reunifier of nature. Remarkable as this early *Scenic Resources* survey was, it never would be repeated as a method of assessing the valley. The sweet melancholy and gentility implicit in the Tennessee Valley surroundings and reflected in the study, really were implicit in the entire landscape of the South, as it reacted to the wastage and suffering of the Civil

75. View of Lake Lure, North Carolina, from Chimney Rock, 1938. This
Tennessee Valley Authority photo suggests the manner in which the
Tennessee Valley Authority planning staff first regarded its demesne. The
staff's was an arcadian, agrarian, all-encompassing way of looking. It was a
world that might have been. There would be no immediate exploitation of
the land for short-term intense govermental purposes (73–74). The initiators
would instead be responsible for the whole landscape and a distant future.
The pair of girls was thought to bring amenity and gentleness to the scene.
Photo reproduced from Draper, *Scenic Resources of the Tennessee Valley*.

76. *Hilltop*, Maxfield Parrish, 1926. Two girls again dominate the river-lake
and mountain landscape beyond with their musings. This idyllic mood,
more typical of the 1890s and Progressivism than of later decades, no longer
was possible after the total loss of innocence in World War II (73–74). That
loss of innocence was roughly equivalent to the loss of environmental idealism
by the Tennessee Valley Authority. Many just-graduated high school girls
were employed at Oak Ridge. Painting reproduced from Coy Ludwig,
Maxfield Parrish (Watson-Guptil, 1973); and Dartmouth College Library
Collection, used with permission.

War and the subsequent boom of northern cities and their factories. In that light, the TVA landscape was an instrument of redemption. The two girls looking out over Lake Lure seem to express a wish that the vista might become a new type of national park, one in which people might dwell inconspicuously, since they had never been allowed to do that in previous parks. Robert M. Howes, one of the authors of the *Scenic Resources* essay, would realize a similar wish when he directed the establishment of TVA's Land Between the Lakes in the 1960s. That latter plot of land looked along an isthmus in a similar manner, between two lakes, but it was more like a national park because it was legislated to have no inhabitants.

Reference to artists who, like Maxfield Parrish, were successful in the 1920s, may give the greatest opportunity for answering the question of whether such a mood and outlook could be perpetuated. Parrish lived in an era that Coy Ludwig has described in his biography of Parrish as "the Golden Age of Illustration."[92] Paintings by him intended for mass reproduction, like *Hilltop* of 1926 (76), show what remained after World War I of the popular hope of a new age of peace and contemplation, and what cultural inspiration held on from Conservationism and Progressivism. The two maidens in the foreground could have come from Burnham's imagined City Beautiful, for they wear semi-Classical garb. Parrish visited the Columbian City Beautiful Exposition of 1893 in Chicago (as did Franklin Roosevelt and Arthur Morgan) and, in a letter to his mother, reported that he had gained a great deal aesthetically from examining its multiplicity, "the conception of the whole thing,"[93] from various visual angles. To the left of the tree under which the *Hilltop* maidens sit is another river-lake, as in the view of Lake Lure (75). Because the maidens are so passive and gentle, their reverie takes in the whole distant prospect of nature, with its meadows and foothills, and the final distance, just as in the TVA picture, is made more intriguing by haze. Parrish himself described the painting in those terms: "It will be of two girls under a big tree at the top of a hill, with a great distance beyond, late afternoon all flooded with golden light, and needless to say, all depending upon the message carried by the figures — their joy or quiet contemplation of the environment."[94] Such was the power of the girls' empathy, Parrish suggested, that their mindset would include and control the look of all the surrounding space. If their view instead had been of the manufacturing plants of Oak Ridge, with their total three-year expendability, could their minds and spirits have been kept from splitting and perishing altogether, faced with such a bleak and cataclysmic prospect? The attitudes and images represented in the book *Scenic Resources* and the painting *Hilltop* could not be perpetuated, because the experiences of World War II and its aftermath had eliminated the possibility of innocence and protracted control of the environment by reverie, such as had turned up in the photograph and the painting. The possibility of Franklin Roosevelt's synthesizing, roseate dreams had passed. Innocent people now looked naive. Idealism had become too confining, a much less reliable commodity. No longer could any region of the country possibly be considered a national template. Rather, *Realpolitik* was in the cards. The Cold War made the aim of generating a more congenial internal environment for the nation under the guidance of a newer technology seem ever less likely to be realized. In 1898, Parrish moved away from his native city of Philadelphia to Cornish, New Hampshire, because he wanted more time to think and to create. The landscapes of New Hampshire and Vermont, still unspoiled, appealed to him. His ambition was to become an undistracted landscape painter.[95] He was searching then for a land and an atmosphere of arcadian innocence, just as, during the 1930s, others would look around Tennessee for the same alternative image. Oak Ridge made something quite different out of the same basic land forms, those rimming foothills, knobs, and ridges.

After World War II, visitors from abroad, with their prewar memories of American idealism and pacifism, still tended to gravitate toward the TVA territory, examining it as a comprehensive, long-term plan and as a key to the American character. Expert English planner Brian Hackett observed, based on a 1950 visit, that while the TVA might have set out to create an ecological balance, always within the agency the "aesthetic approach has been secondary to the economic or productive approach." In Britain, on the other hand, the rural landscape had always been programmed to produce both more food *and* more beauty. In America, he told his British readers, the pronounced variety in "topography, soil, climate, and mechanized techniques" suggested the need for a more careful analysis and balanced adjustment of the land. The greater abundance and availability of raw land in the United States tended to draw attention away from the general topographical condition and toward "special projects like the national parks and the parkways" (even unto the TVA?), he said. However, in the end, "what you will observe in

TVA's site planning is a respect for the topography and landscaping that exists; buildings and roads are neatly fitted in to the existing pattern, and, when the aim is that a landscape should form the major design element, there is little wrong with this approach."[96] *He* interpreted the TVA as having put a first priority on landscape design. The irony, of course, is that in this same year, 1950, the Korean War broke out. From that war followed, during the 1950s decade, a renewed buildup in TVA territory of steam-driven generating plants and consequent strip-mining activity in the eastern Kentucky-Tennessee foothills and mountains. These developments implied less and less respect for and integration with the landscape.

5 | Landscapes and Villages of Ephemeral Display

What Was the Drive Really Toward?

The chronicle of the Tennessee Valley Authority reverberates throughout with altering conditions and new directions. Was its authentic inner drive toward the long or the short term? Had too many priorities, too many ambitions, been crowded into one package, that subsequently clamored to be let out? Only one concern lasted the distance — electricity. Scenic and situational landscaping, communal planning, housing, ecological conservation, forestry, and even agriculture had to play mainly supportive and secondary roles, despite the fact that each had its own partisans in its own time. Recreation was well thought of, but before the war, except with the first two parks at Norris Lake, it tended to be fitted in as was convenient. After the war, the recreational concern came out most strongly at Land Between the Lakes, because unions had secured shorter work hours and there was a feeling that the stringencies of the Depression and World War II were over. With time and affluence, a new "lifestyle" evolved. No matter what the earlier themes were, they emerged in two forms: first, as a technological drive for unlimited power production and distribution, expressed in major construction campaigns largely centered on dams; and second, as the visualization, at least, of new residential arrangements that were intended, like the high-tension wires, to roll out over the land and generate fresh patterns of living and working. At Norris Village, small industry was simultaneously explored. Later it was often asserted that technology had the most legitimate role to play, but there is evidence that settlement and light industry were more genuine aims, at least for those whose minds ranged farther and encompassed TVA's idealistic heritage.

Norris Village: The Past Was Prologue

The whole story can be read in the history of the first site, Norris Dam
(77) and Village. Frederick Gutheim, in his note on the death of Roland
Wank, concluded that Norris Village had failed Norris Dam, only four
miles away, because the village was so small, a more or less inconsequen-
tial "trifle."[1] The site was fashioned for 1,000 houses on 4,500 acres, but
only 294 houses got built.[2] Really, the manifestation was too tiny to serve
as a basis for any broader judgments. And the exemplar was never to be
repeated in further tests. This model was tucked away within the woods.
However, the circumstance unique at the actual moment of creation was
that, as concepts, Norris Dam and Norris Village carried equal weight
within the mind of at least one person who counted, Arthur E. Morgan.
As far as he was concerned, dam and village both were illuminating
truths, regardless of whether the dam stood forth in the sunshine or the
settlement disappeared into the woods. In fact, in the thoughts of Tracy
Augur, the supervising planner of the village, the settlement had potentially greater significance than the nearby dam:

> In a way it epitomizes the TVA, much more than the Norris or any other
> dam. . . . New towns are much less common, especially when they represent
> government enterprise and contain all the novel and attractive features of Nor-
> ris. By a little careful sign painting it would be possible to get most of the tourist
> trade which now visits the dam to visit the town center also, and in the future
> towns can be made of much greater continuing interest than the dam.[3]

Others put the village on a still higher pedestal. Donald Davidson of
Vanderbilt declared that Norris Village was the only bright light he could
see in the whole TVA![4]

The disagreement over whether "settlement" or "technology" ought
to prevail in the pursuit of TVA purposes sprang from the proselytizing
vehemence of both parties of supporters. Entirely forgotten in that debate
was the fact that labor-saving technology already had been introduced
into the region. A 1933 book by Malcolm Ross, *Machine Age in the Hills*,
had summed up this earlier sequence and deemed it traumatic and
threatening for the mountaineers and miners. Furthermore, the book had
predicted the strip mining that would give the TVA so much unfavorable
publicity after World War II. Even before the TVA formally had been
created, Ross wrote:

All this is technological progress of a high order, yet in human terms the greater efficiency means less work for the men in coal mine villages. . . . The improvement lies in the mechanical hands provided to supplant the weaker human ones—steel cutters which rip through coal faster than a pick, stripping machines which grind hills to pieces . . . The condition was general in the Blue Ridge long before the stock market crash of 1929. There the miners entered the lean years of the depression underfed and forlorn.[5]

This account depicted a melancholy situation, getting worse. A similar situation had affected the farmers of the High Plains during the agricultural slump following on World War I. Technology speeded up change and brought dislocations that left vacuums to fill. Technology could look different in different circumstances, was not automatically good in all circumstances. Nothing in the situation appeared to mesh, Ross said; and this meshing, this blending, this integration of worker and machine ought to be a primary concern in the future:

If the hill people could have remained as they were, I for one would have been content to leave them in their innocence, prejudiced but charming, poor but free from civilization's boredom and irritations. That can no longer be. The machine age has changed their hill virtues into industrial vices. Nationally speaking, we have them on our hands, surplus people, misfits who cannot work peaceably in their one industry [mining] and will not be ready as skilled workmen if new industries should be introduced into the region to save it.[6]

Technique and "efficiency," the latter offered as TVA's compensation in the face of the dislocating Depression, had furnished their own justification in the race for "setting standardized devices into motion." As the final consequence was to be described in Frenchman Jacques Ellul's worldwide analysis in *The Technological Society*, "Purposes drop out of sight and efficiency becomes the central concern."[7]

Buildup of the Crafts Seen as a Viable Substitute

Malcolm Ross definitely was not looking for a power surplus that would attract more and more large factories or federal installations to the valley during the World War II and immediately after. His 1933 book looked rather for the realization of Henry Ford's underlying ideal for Muscle Shoals, Detroit, and the rest of America—a smaller form of industry based on multi-streamed mill power: "Lastly, there is water power to be developed on many streams, not for ambitious power proj-

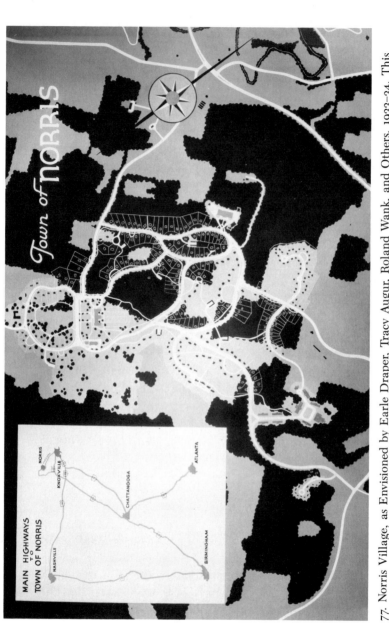

77. Norris Village, as Envisioned by Earle Draper, Tracy Augur, Roland Wank, and Others, 1933–34. This adept and clever plan, with an unprecedented variety of superblocks, loops, and cul-de-sacs, in the final analysis was quantitatively only a small sample, a fugitive item. It passed in and out of the woods. The lot lines actually were not run until the village was about to be sold off. Plan courtesy of Tennessee Valley Authority Archives.

ects but to turn the wheels of small local industries."[8] This "lesser" alternative obviously was *not* the one that the TVA ultimately would pursue. In it, small-scale instead of big-scale dispersal was called for; melding instead of standing forth. This pre-TVA program would have been more modest, interspersed within the countryside at frequent intervals, so as to provide a lifestyle particularly congenial to family and personal stability and dignity.

Who would sponsor this approach, according to Ross? "The Quakers proved that the people turn naturally to handicrafts,"[9] and in this area the people themselves had partially revived the hand arts in order to raise money for more food. Who might be a Quaker? President Hoover was one (at this pre-TVA period, Roosevelt was not yet president), and Ross explained that Hoover had called the American Friends Service Committee to Washington in fall 1931 and asked it to undertake the feeding of miners' children in the Blue Ridge. "Mr. Hoover mentioned that Grace Abbott, in charge of the Children's Bureau, had reported grave conditions there. A fund of $225,000 would be available from the American Relief Administration reserves, surplus money collected originally to feed children in post-war Europe."[10] It was through raising ARA money for Belgian relief after World War I that Hoover first had made his popular mark as a manager of public interests. But the Depression had caused philanthropic attention to swing toward Appalachia, just as following on World War II it switched back again to Europe, this time under the auspices of the Marshall rather than the Hoover Plan. Was the TVA entirely a New Deal inspiration? Do we have to decide that it was either Liberal or Conservative? The times were creating issues and solutions ahead of the deliberate actions of political parties and their leaders. There was even a hint of ecumenical collaboration and convergence. Allen Eaton, in his 1937 history, *Handicrafts of the Southern Highlands*, reported that "Mrs. Coolidge, Mrs. Hoover, and Mrs. Roosevelt have worn dresses of cloth woven on looms in the Highlands."[11] Moreover, in 1933 there had been a Southern Mountain Handicraft Guild exhibit at the Corcoran Gallery in Washington, D.C., sponsored by the same three first ladies, "each of whom has had a special interest in the Southern Highland people and their handicrafts."[12] One of them gave particular impetus to the handicraft movement in that very same year, although it was as the wife of the governor of New York State and not yet as the first lady she was about to become:

> The Homes Crafts League of New York State was for a time one of the agencies promoting the handicraft movement. Organized by the State Bureau of Industrial Education in 1933 as the result of an initial gift by Mrs. Franklin D. Roosevelt through the state Temporary Emergency Relief Administration, a program of rural home craft work for the state was set up.

That retraining, even in New York State, which is not ordinarily considered a bastion of ancient folkways, was to be in "basketry, sewing and embroidery, wood carving, woodworking and carpentry, crochet and knit goods, hooking and braiding, metal craft, pottery, and weaving."[13] For Mrs. Roosevelt, the expression of this interest had begun with her Val-Kill crafts industries, adjacent to Hyde Park and centered at the cottage her husband built for her there in 1924. A national eagerness officially to foster crafts was accumulating. In 1931, the New Hampshire Arts and Crafts League became the first group to be seconded by a state,[14] but the TVA also was becoming a mecca for such products and their exhibition. "Increasing numbers of possible purchasers are being brought into the region through the building of many fine highways, the establishing of the Great Smoky Mountains and the Shenandoah National Parks, the Tennessee Valley Authority developments, and other influences."[15] At the same time, as with so many TVA involvements, there was a hyphenated connection with New York City. "The handicraft salesroom, opened in 1935 at Norris Dam, is well situated for the convenience of innumerable visitors, and with the permanent exhibit and salesroom at International Building, Rockefeller Center, New York City; at the Patten Hotel, Chattanooga; and at Chickamauga Dam, Tennessee, it forms an important sales outlet."[16] This Norris shop could not have been set up had it not been for Wank's previous proposal for a visitors' outlook terrace there. The handmade object took its proper place in the precinct of the highest technology, around the dam, under the encouragement of Wank and A.E. Morgan, who were seeking a transitional work ethic. Their sense of conviction would spread out of the original compound to Norris Village and its houses:

> The Tennessee Valley Authority has also encouraged handicraft work as a part of the training and educational program in the town of Norris, where well-equipped shops give excellent opportunity for residents to carry on work in wood, metal, and other materials as they cannot do in their own homes where equipment is more limited. A number of newly built homes, however, have some facilities for such work, and as a result several residents have made their own furniture and others have developed unusual skill in iron and woodwork-

ing. These shops mark the beginning of a social experiment which it is hoped will be emulated in other communities.[17]

The craft approach, joined with the technological, thus was included at the beginning. It seemed an appropriate association then, whatever it looked like thereafter.

The Craft Ideology Is Embedded Further in the Model Community

While many harbored legitimate doubts as to whether Norris Village was a necessary accompaniment to other architectural and engineering expressions, plainly it was valuable in splicing of value systems — from dam to craft object and on to native house (78) and improved community. Moreover, when David Lilienthal tried to label this patronage of crafts as a crank gesture by Arthur Morgan acting alone,[18] he failed to acknowledge sufficently its pre-Morganic existence, as well as the status, persistence, and size of an earlier constituency that had included Mrs. Coolidge, Mrs. Hoover, and Mrs. Roosevelt. Ross described the home with its crafts as a psychological haven for the hill people: "In the old log cabin there were hooks in the ceiling to which the quilting frame was drawn up each evening to make room for hickory chairs and the supper table. These homely and satisfying objects the miner people can make once more, to add warmth to their drab huts and to appease their restless hands."[19] The poignancy came from the diminutive size of some of the craft objects; contrasts of scale, texture, and size always enlivened the TVA scene. This return to fundamental crafts was offered in compensation for other hardships suffered, particularly the absence of appropriate employment. The substitution would proceed from quilts or hickory chairs to model dwellings (79) and then to the revamped village or town — to be seen as a proper and cohesive community at long last. One reads a heightening expectancy in what Ross had to say in 1933 about the indigenous hill towns: "The fundamental trouble with mine housing is the persistence of the idea that they are camps to shelter roving workmen instead of what they might be — country villages for sane and cheerful living."[20] It was the lack of permanence and fixity, the lack of memory, and the total absence of monumentality and formality that were making America look so temporary and crude. The craft objects would give reassurance that there would be a continuity to be proud of, a carry-over of quality from a past. The dream of a newer type of village, bound to a renewed hope for the future — this Norris Village was supposed to fulfill in haste, out of a long tradition. Also, however, the village was an attempt to rescind the wide-open construction camp. A feel for the range of American life was there. Ross said just before the TVA was founded, "There is an aspiration in America at large to decentralize the cities, to place people in small communities where they may have country life and at the same time enjoy the advantages of good roads, schools and modern homes."[21] Ford had believed in that aspiration, and so did Arthur Morgan and Roland Wank. The dwelling itself became the key emblem for the improved life of the whole community. There was to be no abrupt, abstract postulation of high-density, high-rise public housing, such as became the standard governmental formula for the urban dispossessed after 1948.

78. Mountain Cabin Near Norris Village. Much of the inspiration for Norris Village houses came from local models like this one. Such designs were tied in with cultural notions concerning the homestead and the importance of local crafts. Photo courtesy of Earle S. Draper, Jr.

If Franklin Roosevelt had lived on to a fourth or fifth term after the war, and if Robert Taft had not become such an important influence on public housing, would any high-density houses at all have been built for the dislocated? Roosevelt's own architectural views always seemed arcadian and colonialized, and he was recurringly drawn toward the rural community. Roosevelt's Labor Secretary, Frances Perkins, mentioned that, although Roosevelt might lack a deeper commitment to or feeling for the visual arts, he nevertheless believed strongly that a Missouri Valley Authority or a Tennessee Valley Authority "would make the desert blossom and that more people could find happy, comfortable homes in those areas."[22] The house, the home, the homeplace, the cabin with a family in it was his final device. She "had not been in Washington a month be-

79. Houses in Norris Village. The Norris Village house models followed the examples of vernacular architecture (78), except that, to accommodate new utilities and electrical appliances, the chimney was moved to the middle of the structure; the walls were thinned; and casement sashes were made of steel. Thus it became harder to convey an impression of real solidity and permanence. Photo courtesy of Earle S. Draper, Jr.

fore the President asked her to go down into the southern Appalachian region, from which he had had pathetic letters, to see what the problems were and what could be done."[23] What often has been forgotten in evaluations of the TVA is that a primary hope of all leaders at the time was to relieve a despairing mood, to "make the desert blossom" in an ambiance that had worn painfully thin yet at the same time was taken as an extreme microcosm of the rest of America. Thus, for these leaders, the precise model mattered less than the symbolic implication. What might look to later eyes too bare, too retrospective, too contrived in the domestic architecture of Norris Village actually characterized 1930s architecture at large, whether domestic or monumental. Adaptation, economy, and improvisation were constantly in the wind. The housing was minimally financed. This deflated reality comes out more clearly if Norris Village is compared to the ultimate model of the English Garden City of Letchworth (1906–16). Even at its most diagrammatic stages, Letchworth was always programmed for 30,000 to 32,000 people.[24] Arthur Morgan, by way of contrast, fastened on 5,000 for his village and, by reason of fiscal cuts, had to be content with a population of 1,500.[25]

Another reason for Norris Village's subsequent poor repute in the national housing literature, as William Jordy has pointed out in an unpublished paper, was that Mumford and Stein were greatly disappointed that they had not been called on more for advice about the whole TVA. Had Frederic Delano resented their assaults on the Russell Sage Plan for New York? Neither Mumford nor Stein discussed the TVA at any length in their books, whether Mumford's vast 1938 survey, *The Culture of Cities*, published just when the TVA was ripening; or Stein's *Toward New Towns for America* (1951). As Jordy recalls, Mumford wrote to his counterpart in Britain, Frederick Osborn, the prolific defender of the Garden City and New Town interests, that "the things I worked for during the twenties came to a head in Washington and the Tennessee Valley, without myself or any of my colleagues being given a chance to work in any of the strategic positions."[26] Of course, both Benton MacKaye and Tracy Augur, who had been members of Mumford's Regional Planning Association, were employed by the TVA, and Gutheim, who kept in close touch with it, had an actual part in formulating the TVA law. But Arthur Morgan appears to have been more interested in talents matching the specific territory he had been made responsible for, and so he hired Earle Draper, the designer of mill towns along the Piedmont, to furnish the substance for a congenial southern pattern. Moreover, Morgan's own individuality was likely to get in the way of his endorsing any "cooperative" (Mumford's term) movement originating as recently as the 1920s, since on that score his own sympathy lay more with the utopians of the nineteenth century. The 1930s was a curious decade, at once calling for cooperation and group-think, and yet craving colorful personalities it could readily identify with. The TVA groups and divisions most often came to be known by their leaders, and so did the whole enterprise. Aelred J. Gray, former chief of the regional planning staff, in his 1974 essay on Norris Village brought out this ready identification of TVA undertakings with only one personality. On his very first page, Gray recalled that "the idea of a new town in the Tennessee Valley actually originated with the then newly elected President, Franklin D. Roosevelt."[27]

Norris Village did not offer the "efficiency" and "modernity" expected of the dams, but it substituted another value often yearned for then, "renewed opportunity." Renewed hope — and its manifest fertilization by surrounding nature, a longstanding romantic desire — affected the village's slightly mannered integration with its site, its refreshing diversity of appearance, its nostalgia not to lose its cultural way, and, because of an uncompleted plan, its eventual disproportion. Like the dam, the village, since it was intended as a public showplace, was supposed to indicate intelligent thinking. Its dollhouse or tiny town (77) character was only accentuated by that mission. The American experts could not easily describe what they were seeing, due to the fact that the cue-giving English Garden City motifs of the cul-de-sac, superblock, and greenbelt originally had been developed for relationships larger in scale. Although exquisitely drawn, the attenuation, in response to the irregular ground contours, of the few cul-de-sacs and superblocks that were achieved was great enough to leave an impression of minimal exaggeration. It was as if two other of Norris Village's major models — Stein and Wright's Radburn Village in New Jersey, begun in 1929; or, before that, Unwin's Hampstead Garden Suburb near London, begun in 1905, where Stein used to spend summers in Unwin's house Wyldes — suddenly had become accelerated, fragmented, and cut loose from a larger body of thought. Too great expectation may have placed too great a burden of performance on the little village. Norris was a large model interpreted in a small way. On a map (77) it appeared free willed; surely this was democracy in action. But the democratic freedom constituted a birdshot effect. The plan (77) had an immediacy and an intensity greater than those of any earlier ideal town, and that fact, too, could have reduced the enthusiasm of the New York group. A planning novelty was turned loose in the remoter, not easily commutable, distance from New York, and that can have reduced its magnetism.

The vivacity and animation of Norris Village may be understood better if the development of Progressive ideology since 1890 can be kept in mind. Morgan had had the dream of a model village in a functioning democracy in his mind for many years. Communal reshaping was part of the Progressive aim. Morgan wrote several books on the subject; the one with the most revealing title is *The Small Community, Foundation of a Democratic Life: What It Is and How to Achieve It*.[28] For him, with the villages there was bound to be a release of pent-up feeling. With it, the adjoining Norris Dam could become a demonstration of how a democracy might set up its working monuments in a yin/yang relationship. Comparative size and technology would not matter as much then. As Wank emphasized, the ultimate benefit from the dam, too, was to be domestic. Thus the contrast of actual scale and material in the two entities could not matter as much as the comparative significance of their political and ideological valences. The house would be the breeding center for, the heart of, this idealistic reassertion. What put hopes for domestic progress in America at an apparent disadvantage during the 1920s was that the country had no iconography of housing types closely enough identified with the public weal — like the British semi-detached type for the "Homes for Heroes" Movement, or the German *siedlungen* of the row apartment open to the east and west sun. So the American aspiration was toward a vaguer image of a homestead standing completely free. That step would be taken in order to give individuality a last stand. In an interview with the *Washington Post* of 18 February 1934, entitled "TVA Foresees Utopia Come True in Valley," Morgan indicated his overall design. The inhabitant of the newly-founded "utopian" village, much as in Henry Ford's or Franklin Roosevelt's prescription, would divide his "time between scientific farm and decentralized industry," preferably light industry. By such means American individuality could be reinstated:

> The mountain regions of the South are the last great bulwarks of individuality in America. . . . "The Southern Highlander" is different. He likes rural

life. His income has been and is very small. Agriculture alone will not support him. Today great industries are settling in his midst. Some of these factories tend to destroy his type of civilization. The Southern Highlander is often regarded merely as cheap labor to be exploited, rather than as the representative of a valuable type of culture to be encouraged and protected.

The highlander would need to have his homeplace redeveloped with smaller industries placed around. In Kingsport, Tennessee, where Nolen and Draper had brought the white Colonial house to the fore, it had been customary for printing plant workers to carry on part of their income-producing time in farming and forestry.

In addition to the need to pick out more significant past buildings to use as house models in this redesignated region, it was necessary to keep the surroundings loose, bucolic, pastoral, running free, because, as Morgan had said in the *Washington Post* interview, "We have become [too] used to living in crowds." So in Norris Valley the actual contours drift by (80) from northeast to southwest on wooded ridges, repeating the larger forest pattern of eastern Tennessee as it slopes down from the mountains. Part of this viability on the land, as recognized and delineated around the central green at Norris, came from the tree-surrounded Olmstedian convention of planning a clearing or meadow in the forest — a convention introduced for the first time in the South in 1915, when Warren Manning, an associate in the Olmsted firm, had laid out the golf-playing community of Pinehurst, North Carolina.[29] In his 1918–24 treatment of mill villages on the Piedmont ridge of North Carolina, South Carolina, and Georgia, Draper had observed that "a certain percentage of wooded area is always important in the South."[30] On a smaller scale, this appeared to correspond exactly to what A.E. Morgan and others had envisioned for the whole valley, according to the *Washington Post* interview. "Dr. Arthur E. Morgan and his associates believe their efforts will turn the Tennessee Valley into 40,000 square miles of pastoral, prosperous land." Thus for the landscape of pastoral content, Norris Village had to represent a microcosm of what the entire valley itself might someday become.

The House as Archetype

In order to gain authenticity and distinction, to distill his imagery, Arthur Morgan appears to have wanted to keep the southern district apart

80. Looking South on F Street, Norris Village, 22 April 1935. Tracy Augur, the village's planner, said that he wanted the houses "to nestle amid forest trees and into slopes as though they had grown up together." As few trees as possible were cut down on the home sites. The impression of a converging drift down from the hills through trees was typical of this semi-mountainous area. The final morale-building effect was similar to what Arthur Morgan envisioned for the whole Tennessee Valley — the impression of "a pastoral, prosperous land." Morgan, however, was not given enough money, even for this pilot model, to achieve that goal. Photo courtesy of Tennessee Valley Authority Archives.

from the Northeast. Draper observed to Morgan, during the job interview in Washington, D.C. that led to his earliest appointment as chief
of regional planning on 16 June 1933, that the TVA probably would fare
better with some well-known landscape firm from the Northeast, since
most of his own work had been only in the rural South. Morgan replied
with vehemence,

> That is the kind of man I don't want . . . I don't want anybody coming in
> with an alien philosophy to try to tell us what to do, and what interests me
> is that I talked with (I think he mentioned Fred Johnson down in Kingsport,
> Tennessee), and I think you are the kind of man who knows the South, has
> lived in the South, who can understand local conditions and won't be domi-
> nated by a clique or a group that is preconceived in the notions of what has
> to be done down there.[31]

The proximity to real crises and desperation etched the imagery deeper.
During 1932 and 1933, a hundred families had to be transported out from
the violence and actual starvation of Wilder, Tennessee, during a coal
miners' strike. They were settled in Norris and placed at jobs on the dam.
Sixty-five young men also were driven in three state trucks to La Follette,
Tennessee, where they were provided jobs in the Civilian Conservation
Corps camp.[32] The desire to implement the highest aspirations, in the
face of the harshest realities, was the source of much of the unrelenting
drama in the TVA throughout the Depression and war periods. The tension made for a distracting atmosphere of breathlessness, strain, over-
compensation, and temporizing even amid the most serious efforts.

A wish to respond to a spectrum of needs transformed the houses at
Norris into a microscopic but highly ambitious expression of the whole
TVA. The strict economizing helped to set the tone. Great expectations
were channeled through the fewest means. An outwardly old-fashioned
character, if that was what it was, was derived mainly from the "Back
to the Land" movement of the 1930s, but it displayed the keenest interest
in new building materials. This dual direction or split was to be detected
in one of Draper's early talks: "The TVA is not attempting to impose a
brand-new way of living upon the people of the Valley. Rather we are
attempting to blend modern forms with the long-existing living habits
and social customs of the locality."[33] The stylistic clock was being turned
back in an unusual way that produced the most varied solutions. Thirty
different plans were developed for 294 houses, probably a number of ar-
chetypes unprecedented except perhaps for the similar Depression-

generated set of Usonian houses conceived for the Broadacre City model
(1932–34) of Frank Lloyd Wright. The Depression brought a passion to
escape its drabness and uniformity. In Norris the ends of the houses were
left as open as possible for light and living space, to be cooled by cross-
ventilation. This openness was made more feasible by running a denser,
more specialized utility strip through the middle of the house, via the
kitchen and bath. One or two porches were added, usually at the side
or back for privacy. Another potentially different living space was the
unfinished attic. This permitted the eventual development of bedrooms
when children came. Later, this feature became a finished attic in the
ampler brick-veneered houses of Wheeler Dam and Pickwick Dam villages. The houses built at Wheeler and Pickwick were even fewer in
number — fifteen apiece.[34] An interesting prototype of the split-level house
came into being when it was realized that if the hillsides were excavated,
water would infiltrate the house. Hence a type with staggered floors was
evolved, to be built without excavation up the side of hills, accompanied
by an asymmetrical roof pitch. On the outside of the regular houses, a
wider front door with sidelights and a porch constituted an innocent
afterimage of the dog run (81) of the old log cabin. The historic vernacular house did not require a porch, because the dog run in itself was an
aperture passing through the middle of the house. Now the utilities were
there. Norris exteriors also reflected the past by being left unpainted or
being only stained or whitewashed, reflecting the old southern saying,
"too poor to paint, too proud to whitewash." The outside wall surfaces
were mostly of board and batten, or of shakes (81). The latter were split
by hand. Looked upon as a luxury item today, shakes then were acquired
in thousand-batch lots in order to give winter employment to the mountaineers nearby.[35]

Historically, time typically was passing and fluid, as social status was
uncertain in Depression-time America. Inside the Norris houses, in contrast, everything had a brand-new, synthetic character. Plywood, used
on the walls, contrasted with the hand-hewn shingles outside. The ceiling height was mostly seven feet, six inches, with the ceiling composed
of insulation fiber board. The wall insulation was rock or mineral wool.
The insulation of the floor was aluminum foil.[36] Steel casement windows
were utilized, as elsewhere in that decade, only to be abandoned in later
decades because they rusted through and proved unusually difficult to
keep closed if the house settled. Reflecting this same fascination with

new materials, one experimental house of Armco steel was erected, but it was found not to be repeatable because of corrosion and high square-foot construction costs. The cheapest houses, renting for $14 a month and costing $2,000 apiece, were constructed of cinder blocks (82). This way of turning technology inside out struck some as too stark, especially compared with the more rustic cabins, but Arthur Morgan felt that every avenue of future construction and association with culture ought to be explored.[37] Anticipating the attitudes of 1960s and 1970s Post-Modernism, there was a desire to juxtapose unlike materials and temporal motifs. The floors were of concrete slabs with precast beams. The roofs were of sheet metal, painted to harmonize with the surroundings; blocks were finished in "a smooth, colored surface similar to tile."[38] Into these concrete houses the refugee miners from Wilder, Tennessee, were settled.[39] There were even solid stone houses, built by local masons brought in by Arthur Morgan to see what they could do.[40] It was not designs or styles or even building materials that he primarily wanted to investigate, as much as it was the human condition when put into such units. The experimentation was more with process and content than with style.

This preference for promoting every sort of human potential by means of a functional eclecticism is particularly demonstrated by the electrical equipment of the first 152 houses, concentrating on the kitchen. The TVA facilitated the founding in 1935-37 of the Rural Electrification Administration, which was to mean so much to farmers all over the nation.[41] The Norris kitchens had electric ranges, refrigerator, and water heater, all automatic in operation.[42] Electrical heating was intended. Such features were not looked upon so much as technical triumphs as, once again, a means to upgrade the human psyche. "Women who have lived all their lives in cabins on run-down farms will be moved into modern houses on fertile land and these houses will be equipped with every comfort that cheap electricity can provide." Women would be liberated by electricity. Relieved of drudgery, they *could* bloom. "The people of these mountains would be the physical and mental equal of any in the land if they had the opportunities. Their environment is all that has blocked their progress."[43] The whole TVA effort derived from a democratic conviction that everyone in the nation eventually should be equal in status, condition, and capacity, but always on a comparatively high material, technical, and — at Norris at least — cultural level. Crowding so many devices at once into the houses was naive, of course, and thoroughly American, but also optimistic and upbeat, part of the original TVA intention.

The Multi-Phased Scene, Bursting with Activity

The bucolic mood of Norris resembled that of a southern summer camp meeting in a grove beside a lake, of which Tennessee had had a number,

81. House in Norris Village. The dormers on a low-sloped roof show an effort to utilize every foot of inner space. The allegation that Norris houses were expensive was not true. The walls were covered with hand-split shakes, but the roof was of commercial asphalt shingles, juxtaposing old and new. The type was called a "Dog-Trot, D-2," again combining the terminologies of log cabin tradition with modern efficiency. Sidelight windows on either side of the front door signify where the dog trot would have gone if this were a log cabin. Photo courtesy of Tennessee Valley Authority Archives.

beginning even before the Civil War.[44] Such camp meetings then were called "summer assembly grounds." The image of this fragile revival camp was seen as spiritually renewing because it was away in nature, and had its own inner forms of architectural renewal. Juxtaposed with this revival camp image was that of the reconstituted mining town, straight out of the rip-roaring Old West. Arthur Morgan did not like the atmosphere of the dam construction towns he had known, with their fighting, gambling, drinking, and prostitution, and previously he had reformed those of his Miami River flood-control project in Ohio with five examples containing stores, gardens, and schools. Draper reported, "We were determined that the community should depart from the usual squalid conditions around dams that had existed under the Army and the Department of Interior and the various dams that had been built."[45] The immediate focus of this type of reform in Norris was the southwest sector of town, where there was a quadrangle of temporary wooden buildings for construction workers around a tiny green. That quadrangle later became an administrative center for the TVA. Included in this first matrix were six men's and one women's barracks. The four along Ridgeway Road later were torn down and replaced by houses. In addition, in lieu of another

82. Cinder Block Houses, Norris. The inspiration for the cinder block houses came from Arthur Morgan, who was always eager to experiment with new materials. The houses cost $2,000 apiece. Photo courtesy of Tennessee Valley Authority Archives.

bunkhouse, there were fifteen small cabins at the southwest entrance to the village, adjoining the Norris Parkway, that were later converted into tourist camps.

In another guise, Norris Village was a part of the Subsistence Farms Movement and was likened to the well-known example at nearby Crossville, Tennessee. Norris briefly received financial support from the Subsistence Homesteads Group.[46] Although a dairy farm, a poultry farm, and a creamery were built in the greenbelt for demonstration purposes,[47] the real center of the agricultural initiative was at the second, larger "village green" on the northwest. The first structure built in the civic cluster on that green was the agriculture building, designed to function as a food market "to be supplied, at least in part, by local farmers who would come in with produce and sell from their wagons in the rear of the building. At this time it was also thought that the inhabitants of Norris would engage in part-time agriculture to a considerable extent, and the building was intended to serve as a store for selling farm implements, seed, and other farm items."[48] However, the soil around Norris proved too thin for this kind of intensive, individualized farming (called "truck farming" in the North), so "this idea was subsequently abandoned and substantially the same building, remodeled and added to, now serves as a drug store, food store, telephone exchange, and post office."[49]

At Norris Village, training, largely inspired by Arthur Morgan, filled every nook and cranny. Morgan wanted to produce a whole new lifestyle for each family.

> In order to provide more jobs in a period of severe depression, the working day was reduced to five-and-one-half-hour shifts, with two shifts for each part of the work. Thus there was time to train people in many forms of household skills and activities. . . . Whole families made it a practice to work and learn together. The contempt some people higher up in the TVA expressed for such activities was misplaced. When President Roosevelt had told me of the changes he hoped to bring about in the lives of the mountain people, including those parts of their day that otherwise would have been largely idle, I felt that he was thinking and speaking intelligently. For parents and children to live and work or play together — sharing household interests and recreation and making household belongings — was better than being in a camp surrounded by gambling houses. It was well to meet the needs and wishes of workers for both work and leisure and not to assume that the TVA was interested in providing only work and wages.[50]

For Morgan the effort was all of a piece. In fact, he felt, bringing still another metaphor into play, that "the construction camp resembled a simple college campus."[51]

One of the houses was used as a home demonstration center, with classes including cooking, child care, home furnishings, and family budgeting.[52] On the east side of the village, constituting a third center, barnlike trade shops were built to provide training in electrical, mechanical, and woodworking skills.

> At the trades training center there were classes in machine tooling, welding, in wrought iron blacksmithing, in electrical work, in carpentry and in more refined kinds of woodworking. . . . For trainers we employed master craftsmen — one from the Highpoint, North Carolina, furniture factory and a couple of teachers from the Berea [College], Kentucky, craft shops. . . . For one phase of it they explored locally in the valley samples of early American furniture that was good looking, of good quality and useful in the household. . . . The Norris trade shop, where this craft work was done, became a prominent social center for the town. Men, women and their children would come in and share in the craft making.[53]

There were three basic branches of instruction here — agriculture, trades, and engineering. James Agee, when he inspected the town, was fascinated that auto mechanics, aviation mechanics, plumbing, and wrought-iron working were taught.[54] Arthur Morgan's ultimate hope was, by these means, to bring unskilled employees up through the ranks to become foremen. There was a future in that, because other dams would be built. At the community building there also would be "cultural" courses in natural science, mathematics, history, current events, and music appreciation. But, as with other later programs, there was already an inhibiting effect coming from high TVA officials, according to Arthur Morgan: "The legal counsel, Mr. Lilienthal, contended that personnel training was not justifiable under the TVA Act. I think this is an accurate and fair statement of the opposition within the board, which had a somewhat dampening effect on the fulfillment of the training possibilities."[55] In the end, Norris Village became, as far as social experiment was concerned, a deflated settlement, a commuting suburb, a bedroom city: "Oak Ridge was not as pleasant a place to live in as Norris was and the commuting distance was not too great."[56] In terms of planning accomplishment, Norris Village was perhaps less than a great achievement, as has been said ever since it was built. As a contrast with the first TVA dam nearby, however, it was most effective as a foil, its in-ground tradition contrasting boldly with the up-rising construction and technology of the dam.

The house and town plans were ingenious. Their professed intent was a novel one in America — to employ architecture, planning, and landscape architecture with such a degree of dedication that an improved quality of life would result. While these arts were being exhibited with an emphasis on minimalism, due to the inadequate budget, there was also an aspiration toward utopian perfectionism, toward becoming happy and effective in one's work and in family and community relationships, that surpassed that in all previous demonstration attempts. In the one industry there that got well enough under way, the porcelain factory, that same aspiration was assiduously pursued. It was typical of Arthur Morgan that he would seek out an exotic material and a thoroughly unusual person to begin the manufacturing at a small scale. It was also characteristic that he would wish to put the best of china on sale at "the five-and-ten-cent stores." Said he:

> It came to my attention through the late S.T. Henry of Spruce Pine, North Carolina, that in the mountains of that state there were large deposits of an especially fine quality of kaolin, or disintegrated feldspar, as natural dikes, or strata, in the mountains. Most kaolins have been deposited under water, and in the process they become slightly mixed with some other mineral, which makes them unfit for the better tableware products, such as fine porcelains. I located an especially capable ceramics engineer, Robert Gould, who wanted to return home from Poland, where he had been developing a fine-porcelain industry [at the Giesche Porcelain Works at Katowice]. After only a few years he developed very fine porcelain products in a pilot plant at Norris, Tennessee, by methods that greatly reduced the labor cost. . . . This project was disapproved and stopped in the TVA area by the other members of the board, but was later renewed by the Bureau of Mines."[57]

The laboratory was established in April 1934 and finally closed in October 1965.[58] The big technical attraction had been the possibility of further development of electric kilns, of course.

Disavowal of the Original Medium

There was a hope that Norris Village could be repeated as a model throughout the Tennessee Valley. Earle Draper wrote, "If we could have carried over [the] Norris town & parkway concept all thru the Valley it

would have had a tremendous impact."[59] This legacy of communal idealism likewise was recognized by the TVA on the fortieth anniversary of
Norris Village by a proposal for more planned villages in the "rural
lifestyle" originally advocated.

> Forty years after the planning of Norris began, TVA planners again are at
> work on "new town" projects now in the conceptual stage. Planned villages
> in areas of lower middle Tennessee and north Alabama would be designed
> to concentrate housing and related services to provide for urban needs while
> maintaining a quiet, rural life-style. A much larger planned community that
> would accommodate an ultimate population up to 50,000 is foreseen on the
> shores of the future Tellico Lake in east Tennessee to accommodate to popula-
> ion growth expected to result in that area.[60]

To Draper we are also indebted for the indirect reason why no more
Norris Villages were to be built.

> After the Board was upheld by a *one vote margin* in the Supreme Court on
> its right to take the power to the consumer — rather than sell it at the power
> house, the other two members told [Arthur Morgan] that TVA could do no
> further experimenting in building towns, parkways, etc. in connection with
> dam construction for fear TVA would suffer from Congressional reaction.
> Hence in the next few dams, Wheeler, etc. we laid out a minimum com-
> munity at each dam & only necessary access roads. The Board became fright-
> ened that Sec. 22 and 23 of the TVA Act would seem to the public [to allow
> socialistic activity] by using public money for activities such as housing that
> should be left to private activity.[61]

The open-endedness of Sections 22 and 23 had come back to haunt the
executors, and fear of a threat to capitalism had cropped up amazingly
early, at least in two of the TVA directors. The unfortunate long-term
result was that there was no further opportunity to demonstrate what
the bucolic, pastoral community, existing on agriculture and light industry,
in the manner first visualized for the valley by capitalist Henry Ford,
might have come to. Nor could we ever see what applying the most
thoughtful engineering and architectural skill to the single-family house
might have achieved. Progress had been made in the evolution of housing and the layout of new towns in Europe between the wars, particularly in Britain and Germany; and Norris Village tried to capitalize upon
that progress. Perhaps the village's greatest flaw, at least in the eyes of
its critics, was that it indulged too thoroughly in the revival of nineteenth-
century American utopian Perfectionism. Such a stance was hardly sur-

prising in Arthur Morgan, biographer of the nineteenth-century American utopian Edward Bellamy. Bellamy had had an important effect on Ebenezer Howard, inventor of the English Garden City. The latest technology of the dams could be celebrated, but not the latest technology of human settlement. Earle Draper, regretting the fact that the funds for housing in Norris Village were cut "to what would normally be spent housing workers" in any dam construction camp in the U.S., noted: "In a way that was a disappointment because we had hoped to make a larger demonstration of town planning that would be — would serve as an example for the country."[62] Norris might not have served as a proper example for the rest of the country, but it might well have been the beginning, at least, of an improved settlement pattern for the Tennessee Valley miner, farmer, or small-town dweller of the time. The sensation of drifting and dreaming, of generalized wishful thinking, appears even in impersonal, objective statistics. Norris had 2.7 families per acre. Raymond Unwin, the chief planner of the English Garden Cities, thought he was achieving a low density when he established a British national ratio of twelve families per acre. On exceptional sites, he was willing to go up to twenty to thirty per acre. Even his entirely reasonable numbers have been described by later opponents, the perennial advocates of high-density, high-rise dwellings, as "wasteful." The official explanation for the extremely low density of Norris was that "an open, rambling, informal sort of lay-out resulted because of the rough topography, which had little in common with the highly organized schemes developed for similar projects in England and at Radburn in this country."[63] Nevertheless, the study numbers and diagrams are close enough to identify Ebenezer Howard's 1898 English Garden City ideal as Norris's ultimate source. Plenty of planning expertise was available at Norris Village, to be sure. But what made the town even more engaging was the warm, cozy way in which it sought to establish itself as a proper setting for the mountain miner with his crafts, his cabin, and his enfolding woods. Norris was a nest for fledglings. A plethora of architectural, planning, and landscape devices gave it a lasting charm and attraction.

The Selloff of Norris

While it continued investing in TVA dams and waterways, after the initial crises the federal government had much less interest in sustaining its stake in model housing and settlements, even though any well-planned community would be bound to appreciate, in cold dollar terms, after the war. As soon as the war was over, there was tremendous national pressure for housing as ex-soldiers wished to start families.[64] Aelred Gray indicated that Norris cost the TVA $1,761,000. It was disposed of in June 1948 to a group of Philadelphia investors, headed by Henry D. Epstein, for $2,107,500. The Epstein group traded off the houses as individual units and then remaindered the rest of its holdings for $280,000 to a local corporation made up of residents.[65] The only bulwark against complete disintegration of the communal spirit was a 1939 state law that allowed the Norris inhabitants to form a planning commission with zoning.[66] The Norris incorporation act of 5 April 1949 conformed to the boundaries of the planning and zoning district, instituted at the first planning commission meeting on 29 May 1948.[67] Norris went the way of the New Deal Greenbelt towns — into private hands. One of those towns, Greenhills, north of Cincinnati, Roland Wank had joined in designing, upon being recommended by Draper.[68]

It was not only the eagerness of Congress to jettison Norris Village, or of the TVA management to "localize" it,[69] that mattered in its dissolution, but also the fact, indicated by Professor Jordy, that Norris had been regarded, even by its creators, as a single-phase "demonstration" town.[70] What was patently illogical was that Norris had not been allocated a sufficient budget even for that initial demonstration. Oak Ridge was allowed to grow to a population of 75,000 during wartime, while Norris was reduced in the Depression from 10,000 to 5,000, and then to 1,500.[71] Oak Ridge was not sold off by the Atomic Energy Commission immediately after the war as Norris was by its agency, the TVA.

The Norris Parks

Like Norris's two village greens, its parks, called "Norris" and "Big Ridge," were binuclear. They stretched along the southern shore of the lake, with the town forest between.[72] They were intended for use by local patrons and were contiguous with the village, but, like the center of the village, they seemed to presage a developmental direction for the whole valley. A popular conviction was that several areas of the valley previously had been misused and abandoned.

The Konnarock Basin in particular has suffered severely from misuse of its forest lands and misdirected agricultural efforts. At the present time, several square miles present a picture only slightly less desolate than that presented by the Ducktown Basin in southeast Tennessee. The original stand, a superb example of the climax forest, was clear cut and the area subsequently burned and heavily pastured. The present appearance of the valley is a combination of badly eroded pasture land, dotted with scarred stumps, a few acres of stunted second growth, and very poor agricultural land. Many years will pass before this basin attains any recreational significance, although recreation may provide an additional inducement for the needed restoration of a new forest cover.[73]

A change of use from marginal farming or forestry to carefully organized recreation would begin to restore the land. The first page of the foreword to *The Scenic Resources of the Tennessee Valley*, published by the TVA in 1938, conveyed an expectation of this sort of rejuvenation:

Early settlers found an almost unbroken forest blanket over the Tennessee Valley, ranging in composition from southern hardwoods in the lower elevations to northern conifers on the mountain peaks. . . . In addition, where undergrowth has not been subjected to periodic burning off, shrubs and wild flowers may be found in extraordinary variety. Dogwoods make unforgettable Aprils. Rhododendron, laurel, azaleas, and other members of the heath family form mass displays of color on the mountains and plateaus during the spring and early summer. Equally colorful autumn blazes with sumac, sourwood, sweet and sour gum, dogwood, oak, and maple. Luxuriant forests, together with the almost ever-present haze, soften mountain contours and cloak them with mystery.[74]

This Garden of Eden should be recapturable everywhere in the long valley. Its great potential should not be further dissipated!

The poignancy of all these captivating effects came from an introversion limited in scope. The northerner might set up his resort to escape from the heat, noise, fever, and dirt of the city in summer. The southerner, on the other hand, being in those days less preoccupied with urbanism, was apt to suffer during the winter months from rural silence. Wilma Dykeman, in her book on the French Broad River, wrote of the "loneliness" of the mountain winter. That isolation was conditioned by the geographical remoteness of eastern Tennessee. So the Norris parks offered an opportunity to come together in the warmer season to socialize. These parks partook in a general way of the atmosphere of the sacred groves belonging to the summer assemblies of the Protestant sects, although, in the Norris area at the time, the religion was more likely Democracy. It was difficult for families to travel any great distance in those days. It was only in the 1950s that anyone began to have an opportunity to drive far for recreation or any other purpose, over interstate highways modeled on the high-speed *autobähnen* that General (now President) Eisenhower had seen in Germany. Moreover, at that time no-one had the money to travel that they were later to acquire as postwar affluence expanded; consequently there was no South Carolina Heritage USA, no Las Vegas, no Disneyland or Epcot Center to which people might gravitate in well-equipped recreational vehicles. In all phases of American life, the horizon had to be kept narrower, relaxations simpler and less expensive. Immediately adjoining Norris Village was Norris Park, with its lodge, twenty cabins, and twenty-two miles of trails. It contained 3,887 acres and, as a sign of 1930s attitudes, an outdoor amphitheater (83) to demonstrate that acculturation of "the people" was underway.[75]

Norris Lake had many inlets surrounded by low hills. These differentiated it from later artificial lakes such as Fontana, with its straighter shores in a tighter valley; or Wheeler and Guntersville lakes, between low banks. The involutions of Norris Lake were reflected statistically in the way 34,000 acres yielded an extensive 705 miles of shore, as contrasted with the biggest TVA lake, Guntersville, where 69,000 acres gave rise to only 962 miles of shoreline. The Norris, Wheeler, and Pickwick parks, which tended to be hidden away, held onto their pioneer atmospheres. Twelve miles up the Norris lake from the main dam, a small inlet was sealed off to provide a pond of 45 acres that could be kept at constant level. Annual drawdowns of water by the engineers were a chronic detriment to recreation in artificial lakes waiting for the snowmelt.[76] The drawdown by opening the sluiceway could be only a few feet in some lakes or 111 feet at Hiwassee, usually in the late summer or early fall. Four secondary dams were built in Norris Lake, and four more were projected, mostly for breeding fish.[77] Benton MacKaye described Big Ridge Park, with 4,592 acres around its dammed inlet as displaying "intensive outing recreation on the basis of a large number of people on a small acreage."[78] Therefore, according to him, the park had to be designed as carefully as possible. From photographs (84), one is persuaded that order and decorum were obtained, despite the high density. The main activity appears to have been people-watching (very popular during the Depression because it cost nothing). Baseball and dancing were other chief pastimes. Apparently the area was the turf of single or recently married

83. Amphitheater, Norris Park, 1930s. The 1930s and 1940s were decades of family upheaval and deprivation, with a great deal of economic and military wandering. Consequently, group gatherings came to be more meaningful for individuals. Greek and Latin were still taught in many schools, so it was not difficult for a group to imagine being in an ancient theater. Unlike an amphitheater built in the 1920s, this Depression example does not attempt colonnades, and there are no marble seats. That this is a more rustic emergency project is shown by the tag-end benches and the road-bond aisles. And yet the youths are attired quite formally, wearing neckties and long dresses. Photo courtesy of Tennessee Valley Authority Archives.

84. Big Ridge Park, Norris Lake. The water was kept at a uniform level by impoundment. The bathing beach was relatively small and crowded, as the white towels on the grass signify. The peninsula shape was a TVA form. The main recreation appears to have been people-watching, favored in the Depression because it cost nothing. There are fewer swimmers and boaters than one would see after the war. Photo courtesy of Tennessee Valley Authority Archives.

youth; any elders had on shirts, ties, watch chains, and sensible, dignified Panama hats. The national dress code evidently had not yet been liberated. Sport shirts began to appear only in the late 1930s. Sun bathing predominated over swimming, probably because few had been taught to swim. An artificial peninsula of grass (the peninsula being a land form ubiquitous in Norris Lake) was elevated and revetted against an outer frame of sandy beach twenty feet wide. Despite the high density, there was no Coney-Island-style shoving and pushing, and no-one appears to have been about to step on anyone else. The pictures of the parks go a long way toward answering a question often asked by later generations about the 1930s, with its reforms: "Why was there no armed insurrection during the Depression [in contrast to the much more affluent reform decade of the 1960s], no trashing and burning, no carnage in the streets?" Part of the answer was that people, particularly in the South, then tended to bridge over social and economic difficulties with great politeness and courtesy, even though social maladjustment and feelings of deprivation ran deep in the South and could erupt at any moment. Dignity was very valuable as social currency in the 1930s, resulting in less aggressive behavior than in the 1960s.

At Big Ridge Park, MacKaye wished to see a "wilderness park" used as "a natural science area." Arthur Morgan had hoped to keep Island F in the midst of Norris Lake as an undisturbed ecological sanctuary for a century.[79] A tented camping and picnic ground was proposed for Beech Island, farther up the lake, but was never built.[80] There was constant hope, largely thwarted, for the extension of recreational or wilderness preserves. The example might be established with considerable ingenuity and effort, but then its lessons would be ignored. Robert M. Howes, chief of recreation in TVA's Regional Studies Department, was already speaking with a lonely voice when, in 1947, he described how "in the early days of the program we leased a number of waterfront sites to cities and counties for municipal parks." But the parks did not get built. Instead, "the so-called waterfront parks which we had leased remained in an undeveloped state except for the boat dock site which the city had sub-leased."[81] Fishing, houseboating, and water skiing pushed the notion of onshore recreation in enhanced settings out of reckoning. The local will was not there. And yet there was constant pressure to turn over more of the TVA land to states and counties for recreational purposes. Big Ridge Park, opened in 1936, was transferred to the State of Tennessee in 1949. Similar reservations were set up early at Wheeler and Pickwick dams, the other first dams, although they encompassed only between one thousand and two thousand acres. The Wheeler site followed the general pattern of Big Ridge Park, with one hundred acres devoted to "intensive recreational use," while the rest of the area was more or less left wild.[82]

The Shoreland Question

After recreation was draped around the shores of Norris, Wheeler, and Pickwick lakes, there was less certainty about who was going to own and administer what type of shoreland. Arthur Morgan recalled, "As far as I was able, I kept title to marginal forest land acquired in the process of purchasing reservoir right-of-way and to a strip of land a quarter of a mile wide around each reservoir. The purpose was to control the development of the area, especially of its recreation facilities. After I left the Authority that policy was largely discontinued and only that land that would be actually submerged by construction of the reservoirs was bought. The board gave away or sold one hundred thousand acres of land that were reserved for recreation."[83] Draper corroborated this statement, reporting: "We had the complete hostility of the agricultural group in the Authority who favored no purchase whatsoever of land other than what was actually needed for the waters of the reservoir, leaving the rest for private ownership in any use and anyway desired." This made it further "difficult for those who were operating and doing the planning and the construction in many of their endeavors."[84] Those who might wish to carry on comprehensive regional planning almost at once encountered a roadblock, for the land could not be controlled out from the lakes even as far as the quarter-mile strip mentioned by Arthur Morgan. The TVA lakes could be multiplied after the initial example of Norris, but the new town of Norris never was duplicated. In like manner, the Norris Village greenbelt never was repeated and was partially sold off.[85]

Divestiture of such motifs as Norris Park, Big Ridge Park, and other forest land about the lake became the first retrograde response. Harry Wiersema, Arthur Morgan's one-time engineering assistant, attributed the reversal to the incoming general counsel of the TVA, Joseph Swidler. Swidler "felt that the Act didn't authorize TVA to engage in recreation,"[86] so the board did little more than encourage the states and counties to use already-established facilities, until, Wiersema noted, interest again increased about 1965, with Land Between the Lakes.

Immediately after the World War II, there was a strong commitment in Britain to national planning. In July 1953, Howard K. Menhinick and Lawrence Durisch published a very comprehensive article on TVA, "Planning in Operation," in the University of Liverpool's *Town Planning Review*, a major international organ for such discussion. In the explanation, the abandonment of land around the dams took on as reasonable a cast as possible. At first a protective strip had been purchased for two objectives: so that recreational access for the public could be sustained, and to avoid inequitable confiscation of land, as with a farm left on a peninsula or left with only an eroding hill and no fertile bottomlands. However, it soon became evident that the TVA would end up with ten thousand miles of shoreland or more. The cost of administering such a shoreline could become prohibitive! "As a result of these factors, more extensive use was made of flowage easements and less of fee purchases in the later projects."[87] The TVA board of directors consequently put forth resolutions that resulted in land-use maps for a great variety of purposes on the ground above pool level at each reservoir. These maps were developed in collaboration with state and local planning commissions and other agencies. The purpose of the new program was divestiture, as rapid as possible:

> Reserved sites are transferred to the appropriate public and semi-public agencies and private enterprisers as rapidly as possible. The remaining lands for which no TVA program use is identified are sold as surplus property at public auction. Thus TVA aims eventually to dispose of its reservoir lands and land rights except those required for the primary purpose of navigation, flood control, and hydro-electric power production.[88]

Most noticeable in this recitation are the assumptions, first, that the TVA should be devoted more or less exclusively to navigation, flood control, and electrical production; and, second, that real estate values created by the establishment of a reservoir should be passed on "as rapidly as possible" to local agencies and private interests. A little later in 1953, Durisch, one of the authors of the TVA article, with a new partner, Robert Lowry, interpreted the same process more bluntly (perhaps because this essay was intended first for American rather than overseas eyes):

> TVA had in fact purchased by 1935 several thousand acres of marginal land near Norris Reservoir, with a combination of willing sellers and promising

opportunities for land rehabilitation as the principal reason for purchase. In a major reversal of policy, TVA has gradually withdrawn from land management except as essential to reservoir operations. Surplus lands have been or are being sold, other lands transferred to public agencies, and housing properties disposed of either by long term lease or by outright sale. . . . By these actions, the TVA board recorded its decision to withdraw from ownership and management as instruments for extensive land redevelopment.[89]

As Durisch and Lowry suggested, the issue was not so much shoreline as fresh plans "for extensive land redevelopment." Some New Deal figures were dissatisfied with the land acquisition policies, or lack of them, for the reservoirs, because they felt that the later steps had been taken in disregard of the original purpose of the TVA, as designated by Roosevelt. Long-term protection had been traded for a series of short-term, opportunistic advantages. The most outspoken critic of this approach for the TVA was Rexford Tugwell, Roosevelt's long-term advisor on agriculture, town planning, and housing. Reviewing Philip Selznick's administrative study of the TVA, *TVA and the Grass Roots*,[90] Tugwell in effect wrote his own evaluation of what had transpired before 1950. He asserted that David Lilienthal's approach to land acquisition had derived from a concession made to the American Farm Bureau Federation in order to gain its support for Lilienthal's rapidly expanding electrical power policies. Among other things, the TVA had abandoned its "early program for public ownership and control of submarginal land" and reversed the policy pioneered at Norris Lake, "so that it did not even acquire a protective strip around its reservoirs."[91] The clue to understanding the misfire, according to Tugwell, lay in the recognition that not all agricultural interests in the valley were identical, and that Lilienthal, in order to retain H.A. Morgan's support against the proposals of A.E. Morgan, began to favor the interests of bigger and more prosperous farmers and local businessmen over those of small-holders and marginal farmers who normally would be assisted by the Farm Security Administration and the Soil Conservation Service. H.A. Morgan set about making sure that the TVA would not generate its own agricultural service but would rather depend more and more upon the land-grant colleges, the experimental farms, and the agricultural extension activities of the seven states in the TVA. Thus "the grass-roots approach led TVA to oppose the entry of SCS into the valley and to take an unfriendly attitude toward FSA, the agency chiefly concerned with low-income farmers."[92] Instead, a second and sometimes a third agricultural extension agent was put on in each county at TVA expense.[93] Tugwell did not believe that all parties in the agricultural constituency had been well enough represented, and he did not like the fact that so much emphasis had been placed on the concerns of state governments as to result in "putting local interests far, far ahead of national ones. As it turned out it is pretty easy to see that while democracy, as represented by the demands of local interests, had its way, democracy by a wider definition retreated until it became invisible before the pressures of vested interests. TVA is more an example of democracy in retreat than democracy on the march [Lilienthal's slogan]."[94] So it happened that "from 1936 on the TVA should have been called the Tennessee Valley Power Production and Flood Control Corporation. That is what it became as a result of Mr. Lilienthal's opposition-conscious policies and Mr. H.A. Morgan's responsiveness to the local interests and the state agricultural hierarchy."[95] Lilienthal's account of this effort's function and purpose was different. Speaking in retrospect about Harcourt Morgan's initiative, Lilienthal observed,

> On the technical side, as an agriculturist I've since known a good many who were far superior. I think he was somewhat less of a great agriculturist than he was really a great preacher and thinker. A man who had sensibilities about nature and man and a good deal of the grass roots philosophy as applied to agriculture. A good many people have cast doubt, for example, on TVA's reliance on the County Agents System and the Extension Service System. He has been criticized a good deal on this score . . . We had a man in our organization who was very critical of it on the grounds that it was a kind of a front for the Farm Bureau, etc. Actually, what Harcourt Morgan was doing, I think, was setting up an alternative to Washington-directed agricultural activities.[96]

According to Lilienthal, it was Harcourt Morgan who

> made it possible for me to function under strange circumstances. . . . If he had not supported me, not alone on the policies, but for my way of going about these things, the conservative — the predominately conservative — feelings of the leaders of the Tennessee Valley clearly would have rejected me. . . . The fact is that he could have pulled the plug definitely by simply not supporting me.[97]

Lilienthal's dependence on Harcourt Morgan for support evidently was considerable. The relationship worked the other way as well, for Lilienthal also remarked that when Harcourt Morgan fled from controversy, he, Lilienthal, stood up to it and shielded them both.

Tugwell believed that in the TVA situation the terms "grass roots" and "democracy" had been misused. The national government's control ought to have remained firmer, so that the "interest of the whole people" and of the "forgotten man" "could be kept secure against too strong a manifestation of local interest."[98] More speculative was Tugwell's thought that Arthur Morgan's position in the TVA began to be undermined when the proposal came up in 1935 to give the direction of Social Security over to the states. That transfer was in line with the "Brandeis-Frankfurter view" that federal administration should be broken down into smaller units, and with the demand of the states, after the worst fright of the Depression was over, that they have more input.[99] Tugwell believed that almost everyone understood, but that the liberals themselves were not yet ready to admit, that the TVA arrow had flown too wide of its original target: "That there was not the expected national advantage from TVA did not escape notice, although liberals were always more or less confused about the main issue and afraid to question any development of policy for assisting the private power companies. For what was traded for democracy in the valley was support of a sort for the power program."[100] Tugwell knew from personal contact that "Mr. Roosevelt had an old interest in public power [Frankfurter had advised him on utilities when he was governor—W.C.], but this was only part of a broad conservation concept which ran beyond local or state powers and reached out toward the conservation of people as well as water, soil, forests, and grass." Roosevelt had arrived at a moment when he had begun to realize "that there must be power to discipline exploiters, to channel national forces into one rehabilitation program, to visualize all the resources of the area and shape them into one organic, functioning whole."[101] What Tugwell resented most was the apparent conviction of Lilienthal and H.A. Morgan that all the problems of the Tennessee Valley could be adjusted easily and exclusively through power plants or local agricultural agencies.

Opposed Heroes: Arthur Morgan and David Lilienthal

The 1930s encouraged heroes to appear like genies from bottles. These were apt to be genies who would think the moment ripe for executing their favorite reforms. The difficulty with having Arthur Morgan or David Lilienthal exercise such prerogatives, by themselves, was that they identified TVA's welfare as an extension of their own silhouettes, like the rabbit's head shaped on the window shade for shadow play by a child's hands. Others shared this pair's views that they alone could direct TVA toward a positive result. Thomas K. McCraw states at the beginning of his book, *Morgan vs. Lilienthal*, that famed historian Henry Steele Commager of Columbia University and Amherst College had judged the TVA to be "probably the greatest peacetime achievement of twentieth-century America," while Eric Goldman of Princeton University thought that one true hero was visible everywhere in that unique national accomplishment — that the "TVA was David Lilienthal."[102] McCraw, for his part, reports that "The chief recipient of credit for TVA's accomplishments has been David E. Lilienthal . . . Lilienthal's blunt, 'tough' brand of liberalism, and his twenty years of exceptional public service, part of it accomplished despite attacks from right-wing witch hunters, have made him a hero to liberals."[103] The liberal hero was celebrated, but could the liberal policy ever be identified and described as well? And if the liberal policy were well described, would it then be possible to show what Roosevelt's own original intentions had been? Did these two heroes, David Lilienthal and Arthur Morgan, understand accurately enough the intentions of the superhero?

From 1931 to 1933 Lilienthal largely had run the Wisconsin Public Service Commission, "where his central point was that a public utility enterprise was a public business."[104] Wisconsin had pioneered in electrical utility regulation in 1907, following on a 1905 railway commission bill originated by Governor La Follette. A good deal of that aggressive attitude Lilienthal brought with him to the Tennessee Valley in 1934, when he had to begin negotiating with Wendell Wilkie's Commonwealth and Southern utility corporation. To Lilienthal, "the advice of Cooke, Frankfurter, Norris, and La Follette . . . was unanimously against any negotiation at all with Wilkie's company."[105] To the liberal establishment, therefore, Arthur Morgan's approach appeared permissive and highly susceptible to corporate enticement, since he had tried to talk with Wilkie. During 1937 and 1938, when the effort to remove Arthur Morgan from his directorship reached its peak, one of the main accusations against him was that he consorted with Wendell Wilkie. Lilienthal testified at the congressional hearing that "Arthur Morgan has collaborated with representatives of private utilities we know not how frequently."[106] In March of the previous year, 1937, Lilienthal and Harcourt Morgan had sent a letter to President Roosevelt complaining that Arthur Morgan was "actively cooperating with Mr. Wendell Wilkie in such a manner as to prevent the Board from carrying out its obligations to the President and Congress."[107] Arthur Morgan, on his part, claimed that as chairman of the TVA board, he had exchanged views with Wilkie only twice and that Wilkie not only was reasonable but also was "full of ideas and progressive policies," highest praise coming from a man who considered himself as having the same attributes.[108]

On the other side, Arthur Morgan's group felt that the Lilienthal alliance dealt too easily with potential exploiters, Roosevelt and Tugwell's anathema. Harry Wiersema, Morgan's former engineering assistant, years later explained in an interview with Dr. Charles Crawford how Arthur Morgan's mind worked in relation to Lilienthal.

> C. Then why was there the difficulty with the other two Board members?
> W. That was a personality clash with Mr. Lilienthal, I think. Lilienthal was an opportunist. . . . In the Berry marble case [over land to be flooded by Norris Lake, owned by a man with "connections"—W.C.] the engineers and the geologists reported that the marble was useless — had no value at all. Berry, however, was a very influential person and demanded a million dollars or so for his marble. It appeared that he would settle it for a nominal sum — maybe fifty thousand or something. . . . Lilienthal was all for settling it out of court. Morgan said that not one cent would they give for something that is not worthwhile. . . . Well, with such a stand, you see, it was impossible for those two to get together. . . .
> C. Mr. Morgan thought a great deal in terms of principle, didn't he?
> W. Absolutely in principle. Extremely ethical and always valued principle over and above expediency of any kind.[109]

The time sense of the two men was very different. Arthur Morgan, in his lifelong Progressive championship of the small community, felt that "this neglect of a basic cultural unit may be one of the primary reasons for the failure of human society to advance with greater surety.[110] His outlook was universal and eternal, integral with the ideas of ethics and of standing up for unpopular causes. Norris Village represented the "small community" ideal for him, that he spoke of from an ultimate moral stance.

> Important as is unity of spirit in community life, that unity must represent common respect for and commitment to the general good, and the common habit of living with integrity, courage, and self-respect, or it is a unity not worth having. Such unity often must be fought for, sometimes for the time being against the general current of community opinion. Seldom does a fine

community come into being unless along the way some of its citizens have
been willing to endure unpopularity in the common interest.[111]

One lead figure of the TVA wished the dream of the small community
to be a more lasting and substantial element in the broad culture, while
the other interpreted the same motif as being of only short-term, tran-
sient value — a view that was increasingly favored by urban intellectuals
during the 1950s, when "suburb" became a condemnatory, pejorative,
nonarchitectural term among them. While lecturing at the Harvard
Graduate School of Public Administration on 22 April 1941, Lilienthal
was asked if the houses at Norris Village weren't too high-priced. The
question of price had been critical in the English Garden Cities, where
better designed and slightly more expensive houses had been built in order
to inaugurate a chain reaction: "A beautiful home in a beautiful garden
in a beautiful city for all," as Parker and Unwin, the major architects
of the Garden Cities, had expressed it.[112] The comely, slightly more expensive house had been viewed as indispensable for Stein and Wright's
Radburn, New Jersey, too, which had had a direct influence on Norris
and in its turn had been greatly encouraged by Garden City models.
Lilienthal too, in his sequence of reasoning, sprang from house to town,
but for the very opposite purpose — deprecating the house in order to
downplay the town. At Harvard, his answer was that "Norris is not typical of our housing. Norris is simply a suburban town." Then a non-
sequitur, having nothing to do with the expense of houses: "At least that
is what we expect it to be in the long run."[113] So Norris housing could
not be good, because "a suburban town" automatically was bad!

Arthur Morgan's manner was said to be putting distance between
himself and the Tennesseans, whereas Harcourt Morgan and David Lilienthal were seen as drawing ever closer to their "grass roots" clientele. Arthur Morgan didn't often sketch out the essential character of his two
colleagues in his writings, but they utilized negative descriptions of him
to show how far he had wandered off the chosen (if undefined) path and
how alienated he was becoming from those he had pledged to serve. He
was said to be scatter-brained,[114] out of touch, intransigent, and, perhaps
worst of all, patronizing, an elitist! By contrast, Lilienthal declared, Harcourt Morgan had "an earthiness about him."[115] In the way the personalities were delineated and then colored in, the struggle became very
much an American psychodrama, with intense idealism and dedication,

or at least the proclamation of them on one side; and expediency and opportunism, or at least the accusation of them, on the other. One who shared Lilienthal's grass roots enthusiasm and outlook put it as follows, emphasizing the need for liberation:

> Yes, I think there was a basic philosophical weakness that [Arthur Morgan] was never conscious of that cursed his approach right from the beginning and, because of it, Dave and H.A. teamed up against him — well, they didn't team up against him; they teamed up for the things that he was against. They both proceeded from the assumption that people, as Thomas Jefferson considered them, have enough intelligence, enough good sense, and rights which add up to a situation in which you've got to trust them to do things. And if they were doing a very bad job, as they were, let's say, in the cotton belt of the western and southern valley, if you gave them the right tools, and the better tools, they'll do better.[116]

The claims that Arthur Morgan was paternalistic and an elitist, and that Lilienthal and H.A. Morgan were closer to the common citizen, the small-town person, and the soil, were repeated frequently.

> A.E. had the feeling that the people handling the TVA were an elite . . . Now, H.A. and Dave were very different from each other — entirely different — but they had one thing in common. They said, "This country is made up of ordinary people. . . . " They wanted to leave the basic power for advance in the hands of ordinary people — the Jeffersonian approach to living. H.A. and Dave were both Jeffersonians.[117]

Arthur Morgan did indeed take exception to the view that, in the southwestern Cotton Belt, all that was needed was "better tools" for improving life. He said he had been thus wrong himself:

> An increase in economic wealth is not enough. . . . Years ago, with my associates at Memphis, we planned and directed the reclamation of many hundreds of thousands of acres of very fertile land. The philosophy of that development was that if you gave people the means for creating wealth and comfort, they will work out the situation without further help. Yet, today, that most fertile land in America is the locus of the most miserable sharecropper tenantry, where poverty and bitterness are general and violence appears.[118]

Here, A.E. Morgan appears, of the three directors, closest to the "roots." By instruction and example, he wanted to revive character and decent behavior among these farmers. The reform of character through environmental change was a longstanding hope among American reformers, especially those of the nineteenth century. Morgan believed that the slower tempo of life and lack of urbanism in the South could be turned to good account in terms of both human dignity and the production of physical goods of a worthwhile kind with a change of mission. This article of faith was behind Morgan's idea for reviving weaving and pottery in Norris Village and elsewhere. Of his hypothetical southerner with a new character, Morgan observed,

> Given the opportunity he will have time to create fine things in furniture, in clothing, in ceramics, in scientific instruments. He can be the individualist in American industrial life. With artistic and scientific guidance, he can make the goods America needs to take the curse off its mass production civilization. Every valley can become the home of some kind of excellence peculiar to itself. . . . Mass production will have a place in the design, but it should be as servant and not master.[119]

George C. Stoney summarized the opposing Lilienthal viewpoint in a 1940 article: "The TVA folk . . . cannot attack the problems of rural overpopulation, farm tenancy, and illiteracy that choke the social progress of the Valley folk." But, indirectly, "through its program of richer lands and conserved waters the TVA can increase, and is increasing, the general prosperity of Valley life of which the poorest ones are a part." The poorest were only a part. There was a better system, a triumphant "new method of administration" coming, and "this method of enlisting cooperation from local people has all but eliminated the local feeling of resentment toward a federal agency."[120] The undermining of Arthur Morgan's strength and stability was accomplished mostly by reiteration that he was unrealistic, intoxicated on nostalgia, and distracted from modern management. In an overture speech of 12 June 1936 to TVA employees, Lilienthal declared, "There is, as I see it, no turning back from the machine . . . I am against 'basket-weaving' and all that implies, except perhaps as a temporary expedient. . . . We cannot confess our failure, we cannot prepare for 'the second coming of Daniel Boone' in a simple handicraft economy."[121] In an article in *The Nation* in October 1936, the same year, a steady supporter of Lilienthal, the managing editor of the *Chattanooga News*, J. Charles Poe, warned about unprofitable distractions from the single value of electricity:

> Dr. Arthur Morgan started off with many plans for the social and economic rehabilitation of the Tennessee Valley which took little account of electricity. He talked of restoring the lost folkways, of dancing and singing, of basket-

weaving, of wood-carving, and other handicrafts. He urged the formation of cooperatives for the barter of goods. He established a land-planning and housing section and spent more than $3,000,000 on model housing at Norris Dam as a demonstration project. Dr. Morgan assumed that his task was to create a new way of life in the valley either by imposing it upon the people by experts or by setting up demonstrations which would be gratefully copied. Lilienthal would give them income and let them order their own lives.[122]

All such ideas, including suggestions for setting up a system of scrip for monetary exchange and reducing the number of counties in Tennessee, "irritated Lilienthal."[123] Scrip and barter systems were invoked fairly often in the Depression because of the shortage of ready cash. Morgan reported that he had set up such a system at Yellow Springs in Ohio, "where it was serving usefully."[124] Hard-up manufacturers were using it to pay employees. Summer theaters such as the Barter Theater near Asheville, North Carolina, took vegetables and canned goods as payment for entry. Arthur Morgan claimed that Harcourt Morgan, after hearing him talk at the University of Tennessee in November 1933, had gone to Washington and told the newspapers that Arthur Morgan "had proposed a separate money system for the Tennessee Valley," creating a sensation in the capital.[125]

The thought of reducing the number of county governments was not original with Arthur Morgan. The idea had been brought up in 1919, long before his arrival, in an effort to trim Tennessee's ninety-five counties down to eleven.[126] Hamilton and James counties actually had been amalgamated then. Morgan merely suggested that the eleven previously named be cut further, to nine. The reduction in number had current significance for TVA efficiency, since TVA continually had to negotiate with county governments. Up to 40 percent of a county's territory might be involved in TVA projects at any one time. Another of Morgan's concerns, equally ridiculed by his critics, was what would occur with real estate development, once the TVA's activities had increased land values. This worry, too, was based on actual past experience, since there had been a boom and bust with land at Muscle Shoals after the rumor spread in the early 1920s that Henry Ford was going to buy it.

Arthur Morgan could be described as an anachronistic Daniel Boone by Lilienthal and his friends, but the difficulty with that label was that it brought Morgan into still better alignment with the Progressives of the 1890s, whose thinking had led to the TVA. The last of them, Gov. Gifford

Pinchot of Pennsylvania, in 1925, eight years before the TVA, had posed the Morgan-like thought that "Giant Power may bring about the decentralization of industry, the restoration of country life, and the upbuilding of the small communities and of the family."[127] So there was a certain loyalty to Progressive tradition in Arthur Morgan's supposedly maverick position. From the more contemporaneous, up-to-date, 1930s aspect, too, he filled a yearning, for he was, after all, an actual, highly capable, imaginative engineer who was taking on nonengineering problems, just as the Technocrats had said engineers should.

Perhaps the greatest reason why Arthur Morgan's views could not be reconciled with those of David Lilienthal or Harcourt Morgan derived from the differences in the ways their respective professions were expected to approach and react to problems. Louis B. Wehle, who had been in close touch with the machinery of Democratic administrations since Woodrow Wilson, perceived the rivalry as inherent in the different approaches: "Morgan was an awkward fighter. His memorandum was not effectively written. In discussion his scrupulous anxiety to be accurate and fair [the engineer's habit] made him hesitant. He had the intense yet diffident manner of an introvert, tending to become inarticulate in man-to-man controversy. He was not skilled in the ways of politics and politicians." Morgan couldn't get the political hang of "going with the flow." On the other hand:

> Lilienthal had a controlled, driving, effective intensity, and was brilliantly skillful in the use of many weapons [he was a lawyer, so knew how to set up an effective adversary relationship—W.C.]. He made and marshaled his allies and converts resourcefully and ceaselessly by letter, telegram, and the easier, more direct ways of interview in person or by phone, or above all by his use of the press, which liked his colorful copy. He seldom left any preventable weakness in his lines [as a good lawyer does not].[128]

Lilienthal, in other words, was an inexorable natural force, like a glacier, that had to be ceded room to move forward.

Should Roosevelt and the Law Be Scrupulously Followed?

Arthur Morgan wrote in one of his first essays that "the chief means must be cooperation with the people of the Tennessee River region"—exactly what Lilienthal and H.A. Morgan later said they were doing and Arthur

Morgan was not. However, the latter tied on an addendum phase they would not approve of: "And of the [people of the] nation."[129] It was in the difference between a local and a more national interpretation that much quibbling over law arose. Arthur Morgan wanted to sell TVA fertilizer across the country, and Lilienthal and H.A. Morgan said that he must adhere to TVA law and could not do it. A. Morgan countered by observing that Harcourt Morgan already had broken the law when he turned Muscle Shoals from nitrate to phosphate fertilizer manufacture. It appears that neither of the parties was legalistically correct.

On such a broad canvas, it was admittedly awkward to get words and concepts to line up. Under acute time pressures, it undoubtedly was difficult to establish authentic communication. But both administrative parties frequently paid lip service to one personal motivation — both claimed absolute allegiance to the underlying purposes of Franklin Roosevelt's vision for the TVA. The president clearly had a broad, general purpose in mind well before any discussions began on the TVA *per se*. An unsigned forty-three-page memorandum on national planning in the National Archives, exhibiting the literary style of FDR's uncle, Frederic Delano, mentions at its start that Roosevelt, speaking to the New York Agricultural Society in 1931, had suggested that, "in the long run, state and national planning is essential to the future prosperity, happiness and existence of the American people."[130] The memo indicates that this feeling led directly to Roosevelt's 10 April 1933 message to Congress proposing the new TVA law, in which he stated that the potential of the Tennessee River, "if envisioned in its entirety, transcends mere power development." That view differs, of course, from J. Charles Poe's view that the business of the TVA ought to be electricity and nothing but electricity. In Rooseveltian terms, the vision "leads logically to national planning for a complete river watershed involving many states and the future lives and welfare of millions. It touches and gives life to all forms of human concerns." The TVA, Roosevelt said, "should be charged with the broadest duty of planning for the proper use, conservation, and development of the natural resources of the Tennessee River drainage basin and its adjoining territory for the general social and economic welfare of the Nation." In his congressional message, Roosevelt said that he wanted the TVA to be charged with "the broadest duty of planning." Under that canon, it appears that Arthur Morgan would be legally entitled to peddle his TVA fertilizer wherever he wished in the country. Neither Roosevelt nor Arthur Morgan seemed to want to be confined by too many legal restraints. There is much corroborating evidence, scattered here and there, that electricity and fertilizer were not Roosevelt's first priorities, that very early on he wished for a much wider result. Rexford Tugwell reported in his diary after an Executive Council meeting of 12 December 1934 that Roosevelt had stated "that the purpose of the TVA was not to make and distribute electricity, that its purpose was really to reconstruct the life of the whole region and that as an incident to that it was necessary to build dams, to control floods, to prevent erosion, and, incidentally, to make power."[131] Arthur Morgan felt that he was in line and on the way with the president's intention when he gave an interview printed in the Asheville (N.C.) *Citizen* on 17 March 1934: "The Tennessee Valley Authority is not primarily a dam building job, a fertilizer job or a power transmission job. When I first went to see President Roosevelt, he talked for an hour about its possibilities and there was scarcely a mention of power or fertilizer. He talked chiefly about a designed and planned social and economic order. That was what was first in his mind." The year 1934 appears to have been a major one for such inspirational statements. Even Harcourt Morgan, later the opponent of anything but agricultural and localized involvement, said in an interview with the *Washington Star* on 26 September 1934 that the president wanted the TVA to be "a laboratory experiment" for "our whole infinitely complex national picture." And the conviction rapidly communicated itself to the actual instigators of such change. Earle Draper, then head of regional planning for the TVA, announced in a speech to the National Conference on City Planning and the American Civic Association in Saint Louis on 22 October:

> To achieve the ultimate of the TVA Act, we must seek not solely the production of cheap power, or low cost fertilizer or any of the other desirable results of our varied program, but through wise planning to ensure these instruments of a better regional economy reaching into the lives of all the people in such a way as to benefit all who live in the region and those outside within the scope of influence.

The TVA was to be a national example "for a better civilization."[132] The term Arthur Morgan was inclined to use instead of "civilization" was "cultural environment," to signify the ultimate purpose for the TVA.

> As I discussed the TVA with President Roosevelt, it seemed to me that for twelve years in Yellow Springs I had been actively engaged in an undertaking

that was almost identical in spirit to the one he outlined. The TVA seemed to offer a chance to create a new cultural environment . . . I was surprised and pleased to find that Roosevelt had much the same outlook: he wanted the country to loosen up and become conscious of a wide variety of economic and cultural interests. He wanted a new breath of life. . . . To his mind, it should be concerned with every aspect of the region's well-being. There were few areas in America with such a poor and narrowly based agriculture and economy as the mountainous parts of the Tennessee Valley, but he believed in the possibility of its rebirth and larger life. President Roosevelt talked to me about his early experiences as a traveler in the undeveloped parts of Georgia and South Carolina, Tennessee and Virginia. He told of his long-standing hope that he might help to give a new life and culture to the long-neglected descendants of those indentured servants who had, before the days of slavery, largely made up the working class of southern agriculture. The principle of the all-round development of life is only gradually emerging to consciousness.[133]

It now appears clearer what Roosevelt himself had in mind for the TVA. Bringing southern culture up from its alienated status even before slavery to national dimension through stimulating enterprise ran as a thread through all his early thinking. It was another mode for making the American dream come true. But Raymond Moley, writing of this earliest New Deal period, gives the impartial reader even further pause. He noticed the "the [TVA] bill provided for three directors to manage the corporation. In making the appointments, Roosevelt followed his habit of seeking to reconcile the irreconcilable."[134] Moley, later the disaffected Brain Truster, believed this behavior typical: "Thus, Roosevelt, when confronted with several objectives, embraced all of them."[135] The president, no matter how deep his original conviction, made the TVA a grab bag. Even more arresting, in regard to who actually exerted most influence in TVA's conception, is Moley's closely associated revelation:

I am not sure that between the election and the inauguration Roosevelt had conferred with Norris, but I learned later that he had talked at some length with Dr. Arthur Morgan. Morgan's ideas of what should be done in the Tennessee Valley extended not only to the Federal development of the power facilities of the Wilson Dam and the nitrate plant there, but to a vast Federal enterprise to improve the economic and social resources of the valley. Thus the plan as it developed in Roosevelt's mind was more of a creation of Morgan than of Norris, whose major concern was public power.[136]

*Or Were Justices Frankfurter and Brandeis
the Real Fountainheads?*

Could it have been that Lilienthal felt, as Moley did, that the breadth
of the TVA really was the (illegitimate) brainchild of Arthur Morgan?
Did Lilienthal consequently dedicate himself solely to the cause of Senator Norris's electrical reform? The answer appears to be no, although
Lilienthal came to cite Norris fairly often. Max Friedman, editor of the
letters between Felix Frankfurter and Roosevelt, flipped the narrative.
According to Friedman, Roosevelt aligned himself directly with his old
utilities advisor, Frankfurter. Roosevelt wanted cheap power and nothing else! "As early as 1933 Frankfurter had warned David E. Lilienthal,
then a TVA director and later its famous chairman, that Morgan's approach was 'fraught with every kind of danger.'" After a few years, "an
inquiry followed, in which Arthur Morgan's fluent but repetitive an-
swers showed that he had departed from Roosevelt's hopes for TVA as
primarily a generator of cheap public power."[137] Frankfurter taught public utilities at Harvard Law School, where Lilienthal studied. Friedman
was unequivocal about the course of events: "At every stage of this prolonged controversy Frankfurter was against Morgan and with Lilienthal."[138] Friedman believed that in the larger flow of events, Morgan's
1938 defeat also occurred because of a tidal shift in history, as the First
New Deal gave way to the Second. His source for that conclusion was
Arthur M. Schlesinger, Jr.[139] Schlesinger attempted to show that Morgan's downfall was symptomatic of a larger changing of the guard within
the New Deal, beginning in 1935. According to Schlesinger:

> The key figures of the First New Deal were Moley, Tugwell, Berle, Rich-
> berg, Johnson. From 1935, their influence steadily declined. The charac-
> teristic figures of the Second New Deal were Frankfurter, Corcoran, Cohen,
> Landis, Eccles, in time William O. Douglas, Leon Henderson, and Lauch-
> lin Currie. The shift in TVA from Arthur E. Morgan, the biographer of Ed-
> ward Bellamy, to David Lilienthal, the protégé of Felix Frankfurter, was
> symptomatic.[140]

The theory works, at least in terms of time, except that among the first
group, too, Berle and Richberg were known to be strong supporters of

Lilienthal. The primary reason for the sudden shift in orientation may have made Morgan more dispensable:

> Fundamentally, perhaps, the First New Deal was destroyed by success. The economic disintegration of 1932 could only be stopped by a concerted national effort and a unified national discipline. The method and approach of the Brandeis school would have been ineffective and irrelevant in 1933. But once the First New Deal had reversed the decline and restored the nation's confidence in itself, then the very sense of crisis which made its discipline acceptable began to recede. The demand for change slackened, the instinct toward inertia grew, the dismal realities of life and mediocrities of aspiration reasserted themselves.[141]

So it was, too, with the TVA. Its effective time spans were so brief — especially the first span, with its brilliant power of rejuvenation. Success had come too well and too quickly. The strength of democracy — its "right" to dissent, take exception, and remain outside the pale and alienated — also was its greatest weakness. The "decentralization" from Washington, D.C., which Lilienthal was so proud of achieving for the TVA (that was the main expository motivation of his "grass roots" thinking), was developed from his reinterpretation of Brandesian themes. The somersault effect of this repositioning is noteworthy. At first Lilienthal wanted the federal government to have even more concentrated power: "It would be a wanton disregard of the people's rights, to let the powers of the federal government be hopelessly outdistanced by the trend to centralized control in industry and commerce and finance."[142] But then he wished to have that recentralized power directed outward, through decentralization, to separate his agency in Knoxville from Washington. One could shift position or rubric fairly easily in the Roosevelt days. Schlesinger, for instance, further observed:

> The leader of the Neo-Brandesians, Frankfurter, had himself been a follower of Theodore Roosevelt and the New Nationalism in 1912. . . . William O. Douglas who in 1933 rejected Brandesianism as obsolescent [and the TVA as obsolescent in the late 1960s, when he protested against its Tellico Dam — W.C.], became in a few years almost its most effective champion, while David Lilienthal, a Brandesian in 1933, ended as a prophet of bigness.[143]

Lilienthal appears always to have searched for tradeoffs, a series of "gives" in one direction so that he and H.A. Morgan could gain "takes" in another. His pronouncements on occasion seemed to teeter on the very edge of plausibility, as when he declared that "in the Tennessee Valley Authority centralized authority is expressed through a decentralized administration."[144] Frankfurter's preoccupation with power questions while he advised Governor Roosevelt undoubtedly tilted Lilienthal in the direction of power for the TVA, but the Brandeis influence may have been even more profound, especially in determining the mode of TVA's operation. Brandeis, through his analysis of J.P. Morgan's manipulation of the finances of the New Haven Railroad, became convinced that "railroads by scientific management could save a million dollars a day."[145] Lilienthal followed in this conviction, seeing himself as the hard-hitting "professional" manager up against capital manipulation. He was a well-educated, well-informed, "scientific" manager. The same picture would, of course, work against the image of Arthur Morgan as a self-made (hence awkward), impulsive maverick, a late-nineteenth-century individualistic reformer. Moreover, Brandeis had "destroyed the common delusion that efficiency results from 'bigness.'"[146] Decentralization from Washington would now promote efficiency in TVA planning through "grass roots," Jeffersonian independence and isolation. Lilienthal's ingrained suspicion of bureaucratic "expertise," and his resistance to large-scale planning except for power systems, made it easy for him to shrug off responsibilities that the TVA previously had thought it had — especially in relation to more comprehensive, organic regional planning that would balance land use between agricultural field and forest, settlement and open space, industry and recreation, craft and mass-produced objects; and that could have created the appropriate infrastructure that would support them all for a greater prosperity. In some respects, Lilienthal's grass roots campaign choose the single-minded aim of Frankfurter and Brandeis over the broader, possibly too pluralistic visions of Roosevelt and Arthur Morgan. The latter, perhaps impractically, was seeking an impossible breadth of environmental uplift all at once and in utopian proportion and perfection. Morgan was looking for what he described, when he spoke of the utopian dreams of the American he most liked to write about, Edward Bellamy, as "a great *pattern* of action."[147]

Brandeis claimed that scientific management would increase the efficiency of almost any enterprise, yielding greater economy as a by-product. Brandeis's writing shows that this conviction derived from the ideas of the odd mechanical engineering genius, Frederick Winslow Taylor, who initially developed theories of "scientific" rate fixing at the Midvale Steel Works in Nicetown, Pennsylvania, in the 1890s. Brandeis's own exposition of "scientific" and "efficient" management first occurred in the Eastern Railroad rate case, considered by Interstate Commerce Commission in 1910-11.[148] "Brandeis created a popular image of scientific management in the years before World War I that seemed public-spirited and humane but also modern, technical, and morally neutral."[149]

Fontana Village and Construction Camps

The bill to begin Fontana Dam was signed ten days before the attack on Pearl Harbor. With the onset of World War II, the theoretical, gradual, localistic approach of the TVA had to be abruptly abandoned. From hope of cultivating a more and more open and cooperative society, the staff had to turn to concern with total secrecy. Fontana Village and the City of Oak Ridge came to the forefront as critical emergency settlements to which many scarce resources should be delegated immediately. It was said that the TVA "built Cherokee, Douglas and Fontana [dams] during the war to get power for just one thing — aluminum to build airplanes."[150] Ironically, the intended use of the vicinity had been, in the main, quite different and a more peaceful one — recreation. The 1938 *Scenic Resources of the Tennessee Valley*, put out by the TVA Department of Regional Planning Studies, explained that "adjacent to the national park, [Fontana Lake] will lie near the center of recreational interests in the Tennessee Valley; and unless special recreational lakes are provided elsewhere, a considerable demand will probably arise for facilities along its shore."[151] This broader and more humane use of the topography now was sacrificed to a much more singleminded, rigid goal. "The gospel of efficiency" was being sent to war.

Norris Village had been a pilot study in how to coax local populations, particularly those from the foothills, out into the mainstream of society. Although the population target originally had been the same for Norris and Fontana villages, 5,000 people each, in the end Fontana grew to 6,000, while Norris dropped to 1,500. Norris had been merely "a little eye-dropper thing," as one later critic expressed it.[152] The hill people of the Fontana district were faced with all the ill-begotten crises of modern society, without the gentler and more considerate preliminaries seen at Norris. The Fontana people were buried much deeper in the coves and up against the mountains than those near Norris or Knoxville. These natives were even more used to isolation and having elbow room among themselves, and between themselves and strangers:

> There weren't many people, but those there loved their meager way of life and didn't want to change. Most farms were worked with primitive hand tools. Some owned an ox or mule. There was not a tractor in all Graham County and only ten in Swain County. Mountain families got staples from "rolling stores." . . . The men went into town on Saturdays but women and children seldom left their homes.[153]

A deferential nod might be given the agricultural existence, but these natives, more primitive yet, were happiest indulging in hunting and fishing.[154] All at once, the apparatus and the noise of the most sophisticated construction equipment in the country were thrust in upon six hundred families in five small communities in the immediate area of the future Fontana Dam. These people could not easily buy more land in the immediate vicinity, since ALCOA and the federal government had already acquired much of it; the government land-acquisition program, begun by the states of North Carolina and Tennessee in the 1920s, had ended temporarily with the establishment of the Great Smoky Mountains National Park in the early 1930s.[155] Planning effort in the district had been directed toward the appreciation and enhancement of an unpopulated, "unspoiled" environment. At first, no military emergency was factored in. One of the displaced hill families — father, mother, and children — took up living in the hollowed-out trunk of a giant fallen poplar. As the family required more rooms, they carved them out of the remaining body of the tree trunk.[156] The first construction crew on the scene slept in huge sections of metal culvert pipe.[157] The contrast between the naturopathic and the technologic in the furnishing of houses in this very remote location was indeed striking and prophetic, especially in regard to how fast Americans, when required, could change their basic living arrangements.

Fontana Village (85) was the first community of prefabricated dwellings in the country, and probably in the world. At Norris Village there had been an attempt to hold onto the image of the old four-roomed cabin of logs and shakes, but here smooth plywood wall panels and flat roofs, employed "to save wartime materials," were displayed proudly on the untouched hillside sites. The flat-roofed, prefabricated building was, in some senses, a peeled-off, shaved-down log cabin, for both examples were really basic, initial, pioneer houses. That TVA interplay between ancient and new symbols is affirmed again today in the Fontana Village center by

85. Two-Cell Trailer Houses, Fontana Dam Village. The houses constitute a stack of units running up a hill, prophetic of the famous Habitat house of the 1967 Montreal World's Fair, except that they are not welded together into a megastructural form in imitation of a denser Mediterranean hill town. But Fontana was not an American prewar suburb, either, since there were no developed lots with sidewalks, no curbs and gutters, no streetlights, no lawns, no foundation plantings or ornamental trees. The "bread slice" construction made it possible to break the houses in half for a quick move. The wooden cover curb in the middle of the roof shows where the joint was for that purpose. The Norris houses (79) stretched time, while these encapsulated it. Photo courtesy of Tennessee Valley Authority Archives.

the restoration of the 1875 Jesse Cornwall Gunter cabin of big logs. In the postwar era, one ccasionally came upon a surplus flattopped house that had been sold to a farmer and then moved off TVA property to sit by an old log cabin. Between the attack on Pearl Harbor and the dropping of the first atom bomb (that, amazingly, had been prepared at Oak Ridge), one-quarter of all American housing was factory-built — almost two hundred thousand units.[158] That event, forthcoming under wartime conditions, had been anticipated by housing experts since the 1920s and early 1930s. During the shortage period immediately after the war, the demand for prefabricated houses rose. It then tapered off as affluence returned, because customers associated prefabrication with deprivation, postponement, regimentation, and inferior materials. Such houses failed to cater to the American craving for individualism and domestic romance. The sense of speed and mobility conveyed by the prefabs was likely to be even less acceptable in the romantic setting of a national park or forest. In a depression or a war, however, prefabrication could be a highly efficient method of responding to recurrent pressures for speed, minimalization and standardization, even while the houses' very temporary look would make later rejection inevitable. The era in which they were built was a more important regulator than the particular construction methodology.

In the TVA prefabrication effort, versatility and adaptability were emphasized to such a degree that a single standard was never perceptible. Portability and mutability, spontaneity and mobility were the desired values that drove designers to spend skill and money on the type. In 1942, Carroll A. Towne, who directed the TVA division of demountable and truckable housing, in the war era already was looking forward to the postwar market. His creation, he said, was "strictly temporary shelter, offering performance at the expense of permanence, but that, like it or not, is what the average American looks for — and gets — in nearly everything he uses. To the manufacturer looking for a market sustained by quick turnover of a product with a short life but high sales appeal, the trailer house, or something like it, may look good. That's what made the automobile industry."[159] The casual phrase "or something like it," the architectonic self-denial, the conclusion that prefabrication was a stopgap measure and should be used only in mobile structures — all were bound to be baffling to the theoreticians of European origin who looked upon prefabrication as a permanent solution to postwar housing shortages.

French theoretical architect Le Corbusier came to the United States in January 1946, immediately after the war, to confer with Albert Einstein, Henry J. Kaiser, and David Lilienthal. He mentioned then that one of his former assistants, Konrad Wachsmann, and Walter Gropius at Harvard were endeavoring to supply "the housing industry with the elements of mass production" while guiding the "enterprise towards a true architectural dignity."[160] Wachsmann's system, destined to be unsuccessful, was based upon "a standard in the form of a chessboard," while Le Corbusier's system for prefabrication would be premised on the "modulor" he had invented, of "three intervals which give rise to a series of golden sections, called the Fibonacci series."[161] None of the European proposals, including that of Wachsmann and Gropius, took trailer mobility enough into account. Elegant, "intellectual" images such as chessboards and golden sections did not jibe with American haste and empiricism. The essential difference was demonstrated in Corbusier's encounter with Kaiser, "the famous constructor of Liberty ships during the war. His latest project had been to construct 10,000 houses a day in the United States. But, he told me, I have changed my mind, I am going to make motor cars instead!"[162] That observation was bound to be discouraging to an enthusiastic European architect who had come all the way across the ocean to look into the standardized houses being made by those who knew how to produce them in quantity. Corbusier consoled himself by suggesting that Kaiser was, after all, mercurial, and that, anyhow, when Americans had built all those prefabs, "the towns will be expanded to enormous size by suburbs, vast, tremendous suburbs," and the streets congested by cars "twice as long as . . . need be."[163] He recognized the American eagerness for proliferation and for transport power, even if he didn't approve of it. Following his eminently unsatisfactory talk in New York with Kaiser, Corbusier pushed south to Knoxville to speak to David Lilienthal, "the guiding spirit of that great harmonious plan, sponsored by President Roosevelt, which built the dams on the Tennessee River and the new towns, rescued American agriculture and gave it new life."[164] Corbusier mentioned "new towns" and dams to Lilienthal, but apparently those ideas didn't evoke even a blip of response. Instead, Lilienthal inspired Corbusier to soar to higher and higher planes of harmony: "Harmony is the aim of all Mr. Lilienthal's work. His face lit up at the delightful thought of establishing a reign of harmony . . . by undertaking the most gigantic works and coordinating the most immense projects: water, motive power, fertilizers, agriculture, transport, industry."[165] There was a citation by Corbusier of many factors within the TVA known to be of longstanding interest to Lilienthal or Harcourt Morgan, but nothing from Lilienthal in return about prefabricated house techniques or the "new towns" Corbusier had come for. Americans were disappointing.

The TVA's whole exercise in prefabrication could be thought a shuffling and redealing of a pack of housing cards, none of which stayed with the same player for long. The Fontana two- and four-celled houses really were reworked Hiwassee Dam types. Back of the Hiwassee originals were demountables devised by Louis Grandgent as early as 1934.[166] The Fontana house was two or three feet longer than the Hiwassee house, with the latter containing only "about two-thirds of the floor space required by [1942] war housing standards."[167] The pinching shoe drew even tighter under wartime stresses. The housing cards were dealt far and wide. Some thirty thousand units were shipped to England to rehouse displaced victims of the London Blitz. And when the Oak Ridge population got ahead of the thirteen-a-day house erection rate at which its more conventional (but still pre-paneled, Celotex-Cemesto) residences were going up, five thousand of the flattops previously exclusive to the TVA were sent over to Oak Ridge. The cost of these units was $1,900 apiece.[168] A premonition of such vigorous migration and substitution had occurred within the TVA itself in 1938, when seventy-two conventionally built cottages of wood from the construction camp at Pickwick Dam had been floated on barges two hundred miles down the Tennessee River to the new camp at the river mouth at Gilbertsville, in preparation for the construction of the Kentucky Dam. Moving had not been on the original agenda for these cottages, and they were damaged somewhat en route. But delivery to the new site, including repairs, cost only $1,280 per house.[169]

That amount was expended to bring the house over two hundred miles — a cost of $6.40 per mile. By 1941, such moves to distances over thirty-four miles cost $280 per house, or thirty cents a mile for the individual cells of the sectional houses. A group of one hundred two-cell houses were trucked from Sheffield, Alabama, one of the earliest housing construction sites, three hundred miles to Murphy, North Carolina, at that price.[170] This thirty-cent figure, although seemingly already low, added impetus to the next step (the form was always evolving): designing a trailer house of one unit. The trailer house was constructed of a strong, light plywood. TVA house types A2 and A3 were adapted into trailer-houses A4 and A5 (86). The latter models were "streamlined" to travel once or twice in a lifetime, over difficult roads.[171] These houses could be pulled for the record lowest cost of fifteen cents a mile. Under the body were wooden wheels made with hardwood cores and plywood flanges that allowed them to be pushed sideways on tracks onto the permanent foundations, once the destination had been reached (87).

The difficulties of getting these houses to the intended site were considerable, especially of reaching the Fontana Dam from Knoxville. One hundred and four sectional houses were brought from as far away as Michigan through Knoxville by three-quarter-ton trucks that appeared hardly powerful enough to pull them. Neither the National Park Service, which had instituted the Great Smoky Mountains National Park adjoining the site to the north, nor the Forest Service, which owned Nantahala National Forest touching the south bank of Fontana Lake, had built any major roads into the area. The chroniclers of Fontana Dam construction reported that, on the existing road which the trailer and sectional houses had to traverse (88), "ten miles an hour was a good speed. Many places were too narrow for passing. It wound and twisted through 90-degree and hairpin curves for 35 miles between Bryson City and Deals Gap. Cut out of the side of the mountain, it overlooked the steep wooded gorge of the Little Tennessee. After the reservoir was cleared to expose slopes, looking down was a terrifying experience."[172] From such firsthand accounts it is possible to see why the larger recreational vehicles and mobile homes had to wait until after the war, when the superhighways were introduced, before penetrating these areas. Some of the packages coming in were of considerable number and size: a demountable bunk house was brought in twenty-two parts, a teachers' residence in eight sections.

A tractor and a few men would push a sectional house off a trailer on metal pipe tracks mounted on wooden beams (87). The two sections then were bolted together, the roof joint was capped with wooden rather than ceramic cover tiles and clamped down by large trunk fasteners, the entrance steps and canopy were added and the utility connections made.[173] The house was ready for occupancy. The cranes often used for postwar prefabricated house assemblage were nowhere to be seen. Thirty-two of the Hiwassee type demountables at Fontana were four-celled, and sixty-eight were two-celled. The two-celled dwelling was divided lengthwise into equal parts which, when pushed together, made a one-bedroom house fifteen by twenty-two feet. The four-cell version had the same

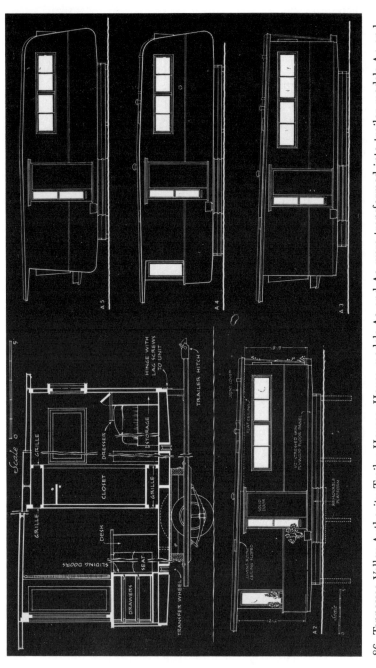

86. Tennessee Valley Authority Trailer Houses. House models A2 and A3 were transformed into trailer models A4 and A5. The latter were "streamlined." For road transport reasons, the roof overhang could only be at the front. The walls were of the "stress-skinned" plywood type, two inches thick. In the section can be seen small wooden transfer wheels for sliding the unit sideways when it arrived at the appropriate site. Drawing reproduced from *New Pencil Points*, July 1942.

module dimension but overall was thirty by twenty-two feet; like the postwar suburban tract house, it had three bedrooms and two baths. Half of the living room was used for a third bedroom.[174]

These cellular units told a great deal about America, about its cumulative triumphs and its longstanding deficiencies. The most striking difference between these houses and the TVA housing product of the 1930s was the lack of romance and rusticity in their appearance, particularly compared to the Norris Village environment and the rustic "vernacular" cabins within its adjoining "people's" parks. At Fontana there was no need to cushion "future shock" because the blow had already landed. The Fontana floor plans (89) were neat, empirical, and too tight, just as we might expect when depression and emergency conditions were operating on the government. In these minimal plans there was none of the expansive feeling that was bound to come after the war. There were no transitional hall or corridor spaces, no "open" planning of the type that overseas is often identified as American, thanks to the earlier distribution of central heating here. What *was* adequately taken care of, and was prophetic of the general American postwar intention (Wank had looked toward that goal at the start of the TVA), was labor-saving equipment designed to assist in tasks then done primarily by women — automatic plumbing, heating, and cooking. For use as summer cottages, which these cellular houses were destined to become after the war, the modules were pretty closed in and unromantic. They did not make any characteristic gestures toward nature, in the American romantic way, but rather just sat there, static. At the same time, there were none of the self-conscious theorizing and rationalization that had typified similar projects abroad, such as Corbusier's 1922 Citrohan houses or 1925 Pessac houses, or the Gropius and Wachsmann attempt at prefabrication for after the war, started in the United States in 1941 for the General Panel Corporation. TVA created an image of a remarkable prefabricated house, but somehow it did not receive the media exposure it might have. Moffett and Wodehouse, in their article on the TVA, "Noble Structures Set in Handsome Parks," exactly caught the drift of aspiration and default that has been so characteristic of large-scale American public undertakings, including the TVA. Ideal images were proclaimed but not always realized concretely:

An elegant set of International-style modular prefabricated houses was designed for the Department of War; it included one-, two-, and three-bedroom units

87. Sliding Half of a Two-Cell Hiwassee Type House onto Foundation Rails at Fontana. The unit next would be pushed left and another half joined to it. The parts were relatively simple, and there were no large boom cranes to put them in place, as would be prevalent in systems building after the war. Boom cranes were used later at Oak Ridge for setting up these TVA flattops. Four of five mountain carpenters could do the fitting as well and maybe better if given adequate tools, as represented by the carpenter's chests in the foreground. Photo courtesy of Tennessee Valley Authority Archives.

88. World War II Trailer House on the Way to Fontana Dam. The houses were made lighter so that they could be towed easily. The tentative streamlining was a little premature, however, since the trucks were not powerful enough to pull the units more quickly than ten miles per hour. The roads were twisting and uneven. The previous norm of the log cabin, with its thick wooden walls and close ties to the land, had begun to be modified with thinner and lighter components at Norris Village, and now at Fontana had been completely dematerialized. Great nervous energy supplanted longstanding stolidity. Photo courtesy of Tennessee Valley Authority Archives.

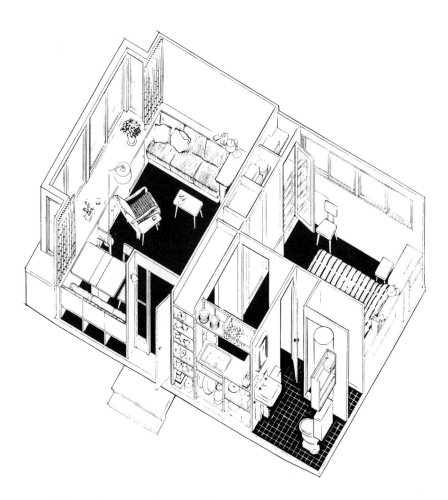

89. Bird's-eye Drawing of a Two-Cell Tennessee Valley Authority House. All the TVA houses were minimal types. Unlike the Norris Village houses, the Hiwassee type gave the impression of being a railroad roomette or ship's cabin. The abundance of storage units was its greatest surprise amenity. Much of the furniture and many of the fixtures came built in. The walls were of plywood, the furniture of pine and maple, and the floors of blue linoleum. No space was wasted on corridors or halls. The living-room bay was added to give a greater sense of space. Under it was even more storage. Drawing from *New Pencil Points*, July 1942.

together with suggested neighborhood groupings, but there is no evidence these were ever produced. As architectural works they represent a far higher standard of design than is generally found in military housing anywhere, their features responsive to efficiency and livability as well as standardization and production ease. Daylight floods major living spaces through generous windows that are shielded by appropriate overhangs to exclude summer sun, while exterior decks extend the necessarily limited interior space.[175]

The attractive 1943 E-2 drawings of these structures for the TVA by Seth Harrison Gurnee show domiciles less tight and efficiently minimal than the one- or two-cell trailer houses. Furnished with Aalto bentwood chairs, the Gurnee designs would have enhanced livability via improved modulation of light coming in through more generous windows. But, shockingly, in these drawings older one- and two-cell prefabricated units, visible through those windows, look as if they had just floated in from outer space, like square flying saucers. Norris Village's attachment to location had entirely vanished. The tentative vista was contrary to fact, since Fontana and Hiwassee literally had "no flat place anywhere, even along the river banks." The drawing takes on the character of a "noplace" place, reminiscent of many of the utopian colonies planted in this region in the nineteenth century. Adding to the sense of detachment created by seeing foreign objects on the broad, sunlit lawn in the Gurnee drawing is the constant awareness that, at any moment, buildings could be shipped out of Fontana Village to anywhere else. Thus the more generous Gurnee image, while superficially serene and placid, is charged with the acute anxieties of a transitory wartime. Such contradictions forecast the "throw-away" or "Kleenex" cities that planners after the war endorsed, out of a continuing conviction that a clean break with the "past" (really, with the basic conditions of the 1930s and 1940s) could and should be made. The American, like the European, saw prefabrication as a means of attaining precision and purity in a less than perfect world. But the American also found it congruent with his more usual fondness for moving on, improvising, making do, and becoming totally immersed in the moment; as well as with his impatience with the task of sorting out and identifying all the working parts of a cumulative culture, or establishing a "new cultural environment," as Arthur Morgan put it.

At Fontana, there was a suspension of the suburban mood (85). There were no lawns to mow and no fences to mend. Ornamental trees and bushes were absent. The single lot was abolished. The dwellings of the village, as distinguished from the dormitories of the construction camp, were organized in free-swinging loops, with a few cul-de-sacs adapting to the rapid rise in the slope of the land. Trailers and cellular houses had their own loops. The basic street motifs were ultimately English in derivation, but their variety and informality here could not be anything but American. The tension of the town plan, its staccato quality, in itself brought a sense of urgency to the scene. Most un-American of all in the photos of the time was the absence of a family car or two at every door, since gasoline and tires—not to mention auto parts—were severely rationed. The twenty-five permanent single-family houses faced on the school green three-quarters of the way from the west end, because the ground was flatter there.

Besides the more uniform prefabricated houses in the village were other categories of shelter that provided still more marginal and temporary effects. The earliest settlements, following almost immediately on the culvert pipes and the giant poplar tree, were streets of four-man tents that appeared in 1942 (90). Camp No. 2, at its maximum, had 216. Camp No. 1 had 13 at first. The structures were called tents, but the term wasn't entirely accurate, for they were really shacks roofed over with canvas. Wooden walls, attached to wooden platforms, were four feet high, lined with building paper, and covered with shiplap siding. They had screen doors for summer, and plywood panels were put in them in the colder months, when the whole interior was supposed to be kept cozy by a "Warm Morning" coal heater. Likewise appearing early in 1942 were the two-man Universal Trailers. There were 127 of them at the peak of village population.

Camp No. 1 was abandoned in September 1943. The best of the tents in that original group then were moved on to Camp No. 2 in the Gold Branch area, where they were upgraded slightly. The trailers were held in the Bee Cove area until September 1942, whereupon they too were moved to the farther, western end of the village, where more permanent services were attached to them. Fifty-five tents then took their place in Bee Cove. In September 1944 Bee Cove was entirely abandoned as a tent colony in favor of houses in the village.[176] A game of musical chairs went on among the various types of accommodation and service units. Even the twenty-by-twenty-foot jail was placed on skids, so that it could be moved according to sociological needs. In the same vein, Fontana house types really represented a gathering, a harvesting, of all the forms that previously had been developed at other TVA dam sites, such as Hiwassee or Pickwick. Even more striking in terms of assemblage was the fact that the standardized houses were not always uniform, because the panel forms sometimes were bought in small lots that would vary slightly in size. At that end of the construction spectrum, the ultimate development was the so-called "slabshack," which could be bought in Bryson City (pop. 1,612) and transported to the Fontana site for an overall cost of $125[177] — a price and a process that might be worthy of notice today by those looking for houses for the homeless.

Something more communal in appearance was a construction camp enclave (91) put up in Gold Branch Cove. There the two dining room and community buildings overlooked a common parking area. The community building was a modification of that at Hiwassee. Above that complex were six (later more) paired dormitories standing against the hill. Their placement was achieved by benching with a bulldozer. At the back, the dormitories were only one story; at the front, two. They were entered from the rear, either by ramps or stairs at the outer ends, or directly into a central washroom pavilion between the two wings. The pavilion acted as a pivot point, like a pintle in a hinge. The "split level" feature not only kept plumbing installation costs lower, but also meant that the wings could be adjusted forward and back: "The points of junction between the wings and the washroom were treated as flexible joints so that the building could be built along the contours of the slope." Hiwassee Dam had been located in a narrow and steep valley, and these dormitories were of the so-called "Hiwassee" type. What is most striking in photographs (91) is their design ingenuity, together with their utter impassivity. They appear black and white, frozen and isolated in time. The architecture was beginning to reflect the character of a calculating superstate, exercising its capacities with greater and greater skill, once the need was perceived clearly enough. They are strangely timeless, like the Manhattan Project nuclear factories, in an era of split-second timing. Yet there never was enough shelter. The category of private trailers was one indication of this fact. Not only did the TVA have to condone, within limits, rent gouging and rental of substandard facilities by private owners (when they were available), but it also informally acquired shacks itself on occasion. Its TVA Tipton Camp held as many as seventy shacks at one moment. Then, in reaction, things tightened up. "The erection of additional shacks was not permitted." Hazel Creek became a "sanitation nightmare" for the

90. Bee Cove Construction Camp, Fontana Dam. Bee Cove was the earliest construction camp. It was also the most transitory, being abandoned within a year. The tent on a box had been used first in American revival meeting camps, as along the New Jersey Coast or on Martha's Vineyard, where it evolved into a summer cottage. Even the ancient Egyptians had the form, which they took with them on expeditions. Photo courtesy of Tennessee Valley Authority Archives.

TVA: "Lax sanitation in isolated mountain spots was not the same as concentrated living in this [place]."[178] Workers had to be placed twenty miles to the east at Bryson City, in order to be ready when the dam flooded into its backwater. A tract was bought there for fifty-two trailer camp sites with absolutely no amenities except one washhouse for men and another for women.[179] These settlements held no permanent time, just as they were really nowhere in this permanent setting of a beautiful mountain area.

The buildings had been engineered to last a maximum of six to seven

91. Construction Camp in Gold Branch Cove, Fontana Dam. In this camp a town
square does start to evolve. The cafeteria is in the right foreground, with the
community building and library across the parking lot at the left. Because food was
rationed and every bit of physical energy needed, there was a sign in the dining room,
"Leave Your Plate Clean!" Outside, on the building front, there was another unusual
admonition: "Work or Fight!" In the Armed Forces the parallel message was "Shape
Up or Ship Out!" Against the Yellow Creek Mountains were six (later more) of the
Hiwassee type split-level dormitories, adjusted to the terraces scooped out for them.
Each pair of windows in them represented a 9 x 9 foot cubicle for two workers. The
dormitories were neutral in expression. America looked different and without choice in
wartime. Photo courtesy of Tennessee Valley Authority Archives.

years, as long as the war might last. Instead, many of them seemed to
endure forever. The village itself was leased in 1946, as soon as the war
was over, by a non-profit private distributing organization called Government Services, headed by Gen. U.S. Grant III, to become a lower-cost
people's resort.[180] But the reshuffling kept on: seventy-five houses were

shipped out to the Watauga Dam site at that time. The three hundred prefabricated houses still there began to be refurbished in 1957. In 1976, a handsome, eighty-room inn was put into service.[181] This kind of physical persistence in a choice location was less out of character than might be thought, despite the clear shift into a peacetime situation. The resort actually was the fifth camp on the site — having been preceded by the earlier war construction camp, and before that by two lumber camps, and before them by a copper-mining village.

Oak Ridge

Oak Ridge exhibited as much variety in habitation as Fontana Village, if not more. It drew in a number of manufactured housing types on very short notice. Its population was much larger than any village within the TVA. A major reason why the Clinton Engineering Works was located there was that Norris Dam, only sixteen miles to the northeast, had kept the Clinch River that came out of Norris Lake from silting or flooding. The easy-flowing Clinch water was needed for cooling the new processes being developed at Oak Ridge.[182] In addition to the Norris electricity, the Watts Bar TVA site to the southwest had been hastily developed with both steam and hydraulic output. So Oak Ridge was within the sphere of influence of the TVA, whether or not the latter had any inkling of what the Army's sudden demand for power actually was for.

In many ways, the TVA had been a last effort to resolve a dichotomy between the earlier idealistic vision of an earthy, agrarian, frontier society, to be made more prosperous in a particular place; and a succeeding dream of perfecting that same environment through the hygienic influence of a "good" civilian technology. Agrarianism had been used to fight industrialism in the earlier utopias. Regionalism and agrarianism, insofar as they had been preserved from the dissolving effects of the Civil War aftermath, were taken much less seriously after World War II. Despite the fact that it had been the early and lasting hope of the Progressives, the vision of generating hydroelectric power, too, began to fade after the war, in the face of the mind-boggling international implications of the atom. Yet availability of the older kind of energy had made the atomic experiments feasible. Ironically, it became necessary to revive the old-fashioned, heavily-polluting method of providing electricity by means of coal-fired boilers for steam-run turbines at Watts Bar (50). Because power came to Oak Ridge from Fontana Dam as well, the two-hundred- or three-hundred-mile transmission lines of the Tennessee Valley made it possible to reach into infinity with a sudden flash and roar. TVA's ultimate purpose had changed. TVA had been conceived for a well-defined district with local folkways and customs, but the atom and jet plane, in a sense, had made the whole world a village, without borders. Therefore, enriching the indigenous values of the Upper South now seemed much less relevant. To look outside the country, even into outer space, seemed the nobler and more productive cause.

Norris, Wheeler, Pickwick, Fontana, Watts Bar, and the other dam settlements counted as villages, even as hamlets, grudgingly built; whereas in size Oak Ridge became a genuine city, surrounded by barbed-wire fences (92) and watch towers, most unusual in America.[183] The concentration was inward, and that with a vengeance, but at first there still seemed little site planning and road and path delineation. The shops were overgrown houses, especially around Jackson Square, the first shopping center. Everything done in Norris Village could be submitted to press scrutiny and press release, but at Fontana and Oak Ridge publicity was avoided and censored. Because the employees of Oak Ridge included hundreds of young women, just graduated from high school, who performed delicate adjustments on electronic equipment,[184] the flow of life there had a curious juvenility, despite the grimness of the task. Since none of the buoyancy and vivacity could spread beyond the barbed wire, there was a sense of artificial containment. In both the Depression and the war, as well as in the high schools of the 1930s, camaraderie could be intimate and intense, but as at Oak Ridge, it was not the kind that could be expected to endure in a permanent community, because all the immediate events around mediated toward an eventual geographical dispersal. Beginning on 5 August 1943, distinguished scientists J. Robert Oppenheimer, Enrico Fermi, James B. Conant, Vannevar Bush, and Arthur Compton turned up at the nondescript Guest House that looked at, and looked like, the other temporary buildings in the city.[185] Such men could not be flamboyant, as the directors of the TVA occasionally had been. They had to be anonymous, like the buildings of Oak Ridge. Few recognized them as they moved through the endless plants (93). They proceeded under assumed names. Fermi had the entirely inappropriate one of "the farmer." Despite the apparently innocuous, colorless, quiet demeanor of the people

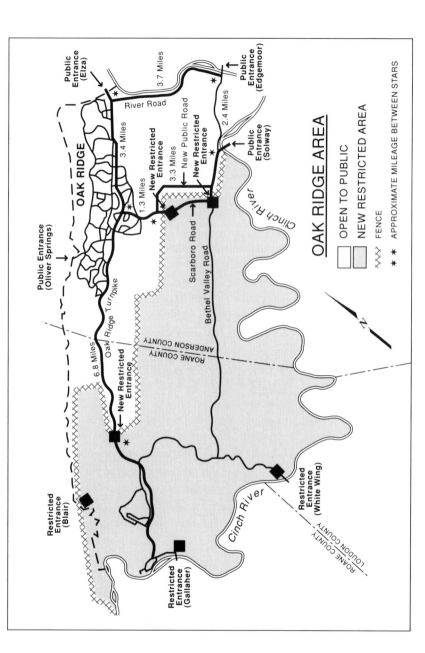

92. Map of Oak Ridge. The settlement is at the upper right for safety. The residential layout, under the guidance of the architectural firm of Skidmore, Owings and Merrill, approached an art form. Its very free manipulation of street motifs, compared to English or European examples, shows it to be indubitably American. Map reproduced courtesy of U.S. Energy Research and Development Administration, redrawn 1990 by University of Tennessee Cartographic Services.

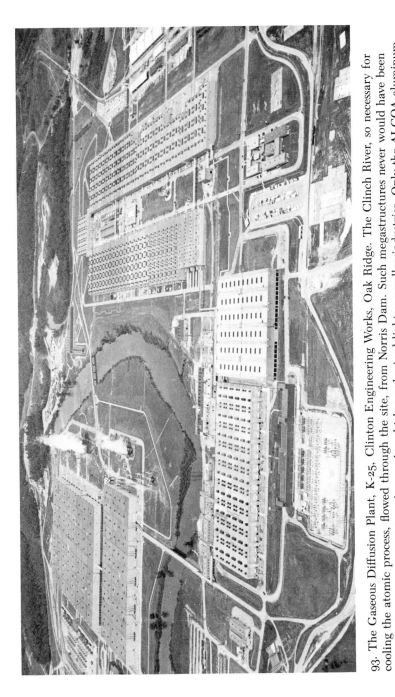

93. The Gaseous Diffusion Plant, K-25, Clinton Engineering Works, Oak Ridge. The Clinch River, so necessary for cooling the atomic process, flowed through the site, from Norris Dam. Such megastructures never would have been envisioned among the TVA's earlier goals, which emphasized lighter, smaller industries. Only the ALCOA aluminum plants previously had approached this scale. Photo courtesy of the U.S. Atomic Energy Commission.

and the buildings at Oak Ridge, there was a constant undercurrent of grimness (94–95). The anxiety felt and seen first at TVA's war-interrupted Watts Bar site grew greater because of the bigger secret being kept at Oak Ridge. Leroy H. Jackson, an early planner of Oak Ridge who later became chief of the Atomic Energy Commission's construction branch, recalled the nervousness beneath the outer layer of reassuring administrative confidence: "The hazard factor was a big one. No one knew too much about what was going to happen at Oak Ridge when it first began. . . . It was wise to have the residential area a couple of ridges away from the plant facilities on the very border of the community."[186] The town remained in the northeast corner. The first gas diffusion plant was in the opposite northwest corner, as far from the town as possible (92). Elbow room was less a desideratum now than the security found in Tennessee isolation. Landscaping appeared at odd outcroppings at Oak Ridge. Anything positive in that vein had to come from individuals acting independently—through flowerbeds or foundation plantings around the simplest cabins or ugliest trailers (96) in what amounted to a peaceful little green insurrection. Unless the protection offered by the ridges can be taken as a form of landscaping, the only other large-scale effect would be from seeding slopes and pruning trees.[187]

Every gesture of refinement was overridden by the general impression of an extensive military camp, although Oak Ridge certainly did not resemble the first of that type, the oh-so-orderly Roman *castrum* with its north and south, east and west streets. "The first impression one gets of the whole area at Clinton Engineer Works, however, is far from prepossessing. It seems to be nothing but a hodge-podge of construction camp, army base and war housing jammed indiscriminately together."[188] The main concessions to the original landscape took place in the preservation of as many trees as possible and in the orientation of the houses on the side of the main ridge so as to catch the best views south over the valley below.[189] The engineer, acting as a total, omniscient leader in the manner advocated by the Technocrats in the early 1930s, came into his own at Oak Ridge in the 1940s. As Johnson and Jackson have pointed out, however, there were in Oak Ridge early reservations about such undiluted authority.[190] In their first designs, Stone and Webster, the renowned Boston engineering firm, was not perceived to be taking advantage of the ground contours in the town area, and their houses were judged dull and routine. District Engineer Col. James C. Marshall consequently turned the townsite and housing responsibilities over to the architectural firm of Skidmore, Owings and Merrill (SOM) of Chicago and New York, with John Merrill the partner in charge; acting in cooperation with the John B. Pierce Housing Foundation and Leon Zach, landscape architect, who acted as landscape consultant. Tracy Augur, the TVA planner who had worked under Earle Draper to carry out the construction of Norris Village, was another consultant. Col. Marshall likewise did all he could to save the trees on the house lots.[191]

The residential area on the hill, with its service compounds below, then took on something of the configuration of an ideal European linear city of the 1920s and 1930s. This was a logical shape for it to assume at the outset, for it soon had to expand to the east and west along the Oak Ridge Turnpike. Oak Ridge was different from any European example, however, because it was so empirical, also incorporating informal English Garden City motifs, and because it actually got built. It ran along the side of an actual ridge rather than on flat paper. It was six miles long and a mile and a half to two miles wide. It was infiltrated with English superblocks and cul-de-sacs on a grand, highly animated, scale. The superblock not only conferred a flexible form to lay over the moving contours of the land, which varied foot by foot, but it also kept the cost of roads and utilities down, in the British manner, since the superblock pattern with cul-de-sacs requires less road for the same number of houses than a grid plan. The superblock scheme also was used so that the hill could be ascended as quickly and easily as possible without heavy regrading. The planners wanted roads with no slope greater than 10 percent.[192] With a grid, that would not have been possible, as San Francisco demonstrates. The cul-de-sacs were longer than they would have been in Britain, coming out like calligraphic sweeps of the pen. The architects at Oak Ridge were never short of land, so they easily could afford the extra spread of cul-de-sac and superblock, as the British with their land shortage and greater sobriety never could.

Whereas continental Europeans developing such a linear plan would reduce the choice of housing types and structural variety to an absolute minimum, in order to take maximum advantage of the potential of standardization, the American impulse seemed to be to show how many different plans, techniques, and housing types might be introduced through "standardization." The method had a Johnny Appleseed character to it. The prefabricated dwellings for Oak Ridge were manufactured in any

94. War Bond Drive in Front of the K-25 Atomic Works at Oak Ridge. A sea of grim faces looks up as a hero from the war speaks. The implacable, fenestrationless gaseous diffusion building furnishes the only backdrop. In the amorphous, crowded mass of laborers, with their strained faces, there was still an effort to retain individuality in the hats — the felt fedora; the railroad engineer's cap of mattress ticking, a popular item in Kentucky-Tennessee; the fuzzy wool hunter's cap; and even a few early metal hardhats. The war, following the Depression, to some extent submerged the individuality nurtured by the 1920s. Many American reform movements protested the recent repression of American individuality, and in the Tennessee Valley Authority that had been one of Arthur Morgan's greatest concerns. Photo by J.E. Westcott, reproduced courtesy of the U.S. Energy Research and Development Administration.

95. Stiner's Store. Lead Mine Bend. E.H. Elam interviews candidates for employment on Norris Dam, 8 November 1933. These were also grim times (94), at the depth of the Depression, but there is retention of prewar, pre-Depression innocence and individuality, too, in the playful test of arm strength on the scales. This activity reflected the ancient custom of log rolling, in which a "pull-down" to lift a log with a hand stick was the first move. The loops hanging from the rafters are mule collars, indicating that mechanization has yet to take command in the valley. There is no fear of becoming anonymous or swallowed in a crowd. Photo by Lewis Hine, reproduced courtesy of Tennessee Valley Authority Archives.

number of states—Tennessee, Michigan, Indiana, Georgia—and then shipped in. The first group of houses was started on 8 February 1943, with Jackson Square as its service area. Three thousand houses were built then, organized into three neighborhood units with their schools.[193] On 27 July 1943, the first house was occupied. The houses (97) were built at a slower pace at first. They were of sloped-roof design and made up of four-foot-wide panels conceived by SOM in cooperation with the Pierce Housing Foundation. They drew their trade name, "Cemesto," from the material they were made of, Celotex-Cemesto from Chicago, which had been used most recently in war housing for the workers at the Glenn L. Martin airplane factory in Baltimore.[194] The chief reason for a sandwich panel was the wartime shortage of wood, but the practice of using synthetic materials had grown up in the 1930s. In the early 1940s there were still high hopes for plywood and fibre board, as well as for steel sash, glass blocks, terra cotta, black glass, and chromium. Celotex-Cemesto consisted of three layers of fibre board between two layers of asbestos, today considered a lethal material. The five-ply lamination was highly praised for its insulating qualities.

The second building campaign brought a wave of five thousand TVA "flattop" residences, in a drive that peaked in August 1944. By then the rate of construction had picked up to thirty or forty houses a day. The Cemesto houses had a production rate of only one every two hours, so slow that it brought in the TVA flattop houses. The number of construction workers at Oak Ridge by then had peaked at fourteen thousand.[195] The rate of construction was made possible not only by the large work force, but also by its application in the sequential assemblage of prefabricated houses by the team method. Two forms of time saving were joined, pre-assembly and systematic motion. The application of assembly-line methods to housing was first thought up by the Pierce Foundation: "House construction was divided into a series of specialized operations, each manned by a separate crew of workmen. One job was done on a number of houses at a time and when the first crew was ready to move on, another crew took over. On a staggered schedule with short gaps between operations to permit flexibility, this technique proved highly efficient."[196] This was the basic construction method used right after the war in Levittown on Long Island, New York, highly praised then for its originality and ingenuity. The Levitts had learned it while in military service.

The third campaign of house construction (omitting mention of the thirteen thousand dormitory spaces, five thousand trailers, and the sixteen thousand hutments and barracks) began in late December 1944, after a lull in November. "Flattops" could not be brought in fast enough, so an Oak Ridge plant for prefabricating fifteen hundred more units was established. From January 1945 until the first week in August, this factory, crudely copying the TVA model, turned out the V and S flattops

96. Flowers in Front of a Wartime Trailer, Oak Ridge. Individualism, with its fondness for lingering innocence and nature, tried to survive in the atomic city through a miniature green revolution in front of a dark, anonymous, standard trailer house. The wartime minimalization of standardization helped to bring on the strong reaction to regimentation in the afterwar toward the purportedly individualized suburban lot and house, with its trees, bushes, and picturesqueness. Photo courtesy of U.S. Energy Research and Development Administration.

disrespectfully called "V and S Chicken Coops."[197] In early August, after
two atom bombs had been dropped, all further contracts, including one
for 475 V and S flattops, were cancelled. But by September 1945, a city
of seventy-five thousand people had been built.

Oak Ridge carried on the "tradition" of the TVA, in being surrounded
by a special aura suggesting implications for a country or a world far
away. Enormous resources had been committed to short-term objectives
by a democracy at war, while in peacetime in a depression there never
had been enough money to project a long-term goal. Had the long-term
implications of the atomic initiative been realized at the time? Was this
democracy capable, in war or peace, of formulating an imagery adequate and ideal enough to sustain the nation? Had time been allowed
to go by too soon for American institutions?

Timberlake City Project, 1967–75

Timberlake City on the Little Tennessee River at Tellico Dam was projected for fifty thousand inhabitants,[198] ten or more times the population
of Norris or Fontana villages, but also twenty-five thousand less than actually had been achieved at Oak Ridge. As a new town, Timberlake City
would have had extra relevance for the TVA in that it could have been
built near Knoxville (98), where so many other of TVA's "firsts" had been
placed. These included Norris Dam and Village; the Watts Bar Steam
Plant; Oak Ridge; Alcoa, the aluminum manufacturing town; and the
dams of the ALCOA corporation on the Little Tennessee River, which
had culminated in the building of Fontana Dam by the government when
the need for aluminum became critical. Fontana had signified the first
effective cooperation between the TVA and a larger industry. Timberlake
City was another enterprise in which business and the TVA might have
cooperated. Boeing Aircraft Corporation was interested in Timberlake
for some time.[199] As an attempt by the railroads to create new businesses
in the depressed area of northeastern Tennessee just before and after World
War I, Kingsport, Tennessee, had not been so different. The buildings
in Timberlake were to be financed privately. The site for Timberlake,
in Loudon, Monroe, and Blount counties, was seen as a remaining pocket
of poverty, in which there was an income of only 68 percent of the national average.[200] The eastern portion of Tennessee still lagged economi-

cally. Outmigration of the young had resumed. "While the tri-county area has many features which lend themselves to industrial expansion, its economy has been characteristic of much of the Appalachian region of which it is a part—lagging behind the remainder of the state, the Tennessee Valley and the Nation in income and nonfarm employment."[201] Timberlake was a scheme born of past postponements. Tellico Dam and Lake, which the town of Timberlake was supposed to rim, had been proposed as early as 1943 as part of the Alcoa-Fontana Dam complex, but because of the war, materials had been too scarce. In 1958, TVA Chair-

97. Celotex-Cemesto Houses, Oak Ridge. These houses were prefabricated out of experimental materials, including asbestos. The curving streets, with the houses rotated slightly on the lots, and the peaked roofs, were minor concessions to the picturesque prewar suburb. Photo by J.E. Westcott, reproduced courtesy of U.S. Energy Research and Development Administration.

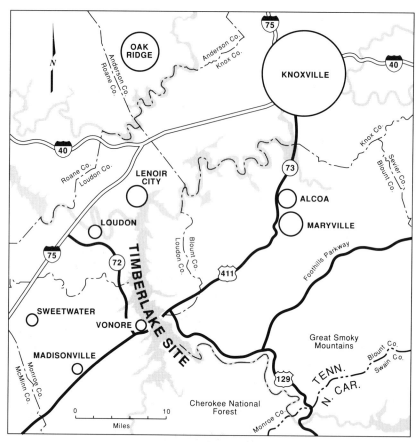

98. The Timberlake Community Site in Relation to Other Key Communities. Much of the previous epic of the Tennessee Valley Authority centered around Knoxville. The linear city contrasts with the concentric cities. Map courtesy of Tennessee Valley Authority Archives, redrawn by University of Tennessee Cartographic Services, 1990.

man Wagner had called for a restudy of the minimal acquisition land policy that had come out of the Jandrey Report of 1944.[202] That move opened the way for visualization of a town on the shore. Norris, the first dam of the TVA, had no nine-foot channel and no boat locks. Knoxville had the channel, but by 1970 its waterfront investment still was less than $4 million, and only 199 people worked on it. At Chattanooga, the investment stood at $318 million, and waterfront employment was 11,800.

The conventional explanation for that unfavorable contrast between Knoxville and Chattanooga was that Knoxville had a "steep and rocky topography close to the river."[203] Muscle Shoals, Guntersville, Decatur, and Scottsboro, all in Alabama, and Calvert City near Paducah in Kentucky, were to become other successful transshipping points. Guntersville had sprung to life on a peninsula with the advent of its new lake.[204] Timberlake was intended to fill the gap for Knoxville, which was only twenty-five miles or a forty-five-minute drive away. Timberlake would be a satellite city for Knoxville (98). The vicinity had offered several demonstrations in the past of how the rest of the country might benefit from a different species of settlement. That sort of demonstration, of course, had been one of the intents of the TVA's originators, mentioned by Draper and others. South of the Ohio River, land was cheaper and less densely inhabited, labor plentiful, and natural resources less developed. Space existed in which TVA might test its mettle. World War II had made this an area more technically proficient.

In that context, Timberlake City could have represented a culmination on the Little Tennessee of what Henry Ford once had proposed for Muscle Shoals — a freely shaped linear city (99) to be a workers' settlement that could have brought about enough amenity and efficiency to support whatever business or industry was next contemplated for the region. At the last dam on the Little Tennessee, just before it enters the big Tennessee, there would have been a greenbelt to separate the city from the traffic of the main river and from Lenoir City. South of that would have been the residential district. A little over halfway down the lake would have come five thousand acres of industrial zoning,[205] put in place because of the existence of the Louisville and Nashville Railroad track and Route 411, as they passed over Tellico Lake from northeast to southwest. That patch of zoning was supposed to provide nine thousand jobs.[206] Small hills or "knobs" were intended to help insulate the residential area from industrial activity. Up to the industrial docks would be as far as barges could go on the entire river system. Farthest south would be the open grounds and recreational spaces, leading up to Chilhowee Dam. Beyond would begin a new scenic gateway to the Great Smoky Mountains National Park and Cherokee National Forest, a background of knobs and mountains reflected in a lake. The locational sequences would take thirty-three miles.

By using the waterfront in some reasonable order, Timberlake City would have fulfilled the principle predicated for the shores of the first

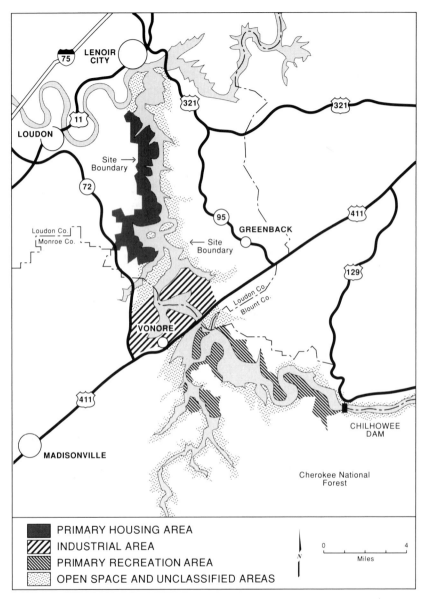

PRIMARY HOUSING AREA
INDUSTRIAL AREA
PRIMARY RECREATION AREA
OPEN SPACE AND UNCLASSIFIED AREAS

99. Spatial Use Concept of the Timberlake Community Site. At its southern tip the city merges into the national park and forest system. The previous existence of the highway and railroad lines caused the industrial zoning to be placed in the center of the site. Map courtesy of the Tennessee Valley Authority Archives, redrawn by University of Tennessee Cartographic Services, 1990.

lake, Norris. The logic and system of the waterfront, protecting the city from sprawl, would have kept the breadth of the settlement from being more than two to five miles. Of a total of 38,000 acres to be acquired, 18,600 would have been covered by water.[207] Tellico Lake would have been a mile wide at the city and narrower as it moved toward the mountains. Tellico also would have resembled Norris Lake in having many inlets and peninsulas. On the promontories, the city itself would have

100. Model of Timberlake City, 1967. The arabesque linear city on a lake had a long provenance in theory in the Tennessee Valley, through the ideas of Henry Ford, Benton MacKaye, and the founders of the recreation resort areas, particularly along Kentucky Lake. Somehow, the concept never came to fruition, even at Timberlake. The most radical reform at Timberlake would have been the careful control of shoreland usage, a sharp turn in previous TVA policy. The "slab in the park" rhythm of high and low rise buildings would probably not be used today. Photo courtesy of the Tennessee Valley Authority Archives.

grown up. Often a disadvantage of such artificial lakes is that the shores look abrupt and untutored (100). The banks at Timberlake would have been carefully modeled, with features such as marinas, restaurants, and stores thoughtfully inserted. James Gober, the planner, brought out the view — not always adhered to elsewhere by the TVA, although the spirit was still there — that "our job is to see to it that this growth is planned and directed in such a way as to contribute to rather than distract from the scenic values nature has provided."[208] Timberlake was intended to have high-rise apartments and group houses (100) as well as single homes. In its linearity, it would have resembled postwar European new towns such as Vällyngby near Stockholm, Cumbernauld near Glasgow, or Thamesmead near London, except that, being American, its distribution would have been looser and less formal. None of those other towns had a long TVA lake to face on or mountains in the distance. Timberlake's scenic values and informal organization probably would have resembled the privately developed Tapiola outside Helsinki, which has handsome views over water, high-rises, and single houses to buy, and the greatest variety of communal amenities.

There were to be a city plaza and two subplazas, each the focus of a village with its own high school, parks, and marinas. The villages would be made up of neighborhoods.[209] Planning devices of all kinds were accumulated to be applied and then to be extended. The linear city was at the heart of European urban idealism; the cul-de-sac and superblock derived from the British Garden City. The neighborhood diagram, with a school at its center, however, came from American sources in the mid-1920s; and the green strips separating neighborhoods and used for bicycle and pedestrian lanes, with their passes beneath the roads, came from the Stein and Wright scheme for Radburn in New Jersey in the late 1920s. Later schemes inevitably attemped to sum up earlier ideograms. Timberlake could have been a significant symbol marking a cumulative effect that could not have occurred in the preceding war and depression years. It also would have reiterated that in this portion of the country utopian colonies had been posited strongly in the past. It would have offered one new town in the South comparable to Reston or Columbia around Washington, D.C. The hope of an American El Dorado, New Jerusalem, or golden dream city had lingered in the American imagination since the Appalachian Mountains had been crossed. The fact that Timberlake was proposed in a bounty time acted against it in its gestation period, 1967-75. From Norris Village on to Fontana Village and Oak Ridge, the evidence was ample that only in times of extreme crisis and disarray would the federal government come forward to support the building of a community, and even then the support would be inadequate. The project might better be undertaken privately, with guidance from TVA. But the difficulty was that the other, earlier communities generally had taken, at most, two or three years to build, while Timberlake was scheduled to require a quarter-century,[210] at a low rate of capital redemption that only a government normally could tolerate. European new towns were financed in that way.

In 1936 the principle of "multiple usage" had been enunciated for the TVA, but in the end that dictum came to apply more to its technology than its settlements. World War II made technology more the final arbiter, because it seemed to have an immediate solution to every problem. And the quick solution was what was sought by the TVA. The other difficulty with the realization of Timberlake came from the potentiality of the recreational area up the Little Tennessee. Some preferred to visualize that as a wilderness, since the area appeared to merge rapidly into picturesque knobs and mountains, heralding the national park and national forest beyond. The thought that the middle of the country was the natural site for the appliqué of dams and other great public works onto the primeval landscape, the machine in the garden, was losing favor. The feeling for untrammeled nature was reasserting itself. After World War II, environmental interest began to spread east from California. In the 1960s, the Vietnam War added to the momentum, along with retreat to communes and religious colonies. Fewer cues would be taken now from the Northeast and New York City. U.S. Supreme Court Justice William O. Douglas, originally from the Northwest, represented the newer, more primitive, all-stops-out way of exerting pressure on established government to save what he believed, apparently on first sight, was a last wilderness stream for fishing for brown trout. Long before there was any thought of Tellico Lake, the trout in the Little Tennessee had had to be restocked constantly. Even with all the ALCOA dams already intervening, however, Douglas preferred to regard the Little Tennessee as a "wild river." He enlarged his objections to include the drama of bureaucracy interfering with the farmer's more worthy lifestyle (the notion Lilienthal and Harcourt Morgan had also used); the flooding of rich bottomlands (an old, down-home type of protest in the South, that H.A. Morgan and

McAmis also had employed); and the fact that Cherokee Indian village sites would be innundated, an objection first raised for archaeological purposes with the great irrigation systems of the West and Southwest. Some of Douglas's companions interpreted the possible building of Tellico Lake and Timberlake City as comparable to the exploitation of the populations of the TVA coal-purchasing districts, which had led to the gashes of strip mining. Had an antitechnological age set in? Had ghosts from the 1960 film *Wild River*, who were also ghosts of earlier defeats, been reincarnated? The heroine of that scenario had liked things "running wild" too. She was "agin dams of any kind." Would America soon see a revival of agrarianism and individualism and be returned to its legitimate wildernesses? The display recalled the 1930s, when an equally long-distance alliance had been shaped between the "Brain Trusters" of Washington and New York and the long-forgotten inhabitants (folk) of the Tennessee Valley. This time, however, the reformists were somehow closer, more immediately on the scene. As McDonald and Muldowny were to put it in 1982:

> In a curiously Hegelian sense, TVA's development guidance, its success as the Valley's educator, and its water management program and recreational advantages have attracted newcomers who have swelled the Valley's population as 'yesterday's people' have left. These technologically skilled outsiders have in many respects formed the nucleus of those who today would halt TVA's continued expansion in the name of environmental and ecological interests.[211]

All well and good, except that it meant that Timberlake City would be increasingly squeezed between two poles, the technological and the arcadian, in a way that was really irrelevant to its being, since it had already tried so hard to offset the disadvantages of each and at the same time capitalize upon their major attributes in a new pattern of settlement. David Lilienthal had said that he saw no future in social "uplift" in the valley. That dismissal had turned into a self-fulfilling prophecy, insofar as housing and new towns were concerned. But the recollection of the original intent was still present and pulsating, although considerably blunted and muted. More important than any social uplift that might have emanated from Timberlake City would have been the sense of closure, of fulfillment, that might have accrued with more consistent land use on the shores. From the main Tennessee River on the north down to the national forest and scenic reserves on the south, Timberlake might have become in actuality the "seamless web" that H.A. Morgan once had advocated for the TVA geography. In a measure, it could have compensated for all the evasions, along the main river and on the artificial lakes, of the onetime chance for optimum development of shoreline amenity. The 33 miles of Tellico Lake shoreline could have made up for other, earlier defaults and divestments along the 680 miles of the whole Tennessee Valley; for settlement potential never capitalized upon at Muscle Shoals in the 1920s, at Norris Lake in the 1930s, and at Kentucky Lake in the 1960s; and even for the failure of Benton MacKaye's double-diagrammed 1933 extension of the TVA, running from Chattanooga to Moosehead Lake, Maine, along his 1921 Appalachian Trail (13–14).

6 | The National Resources Planning Board as a Bigger Image of the TVA

During the 1920s, when Franklin Roosevelt began to obtain status as a political figure, and then, after a long struggle back from an attack of infantile paralysis in 1921, was elected governor of New York in 1929, the conceptual foundations of American life were rapidly crumbling and falling away. Part of American innocence was lost after World War I, diluting the earlier Progressive idealism, and more innocence would be lost in World War II. The 1930s were yet more unusual within the first half of the twentieth century in that they revived visible idealism through public constructs such as the Tennessee Valley Authority and the National Resources Planning Board (NRPB). The latter was not founded before, or after, the former, as is so often thought, but, like Technocracy, the NRPB was created in the same year, 1933. Also the first year of Roosevelt's first presidential term, 1933 marked the low point of the Depression. The fact that the TVA originally took great interest and pride in the comprehensive planning of the Tennessee land and in the intense refinements of its own architecture suggests that the TVA was a last-ditch effort. America normally was not eager or able to push on in such an open, determined way to provide ready definition or self-identification in the visual arts. Aesthetic vision usually had to be imported. In the 1920s and 1930s, design leadership often had gone to architects not originally trained or even born in the United States — Saarinen, Neutra, Schindler, Gropius, van der Rohe, Urban, and TVA's Wank, among a number of others. What was most inconsistent among the many visions of the time was that, although they could be briefly projected as images providing for a far-distant future, they often were implemented as if they were only crash make-work programs, good for three or four years at the most to alleviate immediate deprivations. At TVA, that go-stop psychology peaked during World War II. This split in the motivations and phrasings of various TVA projects led to a later confusion between an evaluation of the TVA dams as monuments for the ages, and the way in which communities, roads, and other support facilities tended — despite the best efforts of some of the TVA architects, landscape architects, and planners — to be treated as *ad hoc*, spur-of-the-moment endeavors. A more ideal outcome would have resulted if those two portions had been treated equally. Emergency ways of seeing and executing tasks, however, became habitual, almost addictive, after a while, and during Roosevelt's administration, the emergencies never ceased. It might be claimed that the TVA and the NRPB together fostered the first sociopolitical ecological overview of the country, based on resource management, both human and natural. But even the TVA itself after awhile became increasingly monocular in its vision and spastic in its actions. Subsequently it became the target of extraneous political and economic pressures. It was a "planned area" entirely without a masterplan. Idealistic young planners coming to work for the TVA after 1938 were taken aback to find that it had no masterplan.[1]

Franklin Roosevelt took problems in at a sweeping glance, starting from the top. But there were also continuities among his improvisations. As early as 1931, two years before he became president, he had declared in a speech to the New York Agricultural Society that national planning was indispensable for the future happiness of the American people.[2] After he became president, his uncle, Frederic Delano, carried on by reporting at the first meeting of the National Planning Board on 30 July 1933, "Fundamentally, intelligent nation-wide planning presupposes as a primary consideration, the thorough study of our land resources, the topography, soil conditions, etc., so as to assign land areas to the uses for which they are best suited for man's benefit."[3] The actual operative statement connecting the TVA and national planning, in theory at least, came in Roosevelt's address to Congress in April 1933, recommending the instigation of the TVA, in which he said, "If we are successful here we can march on, step by step, in a development of other great territorial units within our borders." Visions couldn't get much bigger. Even George Norris, the senatorial initiator of the TVA law, by 1937 evidently had overcome his own preoccupation with electrical rates sufficiently to call, in the grand manner of FDR and his uncle, for "enough TVAs to cover the entire country."[4] Thus national planning was in the minds of several of the creators of the TVA from the outset. The forerunner of the NRPB, the National Resources Committee, in 1935 attempted to fashion a framework to combine the TVA and the eventual NRPB, declaring that "the ultimate test of a regional planning and development scheme is its ready adaptability to national planning."[5] Further, the group with the broadest horizons had the readiest access to the president then. The NRPB had staff-unit status within the executive office of the president. Until Pearl Harbor in December of 1941, Roosevelt met with the NRPB members monthly to go over their reports and agenda.[6] The board tried to devote itself to long-range physical planning in the face of all the emergencies, including those of war. It set up regional offices and encouraged the founding of state planning agencies (the number of which increased from three in 1933 to forty-five in 1935). It studied drainage basins as primary entities.[7]

Roosevelt's intended location for the next valley authority was the Columbia River Basin, which included parts of Oregon, Washington, Idaho, and Montana. He experienced the area firsthand at the dedication of the Bonneville Dam there on 28 September 1937. His speech the same day at the inauguration of Timberline Lodge, the great WPA project on Mount Hood in Oregon, made it evident that he was attracted by the beauty of the locale, and by the fact that so little of it appeared to be occupied. It had possibilities. Like the Tennessee Valley, it was an out-of-the-way, less inhabited region, kept aside by geography and previous history—something of a new, if neglected, hope in terms of the whole continent. The Columbia Valley, in at least one respect, was to become a second Tennessee Valley during World War II, because of the establishment there of the Hanford, Washington, Engineering Works, where the second atomic bomb, dropped on Nagasaki, was manufactured. Because electricity was abundant, heavy aluminum production was intended from there after the war. This installation was a companion to the Clinton Engineering Works at Oak Ridge and was placed on the site because, in its isolation and ready availability of hydroelectric power from new dams, it resembled the Tennessee location. But in 1937, at the Timberline Lodge and Bonneville Dam dedications, Roosevelt could not yet have envisioned an atomic bomb manufactory. That object and the means of its accomplishment were totally unknown. Rather, interest centered on Roosevelt's recollection of how, when he was governor of New York, he had wanted to harness the Saint Lawrence River for electricity and simultaneously prepare for the decentralization of cities. The establishment of the Rural Electrical Authority, taking place in the middle of the decade under Morris Cooke, would provide the final factor required for such a dispersion. A study of the Northwest for regional development purposes was being prepared by Professor Barrows of the University of Chicago.[8] The National Resources Planning Board already had been consulted about "The Columbia River development."[9] Oregon, Washington, Idaho, and Montana had their own Pacific Northwest Regional Planning Commission, as did New England.[10]

River valleys other than the Columbia and the Tennessee faded in and out of view during the Seventy-fourth through the Eightieth Congresses (101). They included the Missouri, the Upper Mississippi, the Arkansas (in many ways the most logical), the Merrimack, and the Wabash in Indiana. The Connecticut was the one most fought over, in an area of high population density and entrenched power interests — further demonstration of why less populous valleys should be sought out for such experiments. The Wabash would have created another lake around Evansville by reaching up the river from there, signaling the penetration of the TVA into a northern state, since, by reason of proximity, it likely would have been attached to the TVA. The Pecos, Platte, Red, and Sacramento rivers were looked into. Eventually the more southern rivers of the Savannah and the Rio Grande were considered.[11] The Tombigee, mostly in Alabama, recently became a canal between the Tennessee and the Gulf.

The number and geographical size of all these proposed river authorities, if ever put into effect, would have required a national plan for protection and utilization of resources of all sorts. Such a plan would have made a different America. Their effectiveness and unity would have depended upon following a single stream and looking first at the single product of surplus electricity. Would this fact have given each enough staying capacity for the future? Although the Tennessee Valley itself had an unusual amount of annual rainfall, after the single-minded technical efforts at the start of World War II it had proved more or less impotent, unable to furnish enough of the "cleaner" power of hydroelectricity, and so had had to resort to coal- and atomic-generated steam, which in turn produced strip mining and acid rain. The TVA experience suggests that such a strategy later might require another technology to solve the problems caused by the first, or rather the heavy demands suddenly placed

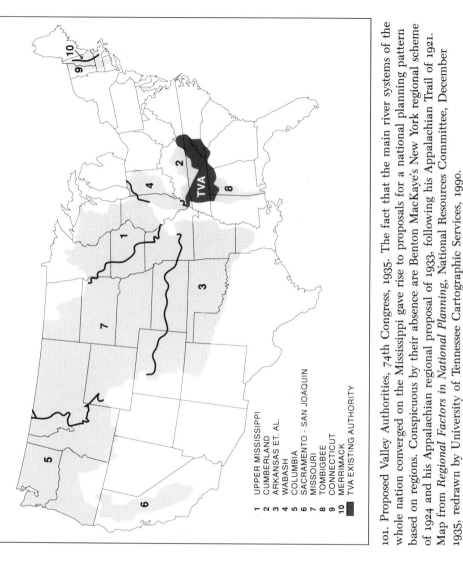

101. Proposed Valley Authorities, 74th Congress, 1935. The fact that the main river systems of the whole nation converged on the Mississippi gave rise to proposals for a national planning pattern based on regions. Conspicuous by their absence are Benton MacKaye's New York regional scheme of 1924 and his Appalachian regional proposal of 1933, following his Appalachian Trail of 1921. Map from *Regional Factors in National Planning*, National Resources Committee, December 1935, redrawn by University of Tennessee Cartographic Services, 1990.

1 UPPER MISSISSIPPI
2 CUMBERLAND
3 ARKANSAS ET. AL.
4 WABASH
5 COLUMBIA
6 SACRAMENTO - SAN JOAQUIN
7 MISSOURI
8 TOMBIGBEE
9 CONNECTICUT
10 MERRIMACK
▮ TVA EXISTING AUTHORITY

upon it. There seemed to be no certitude in technology. Having several states in a row along a river could lead to jurisdictional squabbles, as the 1922 Colorado River compact among states had illustrated. If the ultimate role of the federal government as a guide was not better defined than it had been at the outset for the TVA, and better fulfilled thereafter by yielding less to local interests, would there be any hope for other and later valley developments? The brilliant visions of figures like Roosevelt and Arthur Morgan may not have carried enough weight in the actual execution over time to offset the mores and interests of the Bureau of Reclamation, the Army Engineers, and the Rivers and Harbors Group, all of which had much longer relationships with Congress and therefore were able to block any newer agency's plans for waterways. Roosevelt understood the need for dreams to keep a democracy alive when confronted with the rapid and highly visible successes of fascism and communism abroad, but in a time of severe economic deprivation, he had little cloth to cut his new pattern from. *The Commonweal* summed up the anti-Roosevelt mechanisms working against the TVA and the NRPB simultaneously in an editorial of June 1943, upon the abolition of the NRPB. The editorial identified the move as an act of vengeance by FDR's enemies in Congress, carefully calculated not to harm the war effort. It was as if the 1940s no longer could understand the 1930s: "And so without fanfare or lament dies the present hope of an intelligent, self-disciplined America based on an economy of abundance." This was a point that Thomas Jefferson also had made often — that democracy could not flourish in the absence of the cultivation of both intelligence and self-discipline. The fact was, of course, that the conservatives wanted to restore the "economy of abundance," absent since the Crash of 1929, in their own way; and they correctly guessed that that goal would be in sight with the end of the war. At that moment, however, there was still an "economy of scarcity." The war, in effect, had erased the gnawing memory of the Depression. The businessman began to feel that his lost social priority could be recovered. "Congress has served official notice that it has flunked the exam on the lesson of the Great Depression,"[12] *The Commonweal* asserted. The immediacy of the war crisis had obscured the longer and steadier premonition of the TVA, dating from the Reconstruction era of the 1870s and 1880s and the Progressive Movement of the 1890s.[13] The magazine felt that, with such a noble cause as national planning, the transition back to laissez faire should have been less abrupt and mean-spirited. By 1943 it was beginning to be apparent that the outpouring of guns, planes, and tanks from America would be likely to turn the balance of the war in the Allies' favor. The paralyzing fear of Tojo; Mussolini; Hitler; and Stalin, who at first had joined with the Germans; began to abate. It was significant that the attack on the NRPB was first led by "Mr. Republican," the senator elected from Ohio in 1938, Robert Taft. He was chairman of the special postwar planning committee of the Senate, and as such wanted no continuation of what he considered the soft-headed, sentimental reforms of the New Deal. His first speech to the Senate, on 20 February 1939, was a "vigorous, fact-filled assault against increased appropriations for the Tennessee Valley Authority."[14]

The opposition argument was always couched in those terms: that Roosevelt, like Arthur Morgan, was not hardheaded enough, had a feverish imagination. Such characteristics obviously would eliminate any obligation for genuine long-term concern about the plight of the South since the Civil War. When Taft, to everyone's astonishment, took up the cause of public housing in 1948, it was for statistical and mathematical reasons mostly; he had become convinced that, in the face of postwar inflation, 10 percent of Americans no longer could afford mortgages, but he was not convinced that they lived in squalor damaging to dignity and well-being in a democracy, the New Deal view. The too-seductive (as he saw it) picture of a land and people in distress held no sway in his blueprint of the future. Taft described the NRPB reports as "a combination of hooey and false promises" that would lead the postwar era astray; while his colleague, Senator Tydings of Maryland, observed that the NRPB was derived partly from "socialism and partly the product of a dangerous imagination."[15] Roosevelt, the man with the "dangerous imagination," then made a strategic withdrawal under heavy fire, declaring that he did not care who did the job nationally "so long as provision was made for coordinated planning."[16] In the interest of accomplishing other goals in the midst of a distracting war that the United States might not win (strange as that fear seemed later), Roosevelt abandoned his boldly if somewhat flimsily constructed apparatus for national planning, the "jungle-gyms" of the TVA and NRPB, which previously had been somersaulted upon by his New Deal appointees. The battle over who was to dominate the postwar political scene had been joined, and the conservatives, not the Conservationists, had won out.

The term *multi-purpose usage*, which, when applied to the landscape by individuals such as Gifford Pinchot and Arthur Morgan, had had structural and modeling meanings for the environment, in the hands of later leaders turned into a tool handy in promoting insistent local interests and allowing one TVA program ascendancy over another. Although the TVA dams represented a whole series of carefully-designed monuments of democracy, when the TVA was being threatened most seriously, the same high-calibre consciousness was not always applied to the way in which communities and landscapes were treated. As one drives through the region today, there is little enough display by signage and other means of what the TVA actually is — a continuous network (67) and an unfolding presence. Each dam has to be searched out, whereupon it stands forth largely as an isolated object in a quiet and removed milieu. By now the older dams share some of the romance and nostalgia that attach to ancient southern artifacts, such as the pillared plantation in its lush natural setting. The prospect of there being a continuous, broad ribbon of civil order throughout the length of the Tennessee Valley hardly appears to exist anymore, if it ever did. Yet the continuity of the system, with several variations in the design of the dam complexes (67), was what first fascinated other Americans and foreign visitors. As James Dahir expressed the general impression in 1955: "A number of people who have journeyed to the Valley seeking a paradise have returned to their homes reporting disappointment. Some of the official literature of TVA and other Valley organizations may be in part responsible."[17] The institution had become so locally self-effacing that its actual presence in the milieu was placed in some doubt. In the broader landscape, recollection of how some individual farms had been transformed from erosion-ridden waste to more carefully husbanded property (102–104) was perhaps the major image viewed with respect, although most of that work had been accomplished through the university extension systems. The infrastructure required to draw together and transform the whole 680 miles of the system somehow never got built. The road network in Tennessee had been very poor to begin with. Howard Odum, the southern sociologist who had helped to worry the TVA into existence before 1933, by 1938 had begun to wonder whether it was becoming too narrow and shortsighted in outlook. He noticed that there could be a "possible overemphasis upon economic and material matter and upon engineering and technological processes as against human and cultural fundamentals."[18]

Pinchot's program for forests that would yield profits under scientific

102. Farm Land in Jefferson County, Tennessee, 22 October 1933. There were three types of erosion — sheet, shoestring, and slip. All are present in the photo. The greatest difficulty came from plowing the hillsides, which then caused landslides from row crops. When planted annually with the same crops, such patches would be arable only for a few years. Clearing appears to have been begun on the knob in the near distance. The ubiquity of the desolation would bring on apathy among its population. Photo by Lewis Hine, reproduced courtesy of the Tennessee Valley Authority Archives.

management, and the theme of the national park did have their influence on the TVA, as did the later image of the wilderness preserve. The TVA added two more suppositions to that repertoire: first, that the machine, through careful design, could be fitted into such a naturalistic matrix; and, second, that a people could be adjusted to both ingredients at once. Engagement with electricity could be mixed with disengagement from overuse of the agricultural land. The otherwise disengaged "folk" of the Depression would be found and treated best in Tennessee. Subsequently, the optimum locational conditions and specific testing

103. C.J. Wild Farm, Bush Creek Demonstration Area, Madison, North Carolina, May 1945. Slightly over a decade later than the scene in (102), the land is now better tended, with contour plowing, rotation of differentiated crops, and cover plants such as alfalfa and lespedeza. The steeper hills are not plowed, and less timber has been taken off them. Photo courtesy of Tennessee Valley Authority Archives.

spot for the enormously sophisticated atom bomb and intercontinental missile apparently would be found in Tennessee, too. The place "away" offered an unmatched opportunity to accomplish something out of the ordinary. The TVA normally conducted its affairs in a blaze of publicity (engaged), while the Manhattan Project of Oak Ridge was formed up in a cloak of total secrecy (disengaged). In the Far West, where, with a renewed turn toward the Pacific Basin, the tempo and power of political and cultural life picked up rapidly after World War II, the national park principle appeared to require little or no enhancement or support. The snowy peaks, tall forests, deep canyons, and mirrored lakes made their own unaided statements on colored travel posters, often suggesting a

104. Jesse Vest County Test Demonstration Farm, Russell County, Virginia, May 1957. Twenty-four years later than the scene in (102), magnet farms were flourishing in TVA territory. Usually, there was one magnet farm out of forty farms in a community unit, with twenty communities to a county and one hundred and twenty-five counties in the whole Tennessee Valley. Free fertilizer was produced by the Tennessee Valley Authority. The slopes of these hills also make clear how easily erosion could begin again from a thin soil. Part of the mission was to halt such slippage at its source, but the nurture had to be a close-coupled, vest pocket operation, for this agriculture was enclosed in small valleys and required the most delicate balancing. Photo courtesy of the Tennessee Valley Authority Archives.

landscape without many people but with lots of oranges, surf, and sunshine. It rained more often in Tennessee. Streams moved more slowly there, while the air hung heavier in the summer. The mountains were more rounded and mellowed. Because of the isolation and the setback in the economy dealt by the Civil War, existence could appear more retarded. Time could seem to come to a halt and lock, as it never would in fast-paced California or New York. Hence, in the TVA district it would be more appropriate to intercede with well-placed artificial reinforcements of both nature and time, such as great engineering works and then model settlements, parkways, more and thicker forests, shipping points, smaller landscaped industries, fertile farms (102–104), better schools and hospitals, and the largest possible recreational lakes and "activity parks." In order to control America some wanted it to become increasingly real and pragmatically oriented. But historically the nation also had been controlled by its dreams, and the Tennessee Valley was one place where they began to be embodied by means of well-conceived architecture, engineering, landscaping, and planning. The only chronic difficulty was that "well done" too often changed into overdone or underdone, with too little effort expended in rebalancing or remembering the original aims. Both types of criticism were entered against the TVA by such 1960s environmentalists as Justice William O. Douglas (who actually was himself a New Dealer from the 1930s). He reacted against what seemed to him the stifling artificiality of the Tellico Dam and Lake, because he retained sweet memories of the naturally taller trees, white peaks, "wild rivers," and wilderness of his native Northwest. It is difficult for people from other places to grasp the slow, placid, changeless atmosphere of Kentucky and Tennessee.

The TVA had been dedicated to relieving some of the suffering of the 1930s Depression, at the same moment that it set out to bind up the still-bleeding wounds and humiliations the South retained from the Civil War. Then the TVA immediately had to cast its eyes forward into the 1940s to get ready for a global war. Did that speed things up too much? Did all of America alter its course too often in those times to make any kind of crystallization or synthesis possible? Was its agenda too hectic or varied in purpose to yield an overall plan for another kind of America, regionally or nationally? At first, TVA's effort in the 1940s seemed incredibly energetic; by the same token, it neglected continuity and further amalgamation in its headlong pursuit of electricity. Its speed overcame its reason. Turning out a second, renewable, poised classic landscape to view proved more difficult to accomplish than anyone originally had anticipated, whether the gifted literary Agrarians of Vanderbilt University, the agriculturists of the University of Tennessee, the planning theoreticians of New York or Chicago, or the bureaucratized engineers and lawyers within the TVA itself. Otherwise, the unfolding panoramic vision of the linear city on a manufactured river-lake, first brought up by Henry Ford as a possibility for Muscle Shoals, might have grown up along the Big or Little Tennessee. Linear Timberlake City would have touched the governmentally reserved national forest and mountain park with its southern end. Along both sides of Land Between the Lakes, with Kentucky and Barkley lakes in between, such a settlement also might have formed, except that the image of the national park took too firm hold there. An impromptu linear city did take shape on the western side of Kentucky Lake, in an attempt to fill recreation-retirement needs, its architecture largely "country and western" southern. This linear vision, with its projected smaller industries and craft centers in symbiotic — and, many felt, impossible — relation to the higher technology, became more and more clouded by the interest in serving shorter-term, more sharply-defined ends. People took refuge in specialization. As the culture became more and more electrified, its leaders became more careless about accompanying obligations in other spheres. The hope of sustaining more substantial, better designed architecture and engineering as triumphant symbolic features of the TVA began to fade, too. Thereafter, the TVA did not set the architectural and sociological example for itself and the rest of America, and even for the world, as it once had. For that reason, today it would be more prudent not to expect too much from TVA on the broader front, but rather to strive to recall it as the splendid, bold, gallant, single gesture it once was, in a country that seldom made such gestures visible. The aim had been to accommodate the "forgotten man" more comfortably in a historically valid "shining city," a "distant other place" in America, but this time, at the TVA, beside a shining river-lake (4) rather than on a verdant hill.

7 | Conclusion

Just as the Columbian Exposition of 1893 and the Chicago City Beautiful Movement that came out of it (both prime influences on the TVA, through Frederic Delano and Franklin Roosevelt) were intended to afford a retrospective look, couched in French Classical architectural language, back over the nineteenth century, so in the 1990s the TVA might be used as a similar *fin de siècle* vantage point from which to scan the twentieth century. What the exploit of the TVA appears to underline is that in the twentieth century, Americans have been faced with successive decades, especially in the 1920s, 1930s, and 1940s, that were discontinuous and heavily compartmentalized. The fact that Americans were able to invent anything coherent at all that had a hope of cumulative effect was remarkable enough in itself. For that purpose, Tennessee was a particularly appropriate landscape laboratory, because, during the nineteenth century, all the European, British, and American experiments in utopianism had been implanted in that area of the Upper South and Lower Midwest. In the historical perspective, they and the TVA constituted successive moves toward a better adjustment to the industrial and technological revolutions of the world that had begun in the second half of the eighteenth century. With TVA, the new factor was the sponsorship by a national government rather than by a sect, as with the Shakers or Mormons, or by private philanthropists such as Frances Wright or Robert Owen, who were reacting to swelling and debilitating industrialism in Britain. In many respects, the TVA too was a reaction against first-phase industrialism, as mitigated by nature.

The utopian colonies had been a partial answer to the bafflements of the mushrooming modern city, and so was the TVA. The counteraction this time, in addition to the more usual one of romantic agrarianism, was in the form of a cleaner, quieter technology. This gave the TVA much more physical potential than any of the earlier utopian experiments had had available, as well as a doctrine of secular retrenchment that did not have to be as rigidly held. The TVA's influence could spread out farther into the immediate, bucolic surroundings, which previously had been selected as sites apt for utopianism. The breathing space had been the unique ingredient for both situations.

But in several respects, the time — even in the Depression, when the nation was suffering most from increased industrial and demographic displacement — was not yet ripe. Jointly, World War II of the 1940s and the Korean War of the 1950s, came too soon after TVA's founding in the 1930s, asked too much of over-stressed technology. Because of the wars, the demand for electrical output became so great that the social and ecological programs of the TVA became obscured, if not openly obliterated. In some respects the legend of the TVA is best displayed as encased within the 1930s, as in the cornerstone or time capsule used in that decade at building dedications. Indeed, that may be the safest and easiest place to leave it. How can anyone who lived through the 1930s or 1940s ever believe that those decades belonged to the same century as the 1920s or the 1960s, or that the latter two would have any sympathetic relation? There have been so many Americas! It is, of course, the arduous obligation of historians to try to help their fellow citizens over the barriers of such interdictions by decades. But what may be even more unfortunate in terms of the inhibition of idealism by decades is that the TVA became so distracted by World War II that it was deprived by its government of all of its own strength to affect social evolution, to be acceptable as a genuine long-term plan.

The only means by which such a deprivation evidently could occur was through the overpursuit of an applied speciality. The first hope, of course, had been to bring together and coordinate a number of such specialities. The one speciality became electrical production, while other specialities, such as land-use planning, rural beautification, forestry, and ground occupation and settlement began to be negated as early as 1938, with the departure of Arthur Morgan. So the positive novelty was not so much that the Tennessee Valley Authority progressed into an inadequately guarded future, but rather that it had an illustrious, cumulative body of past thought to draw on, in the form of Progressivist, Conservationist, and City Beautiful principles, now registered in a sylvan setting and newly activated by the ideal of regional planning of the 1920s.

The 1930s gave opportunity for taking exception to the 1920s belief that all telling initiatives for architecture or engineering (or community life) had to come from private or corporate inspiration. With the establishment of the TVA, a clear signal was sent that the federal government intended to be in on the thinking, too, via some unusually knowledgeable and courageous leadership, backed by an unusually talented and eager staff. All the possibilities of improved transport, "scientific" care of forests, mass production of chemical fertilizers, new uses of aluminum and concrete, greater knowledge of how to construct dams, and increased distances that electricity might travel — appeared about to converge in the TVA under the auspices of the federal government. But what the government gave, it could also take away.

The difficulty with the further acceptance of the TVA arose not so much from its novelty as from the fact that the circumstances were extreme, and no convention of procedural discourse had been set up in anticipation of the novelty, despite the repetition of Progressive ideals. Everything seemed to be wanting and waiting for a new pattern. But no *lingua franca* had been sufficiently established, including for the expression of the visual arts. As a result, the manifestation of individual personalities became stronger, and, since the key figures came from such different backgrounds they too quickly fell into confrontation in the directorate. Not enough preparation had been made. A real system and masterplan were not available at first, and the TVA may have been too far geographically from the protective wing of Washington. For any other such agency in the next century, it would be advisable not only to enunciate ideals and personal principles, as the directors of the TVA obviously did, but also to maintain a continuity of sufficiently supported, broader practice and methodology, as well as philosophy.

Architecture without a defined tradition cannot be a very satisfactory architecture. Well-founded, well-designed architecture is not necessary for bodily survival, but it is necessary for an adequately nourished public culture, and the TVA deserved great credit for recognizing that truth. The difficulty was that in the Tennessee Valley there was by now so little left to begin from. Before the Civil War, the valley had had a refined, substantial, well-informed convention of architecture but no economy to nourish and sustain it well enough thereafter. A culture that becomes too impoverished or attenuated cannot sustain its own tradition, no matter how fine that tradition once was, as is demonstrated in the valley's abandoned log cabins or derelict mansions (23–25). The Tennessee Valley's tradition had not been kept sufficiently rejuvenated so that it could be revived easily enough when the appropriate, TVA time came. With Norris Village there was an attempt to revive it, but that was too little, too late. A national deficiency simply was amplified in the valley. As a result, exotic, individualistic artistic expressions had to be introduced instead into the vacuum among all the separate, sequential decades. Now it was overawareness that distorted the image. The desire to sustain novelty and individualism, during a period that could barely afford to entertain either, was another reason why the TVA could not move steadily enough ahead to achieve a lasting visual and policy consensus. The supposed license to pursue one or two specialities, such as electrical production or fertilizer manufacture, to the exclusion of other interests, appeared too early. The lack of discussion and identification of what was worthy to go on with, in terms of building or community making, was not really considered. The nation needed to know better what any of its given regions was doing at any given time. For a long while, scholars hardly helped with that. Their universities were first bent on a national identity and then, after World War II, on an international one.

For the South itself, the problem was less that no adequate dialogue had gone on, for it knew its own tradition fairly well and had respect for architecture as a cultural activity, but rather that it too readily succumbed to the temptation to accept the condition of the romantic ruin, without defining further southern premises for the landscape — although the Agrarians did make a valiant try at such definition. The mellow moods of romanticism could slip too easily into despair and a high tolerance for low maintenance. Undoubtedly that bent was one reason why the TVA had to pay attention to old autos and household appliances abandoned along valley highways. But such a tendency made it even harder for educated people such as the Agrarians to see any good in the antidote of freshness, vitality, and novelty offered by the TVA engineering and architecture, which to them seemed callow. On one side people might cling too long to a displaced past, while on the other paying too much lip service to an overglorified present and future that would need much more by way of implementation than TVA officials appeared willing or able to provide. That gap was the source of some people's deeper disappointment with the TVA, when they actually came to encounter it *in situ*. The extremity of the two positions made the TVA essentially unstable, a situation that originally seems to have attracted Franklin Roosevelt. Continuity as yet had not been structured well enough.

While the autonomy of the decades caused difficulty in passing memories along, the distance between liberals and conservatives, Democrats and Republicans, on occasion was also too exaggerated for the general good. The attacks on the planning programs of the TVA and the National Resources Planning Board by figures such as Senator Taft and President Eisenhower, reflecting their wish to return to "normalcy" in a century that really did not have a hope for it, entirely failed to take into account the fact that wealthy individuals such as George Vanderbilt, Henry Ford, Gifford Pinchot, and even Herbert Hoover, with his own presidential survey of recent social trends, already had pondered, in very similar terms, the need for better understanding and husbanding of natural and human resources on the land, while at the same time trying to put them to new and better uses. In this regard, time itself was coming to a head, encouraging overwrought political partisanship while all the historical signs would have encouraged greater cooperation instead. Regardless of the lack of gratitude from some business and political leaders during and after World War II, Roosevelt refrained from delivering the *coup de grace* to business and banking as he might have done in 1932–33, at the depth of the Depression. Liberals might have thought him weak for that, and conservatives foolish, but he may have been wiser than anyone realized at the time. Maybe America could be set up best with initial cooperation and liason between the political parties and the public and private sectors, such as TVA director Arthur Morgan attempted between himself and Wendell Wilkie of the Commonwealth and Southern Power Company (the Republican presidential candidate against Roosevelt in 1940), for which Morgan was excoriated by his fellow directors when they succeeded in ousting him. Roosevelt himself, in his message to Congress asking for the TVA, said that it should combine the skill and flexibility of a business operation with the sustaining ability of a government.

Roosevelt was genuinely extraordinary mostly in two ways. He wanted, and was able, to take effective action that others could see and note. And he felt that such acts should generate pride in a country that was having great difficulty feeling pride anymore. He wanted to make democracy visible and tangible, by a renewed profession of faith in it, accompanied by experimentation with it. TVA had wanted to provide newly conceived symbols to elevate that pride again in terms of citizenship, but apparently American artists and architects were not ready, were unprepared for this public mission, had not studied their own past sufficiently to come up with convincing civic symbols. Draper and others thought that there had been no continuous cultivation of a visual vocabulary that would sufficiently identify public works as American, although Mumford and Lilienthal were to claim, in connection with the Museum of Modern Art show of TVA architecture in 1941 that what was telling about it was its "Americanism." To obtain sufficient optimism and Modernism, Draper at the outset engaged an architect trained abroad, Roland Wank. As the twentieth century wore on, the instinctive response to increasing political and economic hazards was to take incentives from whatever distance or direction, but such an approach did not lead more readily to synthesis.

What was so daring about the TVA, in addition to its intelligent use of architecture and engineering and its multi-phased engagement with more aspects of American life than are usually encompassed, was the persuasion that human beings rightfully belonged in the midst of their re-conditioned earth. Conditions were supposed to be made supportive enough and attractive enough in regard to such surroundings that people would not be tempted to migrate away any longer. The idea of accommodating people on a seven-state scale was decidedly innovative and was a notion not characteristic of any previous land-reserve policy, such as those for national parks, forests, or wildernesses. It was intended that the TVA people *belong*, according to the older southern preference. That intention recalls a very important postwar issue. Should America be a country prepared for more or less internal migration in the future? Which would be better for its population? The improvement in the speed and convenience of transport following World War II in part unconsciously answered the question by offering greater opportunity.

The misfortune was not only that decades were definitely cut off from each other in the twentieth century, but also that observers, visitors, officials, and natives of the Tennessee Valley too often indulged in prejudices that obscured the interchange values of the chimerical topographic experiments. Thus Benton MacKaye had laid out the Appalachian Trail in 1921, and all wilderness enthusiasts had subsequently praised it as an anachronistic stroke of genius, to be so near the population density of the

East Coast as it rode the windblown, deserted ridge of the Appalachians. But, of course, the founder never meant the trail to stay deserted. It was intended instead to be something else, another kind of settlement of elongated shape, made possible now by long-distance transmission of electricity, by the limited-access highway, and systematic, sequential damming of mountain streams. MacKaye saw the trail as a prime means of relieving population pressure on the East Coast and opening up an American distance to a new system of settlement, from the TVA to the Canadian border (13–14). That concept could not, of course, soon be assimilated by those who were more accustomed to ecological dichotomies, to seeing the landscape in terms of cities *or* wildernesses, high building densities *or* suburbs. Hence, in the late 1960s and early 1970s, young (and older) protesters against Tellico Lake and Timberlake City found it difficult to understand that they were objecting to a proposal that differed considerably from the norms of the national park, forest, or wilderness that they had known previously. In the final analysis, the difference of the TVA concept may have been most useful and valuable because it demanded that the watchers open their eyes and clear their minds. MacKaye, Henry Wright, Clarence Stein, Lewis Mumford, Henry Ford, Stuart Chase, Frederick Gutheim, Donald Davidson, and even Franklin Roosevelt were looking for alternative forms of settlement, other than that of the too rapidly grown, too conventional city. Theoretical and technical means were being accrued to make such change possible, most notably the means of electricity and the new concept of a regional plan.

Does the very fact that a society can be so open, free, individualistic, and diverse as the American one suggest that nothing at all should be coordinated or planned in it? It could be argued that only under the direst conditions and on a most minimal basis should any planning at all be countenanced. Much of the postwar urban public-housing policy was set up on that assumption, and some of the housing had to be dynamited shortly after construction. But other considerations may prove that such self-sufficency and divisiveness are potentially more hazardous than is first supposed. As technology and its appurtenances increased both in size and specialization, it became more difficult for the TVA to handle them, to find satisfactory ways also to include other legitimate human preoccupations, as it had once vowed that it would. The crowding by one technical specialty promised soon to be overwhelming of the others. The TVA remains important for the national future because it began to suggest, although in a somewhat crude and limited way, how factors might be better balanced and adjusted for ongoing life in a particular place. The TVA was the last great refocusing inward on the configurations and habits of the American land and mind. The problems of looking inward may not yet have been solved. The effort, after all, was undertaken in times of acute, successive crises, but it was at least *seen*. Someday the golden age may arrive when the Tennessee Valley and its treatment will be comprehended as an overture to a properly romantic southern landscape, that all Americans, including the Agrarians, would be glad to credit as a final recourse. Americans may someday learn how to unite, then fuse, the absolutely new with the extraordinarily old.

Notes

Abbreviations

FDRL   Franklin D. Roosevelt Library, Hyde Park, N.Y.
MDUVL   Manuscript Department, University of Virginia Library, Charlottesville, Virginia
MSUOHP Oral History Project, The John Willard Brister Library, Memphis State University, Memphis, Tennessee
TVA   Tennessee Valley Authority, Technical Library
USGPO   United States Government Printing Office

Chapter 1

1. Daniel R. Fusfeld, *The Economic Thought of Franklin D. Roosevelt, and the Origins of the New Deal* (New York: Columbia Univ. Press, 1954), 256.
2. Letters between Lilienthal and Delano, 19 Dec. 1934 and 5 Jan. 1935, F.A. Delano Papers, Box 4, FDRL. Delano enclosed in the 19 Dec. 1934 letter an excerpt about the TVA from an anonymous "old friend of mine," an engineer, who had observed on 14 November 1934 that "a board or committee or a commission is a poor form of administration and unless there be within a man who takes and administers the authority, then the whole effort is weak." It is possible that this gave Lilienthal further encouragement to take control away from the two Morgans on the board. The Dec. 1935 *Report on Regional Factors in National Planning* from the National Resources Committee, (Washington, D.C.: USGPO, 1935), 114, chaired by Delano, likewise states that the TVA board system "is of doubtful wisdom," and suggests the appointment of a regular manager beneath the board to direct general affairs. It could be assumed that at this time President Roosevelt would not be too busy to read the report.
3. Frederick A. Gutheim, "Tennessee Valley Authority: A New Phase in Archi-

tecture," *Magazine of Art* 33 (Sept. 1940):520, declared that any TVA architect "is not an individual in control of the work, but one member of a
team. He can operate only to the extent that he can first cooperate."

4. Roderick Nash, "The American Invention of National Parks," *American Quarterly* 22 (1970):727.

5. This anti-Classical attitude was represented by the article on the TVA in *Architectural Forum* 71 (Aug. 1939):83, 98. The immediate target was the Classical cornices on Wilson Dam at Muscle Shoals.

6. R.G. Tugwell and E.C. Banfield, "Grass Roots Democracy—Myth or Reality?", *Public Administration Reviews* 10 (Winter 1950):48. Tugwell reported that he suggested to these first directors that the TVA might "well approximate a new kind of government." It could "supersede the states." It was a time of daring. This image of a Chinese box within a box, a nation within a nation, corresponding to the geographic reality, was invoked fairly often at the start of the TVA.

7. That Roosevelt had the multiplication of TVAs in mind from the start is demonstrated in his message to Congress on 10 April 1933, proposing the TVA. He said, "If we are successful here we can march on step by step, in a like development of other great natural territorial units within our borders."

8. Roland Wank, interview by Decatur (Ala.) *Daily*, 5 Aug. 1965.

9. "Better to Stay at Home," Florence (Ala.) *Times*, 3 Feb. 1934.

10. "TVA," *Architectural Forum* 71 (Aug. 1939):75.

11. Tugwell and Banfield, "Grass Roots Democracy," 47.

12. Arthur Morgan, *The Making of the TVA* (Buffalo, N.Y.: Prometheus Books, 1974), 8.

13. Carl W. Condit, the leading historian of architectural technology, thus gives the TVA the highest marks for execution: "The architectural character of the structures and the enhancement of the surrounding natural beauties (the concept known as total design) represent the highest level of design in any technical-industrial complex of the United States." Condit, *American Building* (Chicago: Univ. of Chicago Press, 1968), 272. The size and refinement of the enterprise were what most impressed the architectural critics.

14. Willis M. Baker, interview by Charles Crawford, 8 Feb. 1970, MSUOHP, 9. In "The New Deal and the Analogue of War," *Change and Continuity in Twentieth-Century America*, eds. John Braeman, Robert H. Bremner, Everett Walters. (Columbus, Ohio: Ohio State Univ. Press, 1964), 109–10, William E. Leuchtenburg made it evident that Franklin Roosevelt did indeed regard the TVA device and condition as something to be handled like a World War I emergency, calling for "war rhetoric," that included the CCC.

15. Donald Davidson, *The Tennessee: The New River: Civil War to TVA* (New York: Rinehart, 1948), 2:176–94.

16. Maj. Lewis H. Watkins, *The Tennessee River and Tributaries*, U.S., 71st Congress, 2d Session, House Doc. 328 (Washington, D.C.: USGPO, 1930).

17. Frank Freidel, *Franklin D. Roosevelt: The Triumph* (Boston: Little, Brown, 1956), 109.

18. Erwin Hargrove, "Introduction," *TVA: Fifty Years of Grassroots Bureaucracy*, ed. Hargrove and Paul Conkin (Urbana: Univ. of Illinois Press, 1983), x.

19. As recounted in Rexford G. Tugwell, *The Democratic Roosevelt* (New York: Doubleday, 1957), 552.

20. Quentin Reynolds, *The Amazing Mr. Doolittle* (New York: Appleton-Century-Crofts, 1953), 169.

Chapter 2

1. Frederick A. Gutheim, "Tennessee Valley Authority: A New Phase in Architecture," *Magazine of Art* 33 (Sept. 1940):527.

2. Attendance at moving pictures greatly affected all other public and popular attitudes in the 1930s. Movies offered big-screen fantasy and escape in an atmosphere of close association, craved at the time. So it was in no way inappropriate that much TVA design would recall that theater atmosphere. Lower-income people and young people of all classes would walk the aisles of five-and-ten-cent stores for hours to keep warm and enjoy the pretty trinkets, especially costume jewelry, sparkling on trays. Those stores had the first "open" displays, in which customers picked up their own items and took them to nearby cash registers. Large-ticket items only rarely could be contemplated. No-one had purchasing or borrowing power. In drugstores, sunburst perfume bottles of vaguely French inspiration, rough-surfaced briar pipes, shiny tie-clasps, cigarette lighters and cases, and key chains competed for attention. In the living room at home, sea shells, postcards, family photos, and summer vacation souvenirs such as model birchbark canoes and balsam pillows served the same purpose — acting as sensory cues. On the challenging open road, when gas could be afforded, the same effect was represented by canvas tops, chrome strips, radiator grilles and ornaments, and chromed headlights standing free, while white sidewall tires and wire wheels offered a bonus as they turned. Technology appeared to promise so much because, ironically enough, business and banking leadership appeared to be failing. For larger objects, such charades are analyzed carefully and illustrated beautifully in the Brooklyn Museum catalog by Richard Guy Wilson, Dianne H. Pilgrim, and Dickran Tashjian, *The Machine Age in America, 1918–1941* (New York: Harry N. Abrams, 1986), which includes a discussion of the TVA on pp. 115–22.

3. Arthur E. Morgan, "Bench-Marks in the Tennessee Valley," *Survey Graphic* 23:5 (May 1934):233–34.

4. That the TVA was considered utopian by its makers is suggested by the titles of some of Chief Engineer Arthur Morgan's books, particularly *No-*

where Was Somewhere: How History Makes Utopias and How Utopias Make History (Chapel Hill: Univ. of North Carolina Press, 1946), and *Plagiarism in Utopia* (Yellow Springs, Ohio: Privately printed, 1944). The *Washington Post* of 18 Feb. 1934 displayed the headline, "TVA Foresees Utopia Come True in the Valley." Arthur A. Ekirch, Jr., included the TVA in his *Ideologies and Utopias: The Impact of the New Deal on American Thought* (Chicago: Quadrangle, 1969).

5. For information on these several colonies, see John Egerton, *Visions of Utopia: Nashoba, Rugby, Ruskin and the "New Communities" in Tennessee's Past* (Knoxville: Univ. of Tennessee Press, 1977).

6. Harold T. Pinkett, *Gifford Pinchot: Private and Public Forester* (Urbana: Univ. of Illinois Press, 1970), 23. Pinchot's successor at Biltmore, Dr. Carl Schenck from Germany, began the first American school of forestry there in 1898. Wilma Dykeman, *The French Broad* (New York: Rinehart, 1955), 218.

7. Frances L. Goodrich, *Mountain Homespun* (New Haven: Yale Univ. Press, 1931), 30–34, and Dykeman, *French Broad*, 216–17.

8. Howard Odum, *Toward the Regional Balance of America: The Way of the South* (New York: Macmillan 1947), 281.

9. Ibid., 333.

10. Letter to the Editor from Earle S. Draper, *Landscape Architecture* 59 (Apr. 1969):196–97.

11. Arthur Morgan, *The Making of the TVA* (Buffalo, N.Y.: Prometheus Books, 1974), 8. Thomas K. McCraw makes the Roosevelt connection, especially with Mrs. Roosevelt, clearer in his *Morgan vs. Roosevelt* (Chicago: Loyola Univ. Press, 1970), 10. Arthur Morgan's reputation as an engineer was already well established before the Roosevelts heard of him. Most striking was his fondness for innovation and experimentation. Between 1917 and 1922, in the Miami Conservancy District, he had instituted self-contracting, very detailed bid specifications, one-union labor, careful technical and historical research, interest in "dynamic design," and "conclusive engineering analysis"— all of which were continued in the TVA. He introduced hydraulic fill and the first dragline for earthen dams. His venturesomeness must have drawn Roosevelt to him. Daniel L. Schodek, *Landmarks in American Civil Engineering* (Cambridge, Mass.: MIT Press, 1987), 227–32. Roy Talbert, Jr., has also pointed out that former governor of Ohio and 1920 running mate of Roosevelt as candidate for president, James M. Cox, also claimed in his autobiography to have "been responsible for Morgan's selection." *FDR's Utopian: Arthur Morgan of the TVA* (Jackson, Miss.: Univ. Press of Mississippi, 1987), 82–83.

12. Harry Wiersema, interview by Walter Creese, Knoxville, Tenn. 26 Apr. 1973. Albert S. Fry, a 1918 graduate of the Univ. of Illinois in civil engineering; and Wiersema, a 1913 graduate of the same institution in architectural and civil engineering, were brought into the TVA from the Morgan firm. Wiersema had been secretary of the Memphis office and its chief

bridge designer. He was liason for the TVA between Knoxville and Denver, the headquarters of the Bureau of Reclamation, which was preparing the plans for Norris Dam because the TVA had not yet organized its own office. TVA continued this practice with the second dam, Wheeler. Wiersema told Creese that he was attracted to Morgan because he happened to hear him preach in the Unitarian church in Memphis. Wiersema eventually became assistant chief engineer of the TVA. He also was responsible for the official reports on the construction of the dams, exemplary books that are models of thoroughness. Harry Tour, although nominally assistant to head architect Roland Wank, also held a degree in architectural engineering from the Univ. of Illinois. That degree was granted by the architecture department rather than the engineering school so, as Tour enjoyed pointing out, he could pass himself off among his TVA colleagues as either an architect or an engineer.

13. Reynold M. Wik, *Henry Ford and Grass-Roots America* (Ann Arbor: Univ. of Michigan Press, 1972), 155–56.

14. Littell McClung, "The Seventy-Five Mile City," *Scientific American* 127:3 (Sept. 1922):156–57; and McClung, "What Can Henry Ford Do with Muscle Shoals?", *Illustrated Word* 37:2 (April 1922):185.

15. Wik, *Henry Ford*, 191.

16. Ibid., 193.

17. Frederic Delano to David Lilienthal, 30 Jan. 1935, F. A. Delano Papers, Box 4, FDRL.

18. Preston J. Hubbard, *Origins of the TVA: The Muscle Shoals Controversy, 1920–1932* (Nashville, Tenn.: Vanderbilt Univ. Press, 1961), 33, 49, 65, 75–76, 90–91, 120. Norris's greatest objection to Ford's acquisition of Muscle Shoals was that the auto manufacturer offered only $5 million for it, when it had cost the government between $125 million and $130 million. Norris regarded the proposed sale as one more giveaway in the long fight to conserve natural resources. George W. Norris, "Shall We Give Muscle Shoals to Henry Ford?", *Saturday Evening Post*, 31 May 1924.

19. Wik, *Henry Ford*, 190. See also Robert Lacey, *Ford: The Men and the Machine* (Boston: Little, Brown, 1986), 228.

20. Wik, *Henry Ford*, 191.

21. Ibid., 190.

22. "Ford Withdraws Offer for Muscle Shoals," *Electrical World* 84:16 (18 Oct. 1924):827.

23. Edward Shapiro, "The Southern Agrarians and the Tennessee Valley Authority," *American Quarterly* 22:4 (Winter 1970):792–93, n. 4.

24. Ibid., 799–801.

25. George F. Milton, "Consumer's View of the TVA," *Atlantic Monthly* 160 (Nov. 1937):654.

26. Arthur E. Morgan, "Bench-Marks in the Tennessee Valley," pt. 4, *Survey Graphic* 23:11 (Nov. 1934):551.

27. Ibid., 576.
28. Arthur Morgan, "Bench-Marks in the Tennessee Valley," pt. 1, *Survey Graphic* 23:1 (Jan. 1934):8.
29. Rexford G. Tugwell, "Conservation Redefined," address for the 50th Anniversary of the Founding of New York's Forest Preserve, Albany, N.Y., 15 May 1935, pp. 2–3; in RGT Speeches, Box 46, 1933–35, FDRL.
30. Paul Ashdown, ed., *James Agee: Selected Journalism* (Knoxville: Univ. of Tennessee Press, 1985), 86.
31. Arthur Morgan, *Making of the TVA*, 63–64.
32. Arthur Morgan, "Bench-Marks in the Tennessee Valley," pt. 1, 9.
33. Samuel P. Hays, *Conservation and the Gospel of Efficiency: The Progressive Conservation Movement, 1890–1920* (Cambridge, Mass.: Harvard University Press, 1959), 271.
34. Gifford Pinchot, *Breaking New Ground* (New York: Harcourt, Brace, 1947), 322. This passage is quoted directly in "The Historical Roots of TVA," *TVA Annual Report, 1953* (Knoxville: TVA Press, 1953), 52.
35. The full title of this "Appalachian Report" was *Message from the President of the United States Transmitting a Report of the Secretary of Agriculture in Relation to the Forests, Rivers, and Mountains of the Southern Appalachian Region* (Washington, D.C.: USGPO, 1902).
36. John Ise, *The United States Forest Policy* (New Haven: Yale Univ. Press, 1920), 207–23.
37. Chicago *Times-Herald*, 24 Dec. 1899.
38. *Message from the President in Relation to Forests*, 26.
39. Ibid., 138.
40. Ibid., 186, 188–89.
41. Strangely, the only report the TVA ever put out on aesthetic values was Earle S. Draper, ed., *The Scenic Resources of the Tennessee Valley: A Descriptive and Pictorial Inventory* (Knoxville: Department of Regional Planning Studies, TVA, 1938). The pictorial enchantment of that survey and its sense of regional character suggest what was missed by not having more such studies.
42. W.J. McGee, "The Proposed Appalachian Forest," *World's Work* 3:1 (Nov. 1901):1386.
43. Frederic A. Delano to Franklin Roosevelt 16 Apr. 1933, Delano Papers, 1933–45, FDRL. Delano obtained the rare copy from Arthur Keith of the U.S. Geological Survey.
44. Frederic A. Delano to James Bryant Conant, president of Harvard University, 30 Oct. 1935, Delano Papers, General Files, 1925–45, FDRL.
45. David Cushman Coyle, "Frederic A. Delano: Catalyst," *Survey Graphic* 35:7 (July 1946):252–53, and Frederic A. Delano, "Railway Terminals and Their Relation to City Planning," *The Western Architect* 15:1 (Jan. 1910):11.
46. Franklin D. Roosevelt, "Extemporaneous Address on Regional Planning," in *The Public Papers and Addresses of Franklin D. Roosevelt*, ed. Samuel Rosenman (New York: Random House, 1938), 1:496. See also Mel Scott,

American City Planning Since 1890 (Berkeley: Univ. of California Press, 1969), 289. Talbert, *FDR's Utopian*, 3–4, says that FDR's father was an alternate commissioner for New York State at the 1893 Chicago Fair, also known as "The City Beautiful" and "The Electric City."

47. John M. Glenn, *The Russell Sage Foundation, 1907–1946* (New York: Russell Sage Foundation, 1947), 2:439.

48. "Correspondence between Howard J. Menhinick and Tracy B. Augur Regarding Origins and History of Sections 22 and 23 of the TVA Act," mimeographed (Knoxville: TVA, Department of Regional Studies, 31 Dec. 1942), 3. About the Virginia meeting, the authors say, "I do not recall that the Tennessee Valley region was ever mentioned by name, but Mr. Odum and other southern regionalists were present and it is possible that the relationship of New York State planning techniques to the problems of the Southern Appalachians were mentioned." Daniel Schaffer also mentions a pre-TVA outline by Roosevelt at Virginia, *Garden Cities for America* (Philadelphia: Temple University Press, 1982), 224.

49. L.M. [Lewis Mumford], "The Regional Planning Association of America: Past and Future," 21 Sept. 1948, 2, Catherine Bauer Papers, Bancroft Library, Univ. of California.

50. Charles W. Eliot II [director of planning, National Capital Park and Planning Commission], "Historical Considerations in Regional Planning," paper presented at the Institute of Public Affairs, Univ. of Virginia, Charlottesville, Va., 7 July 1931, MDUVL, 4–5.

51. Lewis Mumford, "Regionalism," paper presented at the Institute of Public Affairs, Univ. of Virginia, Charlottesville, Va., 8 July 1931, MDUVL, 13.

52. Preston S. Awkright, "Industrial Power Policy for the South," paper presented at the Institute for Public Affairs, Univ. of Virginia, Charlottesville, Va., 12 Aug. 1930, p. 22. MDUVL.

53. Ibid., 11.

54. Gutheim, "Tennessee Valley Authority: A New Phase," 519.

55. *Message from the President in Relation to Forests*, 4.

56. Frank Freidel, *Franklin D. Roosevelt* (Boston: Little, Brown, 1956), 102–103.

57. Ibid., 105–106.

58. James C. Bonbright, "The New Role of Public Ownership in the Power Industry," paper presented at the Institute of Public Affairs, Univ. of Virginia, Charlottesville, Va., 9 Aug. 1930, 6.

59. R.D. McKenzie, "The New Regionalism," paper presented at the Institute of Public Affairs, Univ. of Virginia, Charlottesville, Va., 10 July 1931, 8.

60. Stuart Chase, "The Concept of Planning," paper presented at Institute of Public Affairs, Univ. of Virginia, Charlottesville, Va., 11 July 1931, 11.

61. Ibid., 12.

62. In Newell's estimation, Ford was paralleling ideas promoted by Hugh McRae and his farm colonies in North Carolina. In 1923, with the Farm City Corporation of America, McRae joined Thomas Adams, John Nolen, Elwood Mead, Gifford Pinchot, and Sir Raymond Unwin in the unsuccess-

ful promulgation of those farm colonies. See Paul K. Conkin, *Tomorrow A New World: The New Deal Community Program* (Ithaca, N.Y.: Cornell Univ. Press, 1959), 280; and Glenn, *Russell Sage Foundation*, 2:441.

63. Frederick H. Newell, "The New Industrialism and Power Development in the South," paper presented at the Institute of Public Affairs, Univ. of Virginia, Charlottesville, Va., 9 July 1931, MDUVL, 11.

64. Lacey, *Ford*, 233.

65. Wik, *Henry Ford*, 149.

66. Franklin D. Roosevelt, "Regionalism," paper presented at the Institute of Public Affairs, Univ. of Virginia, Charlottesville, Va., 6 July 1931, MDUVL, 1.

67. Charlottesville (Va.) *Daily Progress*, 6 July 1931, 1.

68. Ibid.

69. Roosevelt, "Regionalism," 4.

70. Ibid., 7.

71. Ibid., 11. Roosevelt's self-proclaimed innocence over what regionalism meant is put in doubt by his 1929 regional survey of New York State in terms of land use and forest resources. Gertrude Almy Slichter, "Franklin D. Roosevelt's Farm Policy as Governor of New York State, 1928–1932," *Agricultural History* 33 (Oct. 1959):171.

72. Franklin D. Roosevelt, "Growing Up by Plan," *Survey Graphic* 67 (1 Feb. 1932):484.

73. Percy Dunsheath, *A History of Electrical Power Engineering* (Cambridge, Mass.: MIT Press, 1962), 354; and T.J. Woofter, Jr., "The Tennessee Basin," *American Journal of Sociology* 39:6 (May 1934):814.

74. The Doctrine of Efficiency arising from Progressivism is explained particularly well in Samuel Haber, *Efficiency and Uplift: Scientific Management in the Progressive Era, 1890–1920* (Chicago: Univ. of Chicago Press, 1964); and Samuel P. Hayes, *Conservation and the Gospel of Efficency: The Progressive Conservation Movement, 1890–1920* (Cambridge, Mass.: Harvard Univ. Press, 1959).

75. Edwin T. Layton, Jr., *The Revolt of the Engineers: Social Responsibility of the American Engineering Profession* (Cleveland, Ohio: Case Western Reserve University, 1971), 143.

76. Ibid., 144, 173, n. 10.

77. Jean Christie, "Giant Power: A Progressive Proposal of the Nineteen-Twenties," *Pennsylvania Magazine of History and Biography* 96 (Oct. 1972):483.

78. Jean Christie,"Morris Lewellyn Cooke: Progressive Engineer" (Ph.D. diss., Columbia Univ., 1963), 72.

79. Morris Lewellyn Cooke, *Giant Power: The Report of the Giant Power Survey Board to the General Assembly with a Message of Transmittal from Gifford Pinchot, Governor* (Harrisburg: Telegraph Printing, 1925), xii. For same subject, see also Christie, "Giant Power," 480–507.

80. Rexford G. Tugwell, *The Democratic Roosevelt* (New York: Doubleday, 1957), 413–14.

81. Nelson Lloyd Dawson, *Louis D. Brandeis, Felix Frankfurter, and the New Deal* (Hamden, Conn.: Archon Books, 1980), 7, 35.

82. Joseph P. Lash, *From the Diaries of Felix Frankfurter* (New York: Norton, 1975), 135, n. 4.

83. Dawson, *Brandeis, Frankfurter, and the New Deal*, 66. Donald R. Richberg, *My Hero: The Indiscreet Memoirs of an Eventful but Unheroic Life* (New York: G.P. Putnam's Sons, 1954), 126–27.

84. Thomas K. McCraw, *Prophets of Regulation* (Cambridge, Mass.: Harvard Univ. Press, 1984), 92–93.

85. Dawson, *Brandeis, Frankfurter, and the New Deal*, 82–83. On p. 215, n. 157, Dawson says that there was no doubt that Frankfurter had "a decisive hand in commending Lilienthal to Roosevelt." Brandeis recommended Lilienthal to Arthur Morgan, too. Arthur Morgan, *Making of TVA*, 22.

86. Clarence S. Stein, Chm., *Report of the Commission of Housing and Regional Planning to Gov. Alfred E. Smith*, 7 May 1926 (Albany, N.Y.: J.B. Lyon, 1926).

87. Benton MacKaye, *The New Exploration* (New York: Harcourt, Brace, 1928), vii. The memo by L. M. [Lewis Mumford], "Regional Planning Association of America," 21 Sept. 1948, 2, reports that MacKaye did the preliminary studies for Wright for the New York Regional Plan.

88. *Report of the Commission of Housing and Regional Planning*, 64.

89. Ibid., 61.

90. Benton MacKaye, "An Appalachian Trail: A Project in Regional Planning," *Journal of the American Institute of Architects*, 9:10 (Oct. 1921):325–30.

91. Benton MacKaye, "Appalachian Power: Servant or Master?", *Survey Graphic* 4:6 (Mar. 1924):618.

92. Ibid., 618–19.

93. Ibid., 618.

94. MacKaye, *New Exploration*, 163.

95. Ibid., 171.

96. Ibid., 150.

97. Lewis Mumford, "The Fourth Migration," *Survey Graphic* 7:2 (May 1925): 133. For Mumford's opposition to Adams, see David A. Johnson, "Regional Planning for the Great American Metropolis: New York between the World Wars," *Two Centuries of American Planning*, ed. Daniel Schaffer (Baltimore: Johns Hopkins University Press, 1988), 178–86.

98. Ibid., 132. Very useful also for further explanation is Carl Sussman, ed., *Planning for the Fourth Migration* (Cambridge, Mass.: MIT Press, 1976).

99. MacKaye, *New Exploration*, 118.

100. Ibid., 143.

101. Ibid., 169.

102. Ibid., 144.

103. Ibid., 138.

104. Roosevelt, "Growing Up by Plan," 506.

105. As quoted in Ellis F. Hartford, ed., *Our Common Mooring* (Athens: Univ. of Georgia Press, 1941), 17, a book for school children, presenting the ideas of H.A. Morgan.

106. Raymond Moley, *Valley Authorities* (New York: American Enterprise Association, 1950), 9, 23. Raymond P. Brandt, "The Valley of Opportunity," in *Today: An Independent National Weekly* [editor, Raymond Moley; publisher, Vincent Astor], 9 Dec.1933, 7, states, with delightful oversimplification "The T.V.A. is patterned after the Port of New York Authority." The year before, Rupert Vance, the southern sociologist of regions, had noted, "There exists also a place for state and federal coordination such as has been established by New Jersey, New York, and the United States in regulation of the Port of the City of New York." Vance, *Human Geography of the South: A Study in Regional Resources and Human Adequacy* (Chapel Hill: Univ. of North Carolina Press, 1932), 492.

107. Douglas Yates, *Bureaucratic Democracy: The Search for Democracy and Efficiency in American Government* (Cambridge, Mass.: Harvard Univ. Press, 1982), 38–39.

108. U.S., National Resources Committee, *Regional Factors in National Planning* (Washington, D.C.: USGPO, 1935), 40.

109. Ibid.

110. Ibid., 41. Frederick Gutheim, "Regional Planning by the Federal Government," *Editorial Research Reports* (Washington, D.C.: Editorial Research Reports,7, 1933): 28–30, lists additional "precedents" for the TVA, including the contentious Colorado River Compact of 1922.

111. Moley, *Valley Authorities*, 23.

112. Richard A. Couto, "The TVA Power Program Since World War II," *TVA: Fifty Years of Grass-Roots Bureaucracy*, ed. Edwin C. Hargrove and Paul K. Conkin (Urbana: Univ. of Illinois Press, 1983), 238.

113. James W. Carey and John J. Quirk, "The Mythos of the Electronic Revolution," *American Scholar* 39:3 (Summer 1970):395. Another article that points up the gap between "social planning and engineering planning" is David Cushman Coyle, "The Twilight of National Planning," *Harper's Monthly Magazine* 171 (Oct. 1935):557–67.

114. Carey and Quirk, "Mythos of the Electronic Revolution," 396.

115. Couto, "TVA Power Program," 238.

116. Carey and Quirk, "Mythos of the Electronic Revolution," 396.

117. Ibid., 404.

118. Ibid., 424. Theodore Roszak, a leading spokesman of the antitechnocratic movement, shared Carey and Quirk's worry in Roszak, *Where the Wasteland Ends* (New York: Doubleday, 1972). He feared that quality would be reduced to quantity, and that "poets, painters, holy men, or social revolutionaries" would be excluded from socially creative roles. However, an irony was that Harold Loeb, the early partner of Howard Scott in Tech-

nocracy, in Loeb, *Life in a Technocracy* (New York: Viking Press, 1933), came out for a number of things that the TVA set out to accomplish in that very first year. He felt that skyscrapers "would have to go, in a technocracy" (139), while "art might become the most important activity in a technocracy" (142). Decentralization of settlements ought to take place. "With the pull toward the centers relaxed, small communities should gradually come to life" (160). The architecture of his Technocracy could be "homogeneous and standardized," since "engineers impervious to aesthetic canons tend to create dignified objects" (118–19). It all sounded much like the desires of the TVA. The year 1933 worked its own amalgamating magic.

119. A succinct explanation of Technocracy is contained in a series of essays in the *New York Herald Tribune*, 15–18 Dec. 1932. Another assessment, frequently cited today, is "Technocracy — Boon, Blight, or Bunk?" *Literary Digest* 114 (31 Dec. 1932):5–6.

120. Richard Lowitt, *George W. Norris: The Persistence of a Progressive* (Urbana: Univ. of Illinois Press, 1971), 563.

121. Editorial, "Scholars and Specialists," *New York World Telegram*, 27 Dec. 1932.

122. Walter Rautenstrauch, "Technological Developments and Social Change," speech to the American Association for the Advancement of Science, Atlantic City, N.J., 28 Dec. 1932, as reported in the *New York Herald Tribune*, 29 Dec. 1932. For further information on Rautenstrauch, see William E. Akin, *Technocracy and the American Dream* (Berkeley: Univ. of California Press, 1977), 46–63.

123. H.G. Wells, *Things to Come* (rpt. ed., Boston: Gregg Press, 1975), 92–93.

124. Henry Elsner, Jr., *The Technocrats* (Syracuse, N.Y.: Syracuse Univ. Press, 1967), 9, 21.

125. Vance, *Human Geography of the South*, 498.

126. Ibid., 511.

127. "Correspondence between Howard K. Menhinick and Tracy B. Augur Regarding Origins and History of Sections 22 and 23 of the TVA Act," 7–8. Department of Regional Studies.

128. Cynthia Stokes Brown, "The Experimental College Revisited," *Wisconsin Magazine of History* 66:2 (Winter 1982–83):97.

129. Ibid., 99.

130. C.K. Bauer to Clarence Stein, 8 Apr. 1933; letter Clarence Stein to Catherine K. Bauer, 10 Apr. 1933; and Catherine Bauer to Frederick Gutheim, 11 Apr. 1933; all in Bauer Collection, Bancroft Library, Univ. of California, Berkeley.

131. Bauer to Stein, April 8, 1933; Stein to Bauer, April 10, 1933; and Bauer to Gutheim, April 11, 1933.

132. Frederick Gutheim, "The Tennessee Valley Regional Plan," memorandum, 14 May 1933, and Frederick Gutheim, "Notes on the Proposed Tennessee

Valley Project," n.d., 1, both in the Bauer Collection, Bancroft Library, Univ. of California, Berkeley.

133. In "Correspondence between Howard K. Menhinick and Tracy B. Augur Regarding Origins and History of Sections 22 and 23 of the TVA Act," 7, Augur reports that John Nolen, Jr., told him, "Early in April Messrs. Gutheim and Nolen submitted an amplified version of these sections to Senator Norris and discussed it with him. This version greatly broadened the concept of Section 22 and introduced the phrases about preparing plans useful to Congress in the guidance of later development." Augur also notes (4) that "Stuart Chase was at that time outlining a regional planning program for the Northwest, along lines very similar to those in the Tennessee Valley."

134. Frederick Gutheim to Catherine Bauer, 9 Apr. 1933, 2, Bauer Collection, Bancroft Library, Univ. of California, Berkeley.

135. Charles W. Eliot II, interview by Walter Creese, 1 Nov. 1972.

136. Charles M. Stephenson, "Report on Section 22 Activities of the TVA," mimeographed confidential report (Knoxville: TVA, Office of the General Manager, 30 Aug. 1940), 35–36. Stephenson goes on to say that these projects were "carried out in a rather hit-or-miss fashion, depending apparently upon the interests and desires of the various department heads and their influence or persuasiveness on the General Manager and the Board of Directors." The dictum for change, if it was indeed contained in section 22, was not fully used.

137. Marguerite Owen, *The Tennessee Valley Authority* (New York: Praeger, 1973), 13.

138. George Norris to Frederic Delano, 5 Oct. 1943, Delano Papers, Box 4, FDRL. See also H.V. Nelles, *The Politics of Development: Forests, Mines and Hydro-Electric Power in Ontario, 1849–1941* (Hamden, Conn.: Archon Books, 1974), 303–306, 474.

139. Harry Slattery, *Rural America Lights Up*, ed. Sherman F. Mittell (Washington, D.C., National Home Library Foundation, 1940), 1.

140. Arthur Morgan, *Making of the TVA*, 22, 183.

141. Slattery, *Rural America Lights Up*, 142.

142. Ibid., 7.

143. Ibid., xi.

144. Tugwell, "Conservation Redefined," 1.

145. Franklin D. Roosevelt, *On Our Way* (New York: John Day, 1934), 54–55.

146. Otis L. Graham, Jr., *An Encore for Reform: The Old Progressives and the New Deal* (New York: Oxford Univ. Press, 1967), 105, 126. Graham looks upon the efforts of Norris for the TVA as another case of the continuity of 1890s Progressivism, and points out that Norris, as an "Old Progressive," helped Roosevelt get elected under the banner of the National Progressive League in 1932 and 1936.

147. Gutheim, "Regional Planning by the Federal Government," 33.

Chapter 3

1. *Tennessee, A Guide to the State: WPA Guide to Tennessee* (rptd. Knoxville: Univ. of Tennessee Press, 1986), 81.
2. Ibid.
3. Ibid., xviii–xix.
4. Ibid., 20.
5. Richard Kilbourne, interview by Charles Crawford, 12 Mar. 1970, MSUOHP, 11.
6. *WPA Guide to Tennessee*, 23–24.
7. Stanley J. Folmsbee, Robert E. Corlew, and Enoch L. Mitchell, *Tennessee: A Short History* (Knoxville: Univ. of Tennessee Press, 1969), 3.
8. Frank Owsley, *Plain Folk of the Old South* (Baton Rouge: Louisiana State Univ. Press, 1949), viii.
9. Ibid., vii.
10. H.L. Mencken, "The South Astir," *Virginia Quarterly Review* 7 (1935):57.
11. Ibid., 59.
12. Ibid.
13. Frank Owsley, "The Pillars of Agrarianism," *American Review* 4 (1934–35):535. *The Junk Car: From Field to Foundry: A Guide for Solving a Community Problem* (Knoxville: TVA, 1972) takes up later manifestations of the abandonment of machinery. In Tennessee, 325,000 cars were estimated abandoned in fields and beside roads in 1972.
14. Ibid., 534.
15. Ibid., 547.
16. Paul Conkin, *The Southern Agrarians* (Knoxville: Univ. of Tennessee Press, 1988), 83.
17. Walter M. Kollmorgen, "Observations on Cultural Islands in Terms of Tennessee Agriculture," *East Tennessee Historical Society's Publications* 16 (1944):71.
18. Ibid., 73.
19. Ibid., 78.
20. David Halberstam, "The End of a Populist," *Harper's*, Jan. 1971, 39.
21. Ibid., 40.
22. Kollmorgen, "Observations on Cultural Islands," 77, fn. 7.
23. Ibid., 69.
24. James R. McCarthy, "The New Deal in Tennessee," *Sewanee Review* 42 (Oct.–Dec. 1934):411–12.
25. Paul Hutchinson, "Revolution by Electricity: The Significance of the Tennessee Valley Experiment," *Scribner's Magazine* 96:4 (Oct. 1934):194.
26. Richard A. Ball, "A Poverty Case: The Analgesic Subculture of the Southern Appalachians," *The Local Area in Large-Scale Society*, ed. Scott Greer and Ann Lennarson Greer (New York: Basic Books, 1974), 124–25. Others

spoke of "the bleak spirit of defeatism and resignation" ("Coeburn—The Town with a Future," mimeographed report, TVA, 13 Mar. 1969, p. 1); deriving from, more poetically, "an environment where weariness is the standard state of being and where the unsophisticated feelings of grief and religious devotion are as real as the change of seasons on the mountainside" (James K. Page, Jr., "Old-time Pickin' and Playin' in Poor Valley," *Smithsonian* 7:1 [Apr. 1976]:107).

27. Ball, "A Poverty Case," 126–27.

28. Harry Caudill, *Night Comes to the Cumberlands* (Boston: Little, Brown, 1963), xi, brings up the concern of John F. Kennedy, while campaigning for president in 1960, over conditions in the coalfields of West Virginia, and how that concern was expressed frequently thereafter in Kennedy's speeches. Kennedy, like Roosevelt, regarded the miners as a "lost" people, to be rescued after he had discovered them.

29. David Cushman Coyle, *Land of Hope: The Way of Life in the Tennessee Valley* (Evanston, Ill.: Row, Peterson, 1941), 12.

30. Odette Keun, "A Foreigner Looks at the TVA," *This Was America*, ed. Oscar Handlin (Cambridge, Mass.: Harvard Univ. Press, 1969), 537.

31. Paul Ashdown, ed., *James Agee: Selected Journalism* (Knoxville: Univ. of Tennessee Press, 1985), 64.

32. Richard Lowitt and Maurine Beasley, eds., *One Third of a Nation* (Urbana: Univ. of Illinois Press, 1984), 272.

33. Louis B. Kalter, "Island Wilderness Biological Survey: Island F Norris Reservoir," mimeographed, Forestry Division, TVA 11 May 1935, 60.

34. Ralph L. Nielsen, "Socio-Economic Readjustment of Farm Families Displaced by the T.V.A.: Land Purchase of the Norris Area," master's thesis, Univ. of Tennessee, Aug. 1940, 136–37.

35. Michael J. McDonald and John Muldowny, "Loyston in the 1930s," *Tennessee Valley Perspective* 2:3 (Spring 1972):26. McDonald and Muldowny, *TVA and the Dispossessed* (Knoxville: Univ. of Tennessee Press, 1982), explains the population removals further.

36. Ibid.

37. The Hawkins fire story is told in the caption of photo 135, Marshall A. Wilson Photo Collection, TVA Library.

38. McDonald and Muldowny, *TVA and the Dispossessed*, 60–61.

39. Jack E. Weller, *Yesterday's People: Life in Contemporary Appalachia* (Lexington: University of Kentucky Press, 1965), 46.

40. James Still, *River of Earth* (New York: Viking, 1940), 10–11.

41. Kalter, "Island Wilderness Biological Survey," fig. 22, 39.

42. W.T. Hunt, "Report of Relocation and Removal of Families from Reservoirs of the Tennessee Valley," mimeographed, Division of Reservoir Properties, TVA, 1953, 25.

43. J.J. Hartzell, "A Study of Population Relocation for Land Between the Lakes," mimeographed, Golden Pond, Ky., Sept. 1970, 2:9.

44. Nielsen, "Socio-Economic Readjustment of Farm Families," 136–37.

45. John C. McAmis, interview by Charles Crawford, 11 Mar. 1970, MSUOHP. Richard Kilbourne, former TVA chief forester, observed about McAmis, "He was afraid that we were going to plant too much of this land in trees and there wouldn't be anything left for the farmers to use." Kilbourne adds, "A farmer couldn't support his family with these little hillside farms, sixty acres, a mule, and a cow. And yet, Mac felt that it was an important way of life that ought to be preserved at all costs." Kilbourne, interview by Crawford, 12 Mar. 1970, MSUOHP, 33–34.

46. "Draft of a Speech before the Power Branch of the TVA by Don McBride, Director," TVA Technical Library, n.d., 2–3.

47. Marian Moffett of the Univ. of Tennessee appears to be the only person who has taken the trouble to investigate these humble churches. She reported that they were usually situated on a rise and, "without exception, they were built during the period between the Civil War and World War I." Moffett, "Some Observations on Vernacular Churches in East Tennessee," *Portfolio* 6 (1982), 61.

48. "Family Case Records," mimeographed, Reservoir Property Management Department, Population Readjustment Divison, TVA, 29 Feb. 1944, 107–11.

49. Ibid., 104–107.

50. Ibid., 68–72.

51. Ibid., 91–97.

52. Ibid., 72–82.

53. "Population Readjustment: Fort Loudoun Area," mimeographed, Reservoir Property Management Department, TVA, 1 Feb. 1943, exhibit 1, p. 1–6.

54. James Patrick, *Architecture in Tennessee* (Knoxville: Univ. of Tennessee Press, 1981), 41.

55. "Family Case Records," 167–71.

56. Jesse Mills, TVA Technical Library, letter to Walter Creese, 2 July 1973, explains that land grants of up to 640 acres originally kept villages of the New England type from appearing, as also had happened in Virginia.

57. George Brown Tindall, *The Emergence of the New South, 1913–1945* (Baton Rouge: Louisiana State Univ. Press, 1967), 576.

58. Rupert B. Vance, *Human Geography of the South: A Study in Regional Resources and Human Adequacy* (Chapel Hill: Univ. of North Carolina Press, 1932), 489–90.

59. Donald Davidson, *The Tennessee: The New River: Civil War to TVA* (New York: Rinehart, 1948), 2:210.

60. Ibid., 21.

61. For a decentralizing trend in the South, see Edward S. Shapiro, "Decentralist Intellectuals and the New Deal," *Journal of American History* 58:4 (Mar. 1972):938.

62. Davidson, *The Tennessee*, 2:211.

63. Ibid., 2:230.

64. Edward S. Shapiro, "The Southern Agrarians and the Tennessee Valley Authority," *American Quarterly*, 22:4 (Winter 1970):797.

65. John L. Stewart, *The Burden of Time: The Fugitives and Agrarians* (Princeton, N.J.: Princeton Univ. Press, 1965), 130.

66. Ibid. However, after World War II, James Dahir did offer a comprehensive view of how the valley might be looked at: "The Valley offers in its rolling landscape, quiet waters, mild climate, and long growing season ideal conditions for a balanced economic and social existence. If it takes its physical environment in hand and moderates its appetite for industry for industry's sake, it may keep what Otte called 'a comparatively low-cost and pleasant home space' in a region of modern farms, well-spaced factories, and human-spaced communities." Dahir, *Region Building* (New York: Harper & Bros., 1955), 111.

67. Robert Penn Warren, *The Flood* (New York: Random House, 1963), 165–66.

68. Ibid., 113.

69. Ibid., 440.

70. Shapiro, "Decentralist Intellectuals and the New Deal," 938.

71. Ibid.

72. Ibid., 948.

73. Map of the Tennessee Valley region, in "TVA Dams and Steam Plants" (Knoxville, Tenn.: TVA, July 1971), 1.

74. "Healing Strip Mine Scars," *A Quality Environment in the Tennessee Valley* (Knoxville: TVA, 1972), 5.

75. Gen. Herbert D. Vogel interview, 9 Jan. 1969, by Charles Crawford, MSUOHP, 18; Marguerite Owen, *The Tennessee Valley Authority* (New York: Praeger, 1973), 111; and Gerald Manners, *The Geography of Energy* (Chicago: Aldine, 1964), 151.

76. "TVA's Coal-Buying Program," mimeographed (Knoxville: TVA, 11 Apr. 1956), 2–3.

77. "Auger May Aid Area Mining," *Knoxville News-Sentinel*, 22 Feb. 1958.

78. "TVA's Coal-Buying Program," 2.

79. Ibid., 7–8. See also "Coal Men Denounce TVA Buying," *Knoxville News-Sentinel*, 9 May 1954.

80. "Lewis Scores U.S. on 'Doghole' Coal," *New York Times*, 2 Feb. 1955. "'Dogholes' Feast on TVA Market," *Knoxville News-Sentinel*, 14 June 1954, notes that "The 'doghole' owners are feasting on the rich TVA market . . . That much coal would make happy all the 'regular' operators if they could grab that market."

81. "Coal Firms Sue for $30 Million," *Knoxville News-Sentinel*, 7 Jan. 1961.

82. "Memo Prepared by Chris Eckl for Dr. Creese," mimeographed, Information Office, TVA, 6 Nov. 1973, 2.

83. "Coal Operator Challenges TVA on 'Pilot' Mine," Nashville (Tenn.) *Tennessean*, 20 Feb. 1961.

84. "Tennessee Valley Authority: The Yardstick with Less than 36 Inches," *Forbes* 115:7 (1 Apr. 1975):27.

85. "Coal Operator Challenges TVA."

86. Bruce Daniel Rogers, "Public Policy and Pollution Abatement: TVA and Strip Mining" (Ph.D. diss., Indiana Univ.), 1973, 58–60, 90–91, 109–10, 129–45. Another model project was proposed for eastern Kentucky, at the Kentucky Oak Mining Company; Rogers, 130.

87. *Reclamation Brings Wildlife Back to Paradise Strip Mine* (Knoxville: Division of Forestry, Fisheries, and Wildlife Development, TVA, 1969), n.p.

88. "Can the Scars be Healed?" *Tennessee Valley Perspective* 1:3 (Spring 1971): 29. For TVA's stronger stance in 1970–71, see Rogers, "Public Policy and Pollution Abatement," 269–77.

89. "Auger May Aid Area Mining." TVA had become very cost-conscious. For that reason, TVA now would bite even the hand that fed it. "Both Power and Purchasing tend to operate on a least cost basis and have conducted cost-benefit analyses relying on costs and benefits in dollar terms. TVA as a whole is a very cost conscious agency. As an example, when President Kennedy came to the Valley he suggested to TVA officials that some new projects be located in Eastern Kentucky, as an area of considerable poverty. TVA's response to him was that projects were not determined on that basis. Rather, costs were the most important factor. Kennedy was not pleased." Rogers, "Public Policy and Pollution Abatement," 82–83.

90. Tom Adkinson, "There's a Spot of Beauty in Them Strip-Mined Hills," *National Observer*, 22 Sept. 1973, 16.

91. "Tennessee Valley Authority: The Yardstick," 25.

92. Ibid.

93. Tom Redburn, "TVA's Perpetual Motion Machine," *Environmental Action* 7:14 (22 Nov. 1975):9. Redburn also observed, 8: "Like the animal caretakers in George Orwell's *Animal Farm*, TVA has come to resemble more and more closely the private utilities it was supposed to supplant." Richard H.K. Victor, *Environmental Politics and the Coal Coalition* (College Station: Texas A & M Univ. Press, 1980), 25–26, came to essentially the same conclusion, that while the TVA had claimed environmental responsibility, in the meantime it had been "leading corporate opposition to effective federal controls of strip mining and air pollution." The TVA directors had lost sight of "environmental engineering."

94. Nat Caldwell, "Aid Hikes Hope of Jobless Area," Nashville *Tennessean*, 19 Feb. 1961.

95. "Wilson Holds that UMW Does Not Violate Laws in Coalfield Operations," Chattanooga *Times*, 5 Mar. 1967.

96. Bernard E. La Berge, "Appalachia: The Lessons of Development," *Exchange* 10:3 (Winter 1975):5. The group, made up of eighteen students from ten developing countries, traveled to Clairfield, Tenn., where underground mining had been replaced by strip mining. The machines used in the latter process had caused the population of Clairfield to shrink from 12,000 to 1,200.

97. Rogers, "Public Policy and Pollution Abatement," 97.

98. Ibid., 95–96.

99. Ibid., 238–39, 288.

100. Ibid., 423.

101. Michael Frome, "The Most Thoroughly Dammed Nation on Earth," *Architectural Forum* 128 (April 1968):80.

102. "Stalking the Law," *Time*, 16 Oct. 1978, 84. "House Floor Vote Could be Dammed by Snail Darter & Proposition 13," *Preservation Action* 3:3 (June/July 1978):1.

103. "Audubon Society Hits Dam," *Knoxville Journal*, 26 Apr. 1965.

104. Carson Brewer, "Little Tennessee Called Top Trout River in East," *Knoxville News-Sentinel*, 17 Nov. 1963.

105. "The Tellico Project," mimeographed, TVA, Dec. 1971, 7.

106. Ibid. There were seven of these industrial terminal areas. In 1970, Chattanooga was employing 11,800 on its riverfront and Knoxville only 199.

107. "Floaters 'Undedicate' Little T's Tellico Dam," *Monroe County Democrat*, 12 May 1971.

108. "New GOP Blow at TVA?" *Tullahoma News and Guardian*, 24 Dec. 1971. See also William B. Wheeler and Michael J. McDonald, *TVA and the Tellico Dam* (Knoxville: Univ. of Tennessee Press, 1986), 142–44.

109. "The Tellico Project," 4.

110. Michael Frome, "Letters," *Architectural Forum* 130 (June 1968):11.

111. "Foresters Ask Halt on Dam," *Knoxville Journal*, 12 Apr. 1965. Cash explained, "I have tried to stay out of politics completely and sing very, very few 'protest' songs, for my business is music. However, half my blood is Cherokee — all my blood is American — and I find myself resenting for various reasons the flooding and destruction of landmarks, buildings and the soil that was my forefathers." His song "Bitter Tears" tells of the confiscation of Indian lands.

112. Walter Amann, Jr., "Cherokee Indians Appeal to Douglas to Save Homeland," *Knoxville Journal*, 5 Apr. 1965, 3.

113. William O. Douglas, in "This Valley Wants to Die," *True for Today's Man*, May 1969, 95, says, "Jarrett B. Blythe, principal chief, was dressed in full tribal costume, and read the following (in part) from the petition." Douglas did not know the Indian was not Blythe.

114. Ibid., 43. For more on the Douglas visits in 1965 and 1969, see Wheeler and McDonald, *TVA and the Tellico Dam*, 65, 83–85, 130–31.

115. "Plea by Cherokees to Save 'Little T' Was Staged Affair," *Knoxville News-Sentinel*, 10 Apr. 1965.

116. Earle S. Draper, ed., *The Scenic Resources of the Tennessee Valley: A Descriptive and Pictorial Inventory* (Knoxville: Department of Regional Planning Studies, TVA, 1938), 133.

117. Probably the best-known Morgan essay in this vein was Arthur Morgan, *dams and other disasters: a century of the army corps of engineers in civil works* (Boston: Porter Sargent, 1971). He was able to be critical of his own trade.

118. Carl Manning, "Carter OKs Tellico Dam Completion," Champaign-Urbana (Ill.) *News-Gazette*, 23 Apr. 1981.

119. Tom Eblen, "Snail Darter Fish Not Rare," Champaign-Urbana (Ill.) *News-Gazette*, 23 Apr. 1981. For Etnier's part, see Wheeler and McDonald, *TVA and the Tellico Dam*, 144–45, 157.

120. "The Tellico Project of the Tennessee Valley Authority," mimeographed, TVA, May 1965, 8.

121. Charles E. Merriam, "The National Resources Planning Board: A Chapter in American Planning Experience," *American Political Science Review* 38: 6, (Dec. 1944):1079–80.

122. Ibid., 1085.

123. "Expanding Outdoor Enjoyment," in *A Quality Environment in the Tennessee Valley* (Knoxville: TVA, Dec. 1972), 9.

124. John Egerton, *The Americanization of Dixie: The Southernization of America* (New York: Harper's Magazine Press, 1974), 73.

125. Phoebe Cutler, *The Public Landscape of the New Deal* (New Haven, Conn.: Yale Univ. Press, 1985), 133.

126. Egerton, *Americanization of Dixie*, 74.

127. Frank E. Smith, *Land Between the Lakes* (Lexington: Univ. of Kentucky Press, 1971), vii.

128. Dero A. Saunders and Sanford S. Parker, "$30 Billion for Fun," *Fortune* 49 (June 1954):226.

129. Ibid., 118.

130. Ibid., 230.

131. Bill Wolf, "These Southerners Just Love Yankees," *Saturday Evening Post* 226 (5 Sept. 1953):42.

132. Ibid., 47.

133. Saunders and Parker, "$30 Billion for Fun," 234.

134. Wolf, "These Southerners Just Love Yankees," 47.

135. "Annual Report," mimeographed, Division of Reservoir Properties, TVA, 1955, 9.

136. "Annual Report," mimeographed, Division of Reservoir Properties, TVA, 1954, 8.

137. "Annual Report," mimeographed, Reservoir Properties Department, TVA, 1947, 17.

138. "Annual Report," mimeographed, Division of Reservoir Properties, TVA, 1948, 8.

139. "Annual Report," Reservoir Properties, TVA, 1954, 10.

140. Ibid., 7.

141. "Annual Report," mimeographed, Division of Reservoir Properties, TVA, 1953, 10–11.

142. As cited in Carl W. Condit, *American Building Art: The Twentieth Century* (New York: Oxford Univ. Press, 1961), 233.

143. See Walter Creese, *The Crowning of the American Landscape* (Princeton, N.J.: Princeton Univ. Press, 1985), 51–52.

144. "Annual Report," mimeographed, Division of Reservoir Properties, TVA, 1950, 11.
145. "Annual Report," Reservoir Properties, TVA, 1947, 20.
146. "Annual Report," mimeographed, Division of Reservoir Properties, TVA, 1949, 16.
147. "Annual Report," Reservoir Properties, TVA, 1955, 10.
148. "Annual Report," Reservoir Properties, TVA, 1953, 5–6.
149. "Annual Report," Reservoir Properties, TVA, 1954, 9.
150. "Annual Report," Reservoir Properties, TVA, 1955, 7.

Chapter 4

1. Earle Draper to Walter Creese, 16 Sept. 1973.
2. William C. Fitts, Jr., interview, 1 Aug. 1969, MSUOHP, p. 20. Mario Bianculli, interview, 4–5 Mar. 1969, MSUOHP, likewise stated that "TVA was fortunate in the misfortune of the country at large because, due to conditions of unemployment elsewhere, TVA was able to gather under one roof, so to speak, the best talent, in many fields — from educational to economical, legal, engineering fields." About the general social atmosphere, Bianculli said, "It was a time of complete stagnation of industrial activities and a complete lack of cheerfulness anywhere."
3. Leo Marx, *The Machine in the Garden: Technology and the Pastoral Ideal in America* (New York: Oxford Univ. Press, 1964), 229.
4. Roland Wank, interview by Gilbert Stewart, Jr., TVA Information Office, 6 Aug. 1965, TVA, 1.
5. Gordon R. Clapp, *The TVA: An Approach to the Development of a Region* (Chicago: Univ. of Chicago Press, 1955), 71.
6. Roy Lubove, *The Urban Community: Housing and Planning in the Progressive Era* (Englewood Cliffs, N.J.: Prentice-Hall, 1967), 15.
7. Samuel P. Hays, *Conservation and the Gospel of Efficiency: The Progressive Conservation Movement, 1890–1920* (Cambridge, Mass.: Harvard University Press, 1959), 41.
8. Earle S. Draper, ed., *The Scenic Resources of the Tennessee Valley: A Descriptive and Pictorial Inventory* (Knoxville: Department of Regional Planning Studies, TVA, 1938).
9. Hays, *Conservation and the Gospel of Efficiency*, 127.
10. Quoted in Richard Kilbourne, TVA forester, interview by Charles W. Crawford, 12 Mar. 1970, MSUOHP, 7–8. In regard to the Harcourt Morgan and McAmis policies, Willis M. Baker, another TVA forester who hosted Pinchot during his 1938 visit to the valley, reported that "there was a great deal of emphasis at that time on the agricultural program, and much of the work was conducted in cooperation with the land-grant colleges and the Agricultural Extension Service. We were pressured to work

in that same pattern." He further noted that in 1938, when Harcourt Morgan became TVA chairman, "TVA started disposing of the forest land." During World War II, the status of the TVA forestry department rose somewhat, "and its work started to become appreciated by the top TVA officials," because of the sudden call for forest products—again a pragmatic, emergency reason. Baker also told about Pinchot's visit. Willis M. Baker, interview by Charles Crawford, 8 Feb. 1970, MSUOHP, 11, 12, 25, 27–29.

11. Edward C.M. Richards, "The Future of TVA Forestry," *Journal of Forestry* 36:7 (July 1938):647.

12. Ibid., 644.

13. Ibid., 650.

14. Willis M. Baker interview, 27–29. When the TVA began, 10 percent of the forest was burned over each year without authorization. Eventually the percentage dropped to .5 percent. The natives were used to reducing litter by light burning, "swingeing." The mountain people started fires to eliminate snakes, ticks, and other insects.

15. *TVA Annual Report*, (Knoxville: TVA, 1953), 69.

16. Harcourt A. Morgan, "Tennessee Valley Authority"—(Washington, D.C.: The Evening Star, The Sunday Star, 26 Sept. 1934), an NBC broadcast, 6.

17. McAmis attitudes from John C. McAmis, interview by Charles Crawford, 11 Mar. 1970, MSUOHP. The McAmis viewpoint is also explained in Norman I. Wengert, *Valley of Tomorrow* (Knoxville: Bureau of Public Administration, Univ. of Tennessee, 1952), 116–17, 123–27.

18. Thomas Bender, "The Rise and Fall of Architectural Journalism," *Design Book Review* 15 (Fall 1988):47.

19. Ibid., 49.

20. Lewis Mumford, "The Architecture of Power," *New Yorker*, 7 June 1941, 60.

21. David E. Lilienthal, "Remarks Made by David E. Lilienthal, Director of the TVA, at the Members' Preview of the Exhibition of TVA Architecture and Design at the Museum of Modern Art, New York City, April 29, 1941," mimeographed, TVA, 2–3.

22. Earle S. Draper to Walter Creese, 10 and 16 Sept. 1973.

23. Also discussed in Walter L. Creese, "The TVA as an Allegory," in *Built for the People of the United States: Fifty Years of TVA Architecture*, ed. Marian Moffett and Lawrence Wodehouse (Knoxville: Art and Architecture Gallery, Univ. of Tennessee, 1983), 61. Roland Wank was undoubtedly affected by German Expressionist design for industry as well. His bibliographies cite such books as Hans Hertlein, *Neue Industriebauten Siemenskonzerns* (Berlin: Wasmuth, 1927), and *Das Grosskraftwerk Klingenberg* (Berlin: Wasmuth, 1928). Marian Moffett and Lawrence Wodehouse, "Roland Wank and Norris Dam," *Tennessee Architect* 3 (Fall 1988):22, puts forth the equally plausible thought that Adolf Loos of Vienna influenced

Wank. A fuller treatment of Wank's life is in Moffett's very recent "Looking to the Future: The Architecture of Roland A. Wank," *Arris* 1 (1989), 5–17.

24. Arthur E. Morgan, "Bench-Marks in the Tennessee Valley," *Survey Graphic* 23:1 (Jan. 1934):42. For another reference to the Danube, see Arthur E. Morgan, *The Small Community: Foundation of Democratic Life* (New York: Harper and Bros., 1942), 74.

25. Walter L. Creese, "American Architecture from 1918 to 1933" (Ph.D. diss., Harvard Univ., 1949), 2, 5–7.

26. Roland Wank, "Nowhere to Go but Forward," *Magazine of Art* 34:1 (Jan. 1941):12.

27. Christiane Crasemann Collins used this German term as a code word in the best summary of recent books about Viennese cultural cohesion in the arts, sociology, and politics: Collins, "Vienna 1900," *Design Book Review* 8, (Winter 1986):28–33.

28. Wank, interview by Gilbert Stewart, 2. Moffett, "Looking to the Future," 12, reports that Col. Theodore Parker of the Army Engineers was also involved in resisting Wank's Norris Dam suggestions.

29. Albert S. Fry, interview by Charles W. Crawford, 17 Oct. 1969, MSUOHP, 23.

30. Earle S. Draper, interview by Charles W. Crawford, 30 Dec. 1969, MSUOHP, 40.

31. Mario Bianculli, architect, interview by Charles Crawford, 25 Mar. 1970, MSUOHP, 36–37), states that Bianculli never had an unpleasant experience with the TVA engineers "except once when the other architectural department — that was regarding the Guntersville Dam — tried to by-pass our designing offices and push their design without even advising us that they were designing it. . . . When they produced the design, of course, we objected. . . . They appointed a big architect from Detroit [Albert Kahn] to judge the two designs after we prepared ours, and ours was selected and built. That was the end of it because it proved that architect expeditors like Mr. [Harry] Tour had better stick to their knitting." Bianculli regarded Tour — who, with his dual training at the Univ. of Illinois, thought of himself as both architect and engineer — as his know-it-all nemesis. Bianculli left TVA in 1945, the year after Wank. Bianculli, interview by Charles Crawford, 20–21 Mar. 1970, MSUOHP.

32. Harry Tour, interview by Charles Crawford, 11 Mar. 1970, MSUOHP, 53.

33. Frederick A. Gutheim, "Tennessee Valley Authority: A New Phase in Architecture," *Magazine of Art*, 33:9 (Sept. 1940):531.

34. Gutheim, "Roland Wank, 1898–1970," *Forum* 133 (Sept. 1970):59.

35. See Walter L. Creese, *The Crowning of the American Landscape* (Princeton: Princeton Univ. Press, 1985), 68–69, 92, for the ways in which Americans regarded their mountains as epics.

36. Paul Zucker, *American Bridges and Dams* (New York: Greystone Press, 1941), 15.

37. Wank, interview by Gilbert Stewart, 2.
38. Arthur Morgan, *dams and other disasters: a century of the army corps of engineers in civil works*, 371.
39. Lilienthal, "Remarks at the Museum of Modern Art," 2.
40. Talbot Hamlin, "Architecture of the TVA," *Pencil Points* 20 (Nov. 1939):45.
41. Ibid.
42. For a picture of the Liberty Bell layout, see William West, *America's Greatest Dam: Muscle Shoals, Alabama* (New York: Frank E. Cooper, 1925), 23, 41.
43. Frances Perkins, *The Roosevelt I Knew* (New York: Viking, 1946), 77.
44. Sibyl Thurman, "Building for the People," *Inside TVA*, 14 June 1983, 5, has a picture of this model in relief.
45. Charles Krutch, interview by Charles Crawford, 10 Nov. 1969, MSUOHP, 2.
46. Talbot Hamlin, ed., *Forms and Functions of Twentieth Century Architecture* (New York: Columbia Univ. Press, 1952), 4, 263.
47. Marian Moffett and Lawrence Wodehouse, "The Art of a Utility: Architecture of the TVA," *University of Tennessee Journal of Architecture* 7 (1983):57.
48. Harry Tour, interview, MSUOHP, 11. For Wank, Tour, and other TVA architects, see letter from Mario Bianculli, *Architectural Forum* 73 (Dec. 1940):15–16.
49. Gutheim, "Tennessee Valley Authority," 522.
50. Stuart Chase, *Rich Land Poor Land* (London: Whittlesey House, 1936), 286–87.
51. Walter L. Creese, "The Even More Extraordinary Jefferson," *Places* 2:3 (1985):3–5.
52. Gutheim, "Tennessee Valley Authority," 524.
53. Creese, "American Architecture from 1918 to 1933," 34–37.
54. Earle S. Draper, "The TVA Freeway," *American City* 49 (Feb. 1934):47–48. Draper also noted that J.C. Bradner, Jr., head of the TVA Section of Engineering under Draper, had been chief engineer of the Taconic State Park System of New York, which included the Taconic Parkway. Draper, "Housing the Workers at Norris Dam," mimeographed, TVA, 25 Nov. 1933, 4.
55. Grant Hildebrand, *The Architecture of Albert Kahn* (Cambridge, Mass.: MIT Press, 1974), 211.
56. TVA, *The Fontana Project*, Technical Report No. 12 (Washington, D.C.: USGPO, 1950), 136.
57. John H. Kyle, *The Building of the TVA: An Illustrated History* (Baton Rouge: Louisiana State Univ. Press, 1958), 114.
58. TVA, *Fontana Project* 502.
59. Ibid., 1.
60. Carson Brewer and Alberta Brewer, untitled manuscript, n.d., TVA, 1.
61. Arthur E. Morgan, *The Making of the TVA* (Buffalo, N.Y.: Prometheus Books, 1974), 113.
62. Ibid., 175.

63. TVA, *Fontana Project* 12.

64. Brewer and Brewer, untitled manuscript, 469.

65. TVA, *The Watts Bar Steam Plant*, Technical Report No. 8 (Washington, D.C.: USGPO, 1949), 49.

66. Ibid.

67. Earle S. Draper to Walter Creese, 16 Sept. 1973; and Draper interview with Crawford, 30 Dec. 1969, MSUOHP, 40. Also "Employment of Albert Kahn," notice of action of the TVA board of directors, 8 Sept. 1937. Kahn then was trying to put distance between his own spare, utilitarian, clean style and the richer effects of the Art Deco and Moderne, previously employed by the TVA, for, in his first communication as consultant to TVA in 1937, he said about the Guntersville power house that "the style had practically been discarded." Kahn, "Architectural Treatment of Guntersville Power House," office memorandum, TVA, 14 Sept. 1937. For the Guntersville power house, Kahn suggested use of gray buff brick and emphasis on volumetric feeling, along with some clarity of horizontal design. Kahn was worried that "decoration" might be sought. His Spartan preference for austerity and economy, in the Midwest mode, showed up in the TVA buildings represented in the 1941 Museum of Modern Art exhibition; for this reason, critics believed the buildings had been influenced by Eastern Internationalism. The hiring of Kahn at this point in 1937 appears to have been at the suggestion of Earle Draper, director of the Department of Regional Studies. Memorandum to J.B. Blandford, Jr., TVA general manager, 30 Aug. 1937; and conference, 1 Sept. 1937.

68. TVA, *Watts Bar Steam Plant*, 142.

69. Knudsen, "said that we should increase our electricity, and he pushed the development of the first steam plants." Krutch, interview by Crawford, 25.

70. John H. Kyle, *Building of TVA*, 125–26.

71. Ibid., 130–33.

72. Krutch, interview by Crawford, 4.

73. Harry Wiersema, interview by Charles W. Crawford, 24 Sept. 1969, MSUOHP, 13.

74. Stanford Research Institute, "Design Manual for an SO2 Emission Limitation Program for the Kingston Steam Plant," vol. 1, mimeographed, TVA, 1976, 3.

75. Richard Guy Wilson, "Massive Deco Monument," *Architecture* 72 (Dec. 1983):47. Wilson's later and fuller discussion is in "Machine-Age Iconography in the American West: The Design of Hoover Dam," *Pacific Historical Review*, Special Issue (Fall 1985):463–93. As Wilson so cogently observed, these dam monuments of the Bureau of Reclamation, Army Engineers, and the TVA have "not yet received substantial scholarly attention." For Boulder (Hoover) Dam he provided the kind of analysis needed.

76. U.S. Federal Power Commission, *Northeast Power Failure, November 9 and 10, 1965: A Report to the President*, 6 Dec. 1965 (Washington, D.C.:

USGPO, 1965), 49. C.P. Almon, Jr.; G.O. Wessenauer; M. Stephen Merritt; and K.E. Hapgood, all from the TVA, were part of the northeast study team.

77. *TVA Handbook*, TVA Technical Library, July 1987, 96.

78. Cecil E. Pearce, "Design of Hiwassee Dam — Engineering Details," *Civil Engineering* 10:7 (July 1940):435.

79. Harry B. Tour, "TVA — Ten Years of Concrete," *Architectural Concrete* 9 (1943):26.

80. Pearce, "Design of Hiwassee Dam," 435.

81. Ibid., 435.

82. Bianculli, interview by Charles W. Crawford, 8.

83. Charles W. Johnson and Charles O. Jackson, *City Behind a Fence: Oak Ridge, Tennessee, 1942–1946* (Knoxville: Univ. of Tennessee Press, 1981), 6.

84. George O. Robinson, Jr., *The Oak Ridge Story* (Kingsport, Tenn.: Southern Publishers, 1950), 81. Three other gaseous diffusion plants were built nearby after the war.

85. Ibid., 84–85.

86. Ibid., 89.

87. Ibid., 42.

88. Ibid., 26.

89. Johnson and Jackson, *City Behind a Fence*, 289–90; and Robinson, *Oak Ridge Story*, illustration opposite 113.

90. Earle S. Draper, ed., *The Scenic Resources of the Tennessee Valley: A Descriptive and Pictorial Inventory* (Knoxville: Department of Regional Planning Studies, TVA, 1938), 36.

91. Ibid., 61–62.

92. Coy Ludwig, *Maxfield Parrish* (New York: Watson-Guptill, 1973), 201.

93. Ibid., 13.

94. Ibid., 144, quoting a letter from Parrish to Stephen Newman of Reinthal and Newman, picture reproducers, 2 Jan. 1926.

95. Ibid., 17, 175.

96. Brian Hackett, "TVA: Creator of Landscape," *Journal of the Town Planning Institute* 37 (Nov. 1950):7–11.

Chapter 5

1. Frederick A. Gutheim, "Roland Wank, 1898–1970," *Forum* 133 (Sept. 1970):59. A reservation similar to Gutheim's came from Talbot Hamlin of Columbia Univ., who reported that Norris Village "has been criticized as too romantic in its labyrinth of curving roads, and too historical in its usual type of house design." Hamlin, "Architecture of the TVA," *Pencil Points*, 20 Nov. 1939, 41.

2. TVA, *The Norris Project*, Technical Report No. 1 (Washington, D.C.: USGPO, 1940), 174, 193. Tracy Augur, "The Planning of the Town of Norris," *Ameri-*

can *Architect* 148, p. 24, gives a slightly different figure of 281 single houses, 10 duplexes, and the reuse of existing farmhouses at Norris, to add up to 350 units.

3. Tracy Augur to F.X. Reynolds, 20 Feb. 1935, Norris Microfilm, TVA Technical Library quoted in William H. Jordy, "A Wholesome Environment Through Plain, Direct Means: The Planning of Norris by the Tennessee Valley Authority," manuscript, 55–56. Professor Jordy was kind enough to share this very valuable and perceptive but unfortunately not yet published manuscript.
4. Donald Davidson, *The Tennessee* (New York: Rinehart, 1948), 2:230.
5. Malcolm Ross, *Machine Age in the Hills* (New York: Macmillan, 1933), 57.
6. Ibid., 235.
7. Jacques Ellul, *The Technological Society* (New York: Knopf, 1964), vi–vii.
8. Ross, *Machine Age in the Hills*, 233.
9. Ibid., 237.
10. Ibid., 204.
11. Allen Eaton, *Handicrafts of the Southern Highlands* (New York: Russell Sage Foundation, 1937), 108.
12. Ibid., 254.
13. Ibid., 299.
14. Ibid., 297.
15. Ibid., 274.
16. Ibid., 294.
17. Ibid. A craft museum was proposed at the west abutment of the dam.
18. David Lilienthal, "Some Observations of the TVA," address to TVA employees, 12 June 1936, in *Speeches and Remarks of Lilienthal*, TVA Technical Library, 1:7–8.
19. Ross, *Machine Age in the Hills*, 237.
20. Ibid., 88.
21. Ibid., 89.
22. Frances Perkins, *The Roosevelt I Knew* (New York: Viking, 1946), 73.
23. Ibid., 69.
24. See Walter L. Creese, *The Search for Environment: The Garden City Before and After* (New Haven, Conn.: Yale Univ. Press, 1966), 218.
25. Arthur E. Morgan and J. Dudley Dawson, interview by Charles Crawford, Yellow Springs, Ohio, 20 June 1969, p. 17, MSUOHP.
26. Cited in Jordy, "A Wholesome Environment," 25–26. Phoebe Cutler, *The Public Landscape of the New Deal* (New Haven, Conn.: Yale Univ. Press, 1985), 134, says, "Not associated with the regional and city planning efforts around New York, Draper was a safe, non-controversial figure to lead an experimental planning venture." Her source was Jordy. Mumford does not appear to have been friendly to Draper and vice versa.
27. Aelred J. Gray, "The Maturing of a Planned New Town: Norris, Tennessee," *Tennessee Planner* 32 (1974):1.

28. Other small town titles by Arthur E. Morgan were *The Community of the Future and the Future of the Community; The Heritage of Community* (Yellow Springs, Ohio: Community Service Books, 1957); and *Bottom Up Democracy: The Affiliation of Small Democratic Units for Common Service* (Yellow Springs, Ohio: Community Service Books, 1954). Otis L. Graham, Jr., *An Encore for Reform: The Old Progressives and the New Deal* (New York: Oxford Univ. Press, 1967), 181, cites George L. Record as the spokesman for those Progressives who wished "to restore the small-town synthesis."

29. Earle Draper to Walter Creese, 10 Sept. 1973.

30. E.S. Draper, "Southern Textile Village Planning," *Landscape Architecture* 18:1 (Oct. 1927):5.

31. Draper, interview, 30 Dec. 1969, MSUOHP, 31.

32. Letter from Jesse Mills to Walter Creese, 14 Aug. 1973.

33. Earle S. Draper, "Applied Home Economics in TVA Houses," *Journal of Home Economics* 27:10 (Dec. 1935):632. That the houses at Norris did not appear more "modern" was due partly to the fact that, although Wank wished to innovate, he was removed early from the village to work on the dam. Earle Sumner Draper, Jr., the son, in a useful recent article, "The TVA's Forgotton Town: Norris, Tennessee," *Landscape Architecture* 78:2 (Mar. 1988):97, remarked, "Needless to say, at a time when American architects were looking to Europe for new directions, this design approach came under criticism. It was not until the noted Finnish-American architect and planner Eliel Saarinen visited Norris and submitted a favorable critique of the building and house designs that the criticism subsided." The inclination was to appeal to a higher court of "advanced" architects of the day, as with Roland Wank and Albert Kahn on the design of dams. Saarinen had begun his own career with an investigation of folk architecture of the Karelian district of Finland. Harry Tour, interview by Charles W. Crawford, 11 Mar. 1970, MSUOHP, 15, likewise reported that some of the designers then wanted "quite modern design . . . but the wiser heads prevailed on the TVA directors to stick with house designs that were more indigenous to that particular area." The sense of drama lay in the contrast of these designs to the smooth sophistication of TVA technology, a contrast that suddenly was lost later, in the "modern" prefabricated portable houses of the war period.

34. Draper, "Applied Home Economics," 636.

35. Draper, interview, MSUOHP, 10.

36. Draper, "Applied Home Economics," 635.

37. "TVA to Back Small Cities," *Florence (Ala.) Times*, 23 Apr. 1934. Rexford Tugwell particularly noted the low cost of these houses; *Knoxville News-Sentinel*, 17 Aug. 1934. James Agee reported on them in "TVA I: Work in the Valley," *Fortune*, 11 May 1935, 145.

38. Draper, "Applied Home Economics," 635.

39. Letter from Jesse Mills to Walter Creese, 14 Aug. 1973.
40. Jordy, "A Wholesome Environment," 46A.
41. Denis W. Brogan, *The Era of Franklin D. Roosevelt* (New Haven, Conn.: Yale Univ. Press, 1950), 261–62.
42. Draper, "Applied Home Economics," 633–34.
43. "TVA to Bring Luxuries to Women of Mountains," *Asheville (N.C.) Citizen*, 17 Mar. 1934. On the other hand, Jesse Mills, in a letter to Walter Creese, 8 June 1973, says, "I remember hearing of one woman (a native) remark after visiting in Norris, 'I wouldn't live way back there in those woods!' Of course she was reacting to her own lifetime of isolation; she wanted to live on a main road." Wank was upset because Mrs. Arthur Morgan wished the houses to "be designed for cooking with wood or coal," in order to avoid inducing culture-shock in the mountain women. She was a considerable student of home economics and domestic arrangement. Wank complained to Lilienthal, who wished to have heating and cooking by electricity. Gilbert Stewart, Jr., "Interview with Roland Wank," 6 Aug. 1965, TVA, 2.
44. Earle Draper, ed., *The Scenic Resources of the Tennessee Valley: A Descriptive and Pictorial Inventory* (Knoxville: Department of Regional Planning Studies, TVA, 1938), 68–69.
45. Draper, interview, MSUOHP, 7.
46. Jordy, "A Wholesome Environment," 15.
47. Arthur Morgan, *The Making of the TVA* (Buffalo, N.Y.: Prometheus Books, 1974), 124.
48. TVA, *The Norris Project*, 210.
49. Ibid., 210–11.
50. Arthur Morgan, *Making of the TVA*, 121.
51. Ibid., 126.
52. Arthur E. Morgan and J. Dudley Dawson, interview, MSUOHP, 13.
53. Ibid., 11–12.
54. Paul Ashdown, ed., *James Agee: Selected Journalism* (Knoxville: Univ. of Tennessee Press, 1985), 76.
55. Morgan, *Making of the TVA*, 129.
56. Harry Tour, interview, 11 Mar. 1970, MSUOHP, 23.
57. Morgan, *Making of the TVA*, 67–68. See also Ashdown, *James Agee*, 78, 94.
58. Morgan, *Making of the TVA*, 69.
59. Letter from Earle Draper to Walter Creese, 16 Sept. 1973.
60. "Early 'New Town' has 40th Birthday," *Weekly Newsletter*, Information Office, TVA, 1 Nov. 1973.
61. Draper to Creese, 16 Sept. 1973.
62. Draper, interview, MSUOHP, 8.
63. TVA, *The Norris Project*, 181.
64. Gray, "Maturing of a Planned New Town," 11, n. 17.

65. Ibid., 12.
66. Tour, interview, MSUOHP, 24–27.
67. Gray, "Maturing of a Planned New Town," 12.
68. Letter from Draper to Creese, 12 June 1974.
69. Gray, "Maturing of a Planned New Town," 11.
70. Jordy, "A Wholesome Environment," 48–49, 54–55.
71. Tracy Augur, interview, *Knoxville Journal*, 16 Mar. 1935.
72. Draper, *Scenic Resources of the Tennessee Valley*, 116.
73. Ibid., 13.
74. Ibid., v, vi.
75. Cutler, *Public Landscape of the New Deal*, 81, speaks of the "amphitheater endemic to the public vacation grounds of the Depression era."
76. Draper, *Scenic Resources of the Tennessee Valley*, 102.
77. Ibid., 116.
78. Benton MacKaye, "Habitability: A Study of the Norris Sub-Region," 22 Oct. 1935, mimeographed, Engineering and Construction Departments, TVA, 12. Complete descriptions and maps of the Norris parks are in Earle S. Draper, "Demonstration Parks in the Tennessee Valley," *Parks and Recreation* 20:8 (Apr. 1937):357–71. Accounts of Wheeler, Wilson, and Pickwick parks; Monte Sano Park near Huntsville, Ala.; and the Norris Freeway are Earle S. Draper, "Demonstration Parks in the Tennessee Valley," Part 2 *Parks and Recreation* 20:9 (May 1937):418–32.
79. Arthur Morgan, "Island Wilderness Preserve," office memorandum to Blandford, 7 May 1934; and Louis B. Kalter, "Island Wilderness Biological Survey," Forestry Division, TVA, 11 May 1935; and letter from Jesse Mills to Walter Creese, 11 Jan. 1974.
80. Draper, *Scenic Resources of the Tennessee Valley*, 131–32.
81. Robert M. Howes, Chief, Recreation Relations Staff, Regional Studies Department, "Recreational Planning," mimeographed, 6 Nov. 1947, TVA, 9.
82. Draper, *Scenic Resources of the Tennessee Valley*, 186–87.
83. Arthur Morgan, *Making of the TVA*, 69.
84. Draper, Interview, MSUOHP, 19–21.
85. Jordy, "A Wholesome Environment," n. 40; and Gray, "Maturing of a Planned New Town," 13–15.
86. Wiersema, interview MSUOHP, 16–17.
87. Howard K. Menhinick and Lawrence L. Durisch, "TVA: Planning in Operation," *Town Planning Review* 24:2 (July 1953):131.
88. Ibid. William B. Wheeler and Michael J. McDonald, *TVA and the Tellico Dam* (Knoxville: Univ. of Tennessee Press, 1986), 10–11, 19–21, recount the evolution of the land purchase policy in detail. They report that Harcourt Morgan and his agricultural group resisted the purchase of land and comprehensive regional development partly because they did not want tourism.
89. Lawrence L. Durisch and Robert E. Lowry, "The Scope and Content of Administrative Decision—The TVA Illustration," *Public Administration Re-*

view 13:4 (Autumn 1953):223. See also Charles J. McCarthy, "Land Acquisition Policies and Proceedings in TVA — A Study of the Role of Land Acquisition in a Regional Agency," *Ohio State Law Journal* 10 (Winter 1949): 46–63.

90. Phillip Selznick, *TVA and the Grass Roots: A Study in the Sociology of Formal Organization* (Berkeley: Univ. of California Press, 1949).

91. R.G. Tugwell and E.C. Banfield, "Grass Roots Democracy — Myth or Reality?", *Public Administration Reviews* 10 (Winter 1950):49.

92. Ibid., 52.

93. Neil Bass, interview by Charles W. Crawford, 9 July 1969, MSUOHP, 11.

94. Tugwell and Banfield, "Grass Roots Democracy," 49.

95. Ibid., 50.

96. David Lilienthal, interview by Charles Crawford, 7 Feb. 1970, MSUOHP, 26.

97. Ibid., 23–24.

98. Tugwell and Banfield, "Grass Roots Democracy," 54.

99. Ibid., 50.

100. Ibid., 49.

101. Ibid.

102. Thomas K. McCraw, *Morgan vs. Lilienthal* (Chicago: Loyola Univ. Press, 1970), ix.

103. Ibid.

104. Anthony Coelho, "David Eli Lilienthal: Pragmatic Liberal at the Crossroads," term paper, History 268, Brown Univ., Prof. John Thomas, 19 Jan. 1973, 8.

105. McCraw, *Morgan vs. Lilienthal*, 40.

106. U.S. 75th Congress, 2d sess., *Hearings Before the Joint Committee on the Investigation of the Tennessee Valley Authority*, May to Dec. 1938, p. 854.

107. Morgan, *Making of the TVA*, 137.

108. Ibid., 141–44.

109. Wiersema, interview, MSUOHP, 35.

110. Arthur E. Morgan, *The Small Community: Foundation of Democratic Life* (New York: Harper & Bros., 107.

111. Ibid., 278.

112. Creese, *Search for Environment*, 292.

113. David Lilienthal, lecture, Seminar on Planning, Harvard Graduate School of Public Administration, 22 Apr. 1941, mimeographed, 23.

114. Lilienthal, interview, MSUOHP, 5–6.

115. Ibid., 6.

116. John P. Ferris, interview by Charles W. Crawford, 7 Dec. 1969, MSUOHP, 8.

117. Ibid., 11. In his interview with Crawford, 13, Lilienthal amplified the idea that Arthur Morgan was "elitist."

118. Arthur E. Morgan, speech at TVA, 27 July 1936, in reply to David Lilienthal, "Some Observations of the TVA." 12 June 1936.

119. Arthur E. Morgan, "Planning in the Tennessee Valley," *Current History* 38, (Aug. 1933):667.

120. George C. Stoney, "A Valley to Hold To," *Survey Graphic* 29:7 (July 1940): 395–99.

121. Lilienthal, "Some Observations of the TVA," 7–8.

122. J. Charles Poe, "The Morgan-Lilienthal Feud," *The Nation*, 143, 14, 3 Oct. 1936, p. 386; see also Jonathan Mitchell, "Utopia, Tennessee Style," *New Republic* 76 (18 Oct. 1933):272, 274.

123. Poe, "The Morgan-Lilienthal Feud," 386.

124. Morgan, *Making of the TVA*, 58.

125. Ibid.

126. Rupert B. Vance, *Human Geography of the South: A Study in Regional Resources and Human Adequacy* (Chapel Hill: Univ. of North Carolina Press, 1932), 492. A partisan of Arthur Morgan, Clarence Lewis Hodge, in *The Tennessee Valley Authority* (Washington, D.C.: American Univ. Press, 1938), 69, supported Morgan's recommendation as an entirely logical one.

127. Gifford Pinchot, "Introduction," in Morris Lewellyn Cooke, *Giant Power: The Report of the Giant Power Survey Board to the General Assembly with a Message of Transmittal from Gifford Pinchot, Governor* (Harrisburg, Pa.: Telegraph Printing, 1925), xii.

128. Louis B. Wehle, *Hidden Threads of History: Wilson Through Roosevelt* (New York: Macmillan, 1953), 164.

129. Arthur E. Morgan, "Bench-Marks in the Tennessee Valley: The Strength of the Hills," *Survey Graphic* 23:1 (Jan. 1934):8.

130. Frederic Delano, "National Planning," introduction, mimeographed, File 450, National Archives.

131. Rexford Tugwell, "Meeting of the Executive Council," Tugwell Diary 1934, Box 14, 12 Dec. 1934, FDRL.

132. E. S. Draper, speech delivered at meeting of National Conference on City Planning and American Civic Association, St. Louis, Mo., 22 Oct. 1934, p. 11.

133. Morgan, *Making of the TVA*, 54–55.

134. Raymond Moley, The First New Deal (New York: Harcourt, Brace and World, 1966), 331.

135. Ibid., 328.

136. Ibid., 327–28.

137. Max Friedman, ed., *Roosevelt and Frankfurter: Their Correspondence 1928–1945* (Boston: Little, Brown, 1967), 448.

138. Ibid.

139. Ibid., 24.

140. Arthur M. Schlesinger, Jr., *The Politics of Upheaval in the Ages of Roosevelt* (Cambridge, Mass.: Riverside Press, 1960), 387.

141. Ibid., 396.

142. David E. Lilienthal, "The TVA and Decentralization," *Survey Graphic* 29: 6 (June 1940):336.

143. Schlesinger, *Politics of Upheaval*, 393.

144. Lilienthal, "TVA and Decentralization," 337.

145. Alpheus T. Mawson, *Brandeis and the Modern State* (Washington, D.C.: National Home Library Foundation, 1936), 91.

146. Ibid., 93.

147. Arthur E. Morgan, *Nowhere Was Somewhere: How History Makes Utopias and How Utopias Make History* (Chapel Hill: Univ. of North Carolina Press, 1946), 4.

148. Samuel Haber, *Efficiency and Uplift: Scientific Management in the Progressive Era, 1890–1920* (Chicago: Univ. of Chicago Press, 1964), 51–55.

149. Ibid., 82. Dewey W. Grantham, "The Ambiguity of American Reform," in *TVA: Fifty Years of Grass-Roots Bureaucracy*, ed. Erwin Hargrove and Paul Conkin (Urbana: Univ. of Illinois Press, 1983), p. 325, says that the TVA "appeared to demonstrate the virtues of decentralization, which was at the heart of the neo-Brandesian approach," but, ironically, the Brandeis message also was aimed toward "scientific management" and "efficiency." The signals were ambiguous.

150. Ferris, interview, MSUOHP, 29.

151. Draper, *Scenic Resources of the Tennessee Valley*, 135.

152. Ferris, interview, MSUOHP, 33.

153. Carson Brewer and Alberta Brewer, untitled manuscript, n.d., TVA Library, p. 454.

154. Ibid., 456–57.

155. Marian Moffett and Lawrence Wodehouse, "Noble Structures Set in Handsome Parks: Public Architecture of the TVA," *Univ. of Virginia Architectural Review, Modulus 17*:1984, 77. The title was based on an observation of the TVA by Raymond Unwin.

156. Brewer and Brewer, untitled manuscript, 462.

157. *Knoxville News-Sentinel*, 23 Aug. 1942; Brewer and Brewer, 459.

158. "The Truth about Prefabricated Houses," *PM Magazine*, 27 Aug. 1945.

159. Carroll A. Towne, "Portable Housing: TVA Leads to Trailer-Houses," *New Pencil Points*, July 1942, p. 49.

160. Le Corbusier, *The Modulor* (Cambridge, Mass.: Harvard Univ. Press, 1954), 54.

161. Ibid., 54–55. Gilbert Herbert, *Dream of the Factory-Made House* (Cambridge, Mass.: MIT Press, 1984), 309, noticed the tendency of Gropius and Wachsmann to string effort and production out for too long because of "a perfectionist attitude," so that no "ad hoc solution of improvised manufacture" could result, as would occur with native Americans. They were too much "delayed at each stage by this search for the ideal . . . and its consequences were disastrous."

162. Ibid., 52.

163. Ibid., 52–53.

164. Ibid., 53.

165. Ibid., 54.
166. Towne, "Portable Housing, 50.
167. Ibid.
168. "The Truth about Prefabricated Houses"; and Moffett and Wodehouse, "Noble Structures," 81.
169. Towne, "Portable Housing," 50.
170. Ibid., 52.
171. Ibid., 55–56.
172. Brewer and Brewer, untitled manuscript, 469.
173. *Knoxville News-Sentinel*, 23 Aug. 1942.
174. TVA, *Fontana Project*, 208.
175. Moffett and Wodehouse, "Noble Structures," 81–82. See also Moffett and Wodehouse, eds., *Built for the People of the United States: Fifty Years of TVA Architecture* (Knoxville: Univ. of Tennessee, 1983), 24–25. This catalogue for the exhibit at TVA's fiftieth anniversary is a serious, very valuable, key document for the physical history.
176. TVA *Fontana Project*, 201–202.
177. Brewer and Brewer, untitled manuscript, 462.
178. Ibid.
179. TVA, *Fontana Project*, 201–202.
180. Lucile Boyden, *The Village of Five Lives* (Fontana Dam, N.C.: Government Services, Inc., 1964), 40–41.
181. "30 Years of Growth — Just a Beginning," *Fontana Villager*, Feb. 1977, 4.
182. *Oak Ridge Operations* (Oak Ridge, Tenn.: U.S. Atomic Energy Commission, 1972), 2.
183. These fences and towers were not permanent, however, and, starting in March 1949, were removed. George A. Sanderson, "America's No. 1. Defense Community: Oak Ridge, Tennessee," *Progressive Architecture* 32 (June 1951):84.
184. George O. Robinson, Jr., *The Oak Ridge Story* (Kingsport, Tenn.: Southern Publishers, 1950), 77–78.
185. Ibid., 51–52.
186. Dick Smyser, "Planners Tried to Avoid Resemblance to Army Camp," *Oak Ridger*, 4 May 1951, pp. 1, 4. Charles W. Johnson and Charles O. Jackson, *City Behind a Fence: Oak Ridge, Tennessee, 1942–1946* (Knoxville: Univ. of Tennessee Press, 1981), 6, makes the same point: "Indeed, the actual terrain at the site was excellent in that operational dangers could be further minimized by placing plants so that they were separated by natural barriers."
187. Ibid., 75.
188. "Atom City," *Architectural Forum* 83 (Oct. 1945):103.
189. Ibid., 105.
190. Johnson and Jackson, *City Behind a Fence*, 12–21.
191. Ibid., 21.

192. "Atom City," 103.

193. "Oak Ridge—A new town planned for destruction," *Town and Country Planning* 13 (Winter 1945–46):183.

194. Dick Smyser, "Planners Tried to Avoid Resemblance to Army Camp," 4. Actually, the first prefabricated houses at Oak Ridge were the Temporary Dwelling Units (TDU) moved from La Porte, Indiana, and Point Pleasant, Virginia, where TNT was manufactured. They were in the West Outer Drive section. The Glenn L. Martin scheme also was designed by Skidmore, Owings and Merrill.

195. Dick Smyser, "Whatever Your Current Housing View, Original Job was Big Task," *Oak Ridger*, 3 May 1951, 2.

196. "Atom City," 105.

197. Ibid., 103; and Smyser, "Whatever Your Current Housing View," 2. In 1959, the city was sold off by the Atomic Energy Commission. Johnson and Jackson, *City Behind a Fence*, 204.

198. "Timberlake—A Planned Environment," *Tennessee Valley Perspective* 1:3 (Spring 1971):17.

199. Withers Adkins, interview by Walter Creese, 27 Apr. 1973, TVA; and "The Tellico Project: A Dam and Reservoir on the Little Tennessee River in Eastern Tennessee," TVA, Dec. 1971, mimeographed, 11; Wheeler and McDonald, *TVA and the Tellico Dam*, 173–83, also discusses the Boeing involvement.

200. Ibid., 7.

201. "The Tellico Project of the Tennessee Valley Authority," May 1965, mimeographed, TVA, 3. Wheeler and McDonald, *TVA and the Tellico Dam*, 59–60, discusses the outmigration from these three counties.

202. "The Tellico Project of the TVA," 2. Wheeler and McDonald, *TVA and the Tellico Dam*, 19–20.

203. "The Tellico Project: A Dam and Reservoir," 3.

204. Donald A. Krueckeberg, ed., *The American Planner* (New York: Methuen, 1983), 317–18, reported that Ladislas Segoe of Cincinnati, a consultant to the TVA during its earliest years, was amazed that Guntersville boomed as a yachting center for Birmingham. In the early years, recreation simply was not thought of as an incipient industry.

205. James L. Gober, chief, Regional Planning Staff, TVA, interview by Walter Creese, Knoxville, 8 May 1973.

206. "Timberlake—a planned environment," 18.

207. Gober, interview by Creese.

208. "Timberlake—a planned environment," 18.

209. Ibid.

210. Ibid., 19.

211. Michael J. McDonald and John Muldowny, *TVA and the Dispossessed* (Knoxville: Univ. of Tennessee Press), 272.

Chapter 6

1. The letdown following disappointment of too-high expectations of the
   TVA is brought out in John Friedmann's *Retracking America: A Theory
   of Transactive Planning* (New York: Anchor, 1973), 5. Friedmann was disturbed to discover when he went to work for TVA in 1952 that it had "not
   ever had a comprehensive plan." An organizational network had been
   built to protect the TVA and the "outside" interests that were helping to
   sustain it, making the TVA "less capable of dreaming big dreams, of ex-
   ploring new solutions, and of influencing the wants of people in the community where the 'poverty of aspirations' limits the horizon of what is
   thought to be possible." See also Friedmann, "The Spatial Structure of the
   Economic Development in the Tennessee Valley," Univ. of Chicago Department of Geography, Research Paper, no. 39, Mar. 1955, 11. Norman I.
   Wengert expressed a similar conclusion. He believed that TVA had not carried out systematic, comprehensive social planning since 1938: "In fact,
   the word 'planning' has been deliberately avoided, and when possible such
   euphemisms as 'forecast,' 'study,' or 'survey' have been substituted." Wengert, *Valley of Tomorrow: The TVA and Agriculture* (Knoxville: Bureau
   of Public Administration, Univ. of Tennessee, 1952), 13.
2. "Introduction to National Planning," mimeographed, file 450, National
   Archives.
3. Frederic A. Delano, "A Memorandum Relating to the Functions of a Proposed Advisory Council for National Planning and Research to Consider
   Comprehensively the Physical, Economic and Social Aspects of the Problem, July 18, 1933." R–G 187, files 103–104, National Archives.
4. Quoted in William E. Leuchtenburg, "Roosevelt, Norris and the 'Seven
   Little TVAs,'" *Journal of Politics* 14:3 (Aug. 1952):418.
5. "National Resources Committee," in *Regional Factors in National Planning and Development* (Washington, D.C.: USGPO, 1943), 15, refers to the
   TVA as the model for the NRPB to watch. Delano was the NRPB "head
   of office" throughout, and Charles W. Eliot II was its executive director.
   Charles E. Merriam and Wesley C. Mitchell were two other prominent
   members of the board.
6. Norman Beckman, "Federal Long-Range Planning: The Heritage of the
   National Resources Planning Board," 23 Dec. 1958, mimeographed (Washington, D.C.: U.S. Bureau of the Budget, 1958), 13–14.
7. Ibid., 12–13, 21–23.
8. Charles W. Eliot II, interview by Walter Creese, Cambridge, Mass., 1 Nov.
   1972.
9. Charles E. Merriam, "The National Reources Planning Board: A Chapter
   in American Planning Experience," *American Political Science Review* 38:
   6 (Dec. 1944):1076. Merriam also made the interesting point that the

Roosevelt national planning board "was substantially the projection of the Advisory Council proposed by President Hoover's Committee on Recent Social Trends in 1933." "The National Resources Planning Board," *Planning for America*, ed. George B. Galloway (New York: Holt, 1941), 493.

10. C. Herman Pritchett, "Organization of Regional Planning," in *In Search of the Regional Balance of America*, ed. Howard W. Odum and Katharine Jocher (Chapel Hill: Univ. of North Carolina Press, 1945), 150.

11. Leuchtenburg, "Roosevelt, Norris and the 'Seven Little TVAs,'" 419ff.

12. "Canning the Planners," *Commonweal*, 11 June 1943, 192. See also Philip L. White, "The Termination of the National Resources Planning Board" (master's thesis, Columbia Univ., 1949), 5.

13. C. Herman Pritchett, "Transplantability of the TVA," *Iowa Law Review* 32 (Jan. 1947):327–38, pointed out that the TVA endured because of a special set of circumstances that included a strong president, a willingness of Congress to care, a readiness of competing agencies, such as the Army Engineers and the Bureau of Reclamation, temporarily to bury their appropriation hatchets, and the fact that the Depression had allayed for the moment the suspicions of southerners about "Yankee-Greeks bearing gifts."

14. James T. Patterson, *Mr. Republican* (Boston: Houghton Mifflin, 1972), 189. It was ironic that Republican Herbert Hoover initiated the trend toward a national assessment of resources, a fact that Senator Taft may not have realized, although the results of the study came out in a two-volume set: President's Research Committee on Social Trends, *Recent Social Trends in the United States* (New York: McGraw-Hill, 1933), 2 vols. Charles E. Merriam and Wesley C. Mitchell of the NRPB board had been members of the Hoover committee. See "Antecedents," in Arlene Inouye and Charles Süsskind, "'Technological Trends and National Policy,' 1937," *Technology and Culture* 18 (Oct. 1977): 596.

15. "Senate Votes Drastic Curb on National Planning Board: Taft and Tydings Denounce Its Proposals," *New York Times*, 28 May 1943. Senators Dirksen, George, and Stennis were also active in eliminating the NRPB. Charles W. Eliot II, interview by Walter Creese, 1 Nov. 1972.

16. Leuchtenberg, "Roosevelt, Norris and the 'Seven Little TVAs'" 434.

17. James Dahir, *Region Building: Community Development Lessons from the Tennessee Valley* (New York: Harper & Bros., 1955), 13.

18. Howard W. Odum and Henry Estill Moore, *American Regionalism: A Cultural-Historical Approach to National Integration* (Gloucester, Mass.: Peter Smith, 1966), 107.

Index

TVA's Public Planning was designed by Dariel Mayer, composed by Lithocraft, Inc., and printed and bound by BookCrafters, Inc. The book is set in Caledonia and printed on Warren's 60-lb. Patina.